WIRELESS
COMMUNICATIONS
SECURITY

WIRELESS COMMUNICATIONS SECURITY

SOLUTIONS FOR THE INTERNET OF THINGS

Jyrki T. J. Penttinen

Giesecke & Devrient, USA

Library of Congress Cataloging-in-Publication data applied for

ISBN: 9781119084396

A catalogue record for this book is available from the British Library.

Set in 10/12pt Times by SPi Global, Pondicherry, India

10 9 8 7 6 5 4 3 2 1

Contents

About the Author

Dr **Jyrki T. J. Penttinen**, the author of this *Wireless Communications Security* book, started working in the mobile communications industry in 1987 evaluating early stage NMT-900, DECT and GSM radio network performance. After having obtained his MSc (EE) grade from Helsinki University of Technology (HUT) in 1994, he continued with Telecom Finland (Sonera and TeliaSonera Finland) and with Xfera Spain (Yoigo) participating in 2G and 3G projects. He also established and managed the consultancy firm Finesstel Ltd in 2002–03 operating in Europe and the Americas, and afterwards he worked with Nokia and Nokia Siemens Networks in Mexico, Spain and the United States in 2004–2013. During his time working with mobile network operators and equipment manufacturers, Dr Penttinen was involved in a wide range of operational and research activities performing system and architectural design, investigation, standardization, training and technical management with special interest in the radio interface of cellular networks and mobile TV such as GSM, GPRS/EDGE, UMTS/HSPA and DVB-H. Since 2014, in his current Program Manager's position with Giesecke & Devrient America, Inc, his focus areas include mobile and IoT security and innovation.

Dr Penttinen obtained his LicSc (Tech) and DSc (Tech) degrees in HUT (currently known as Aalto University, School of Science and Technology) in 1999 and 2011, respectively. In addition to his main work, he is an active lecturer, has written dozens of technical articles and authored telecommunications books, the recent ones being *The LTE-Advanced Deployment Handbook* (Wiley, 2016), *The Telecommunications Handbook* (Wiley, 2015) and *The LTE/SAE Deployment Handbook* (Wiley, 2011). More information about his publications can be found at www.tlt.fi.

Preface

This *Wireless Communications Security* book summarizes key aspects related to radio access network security solutions and protection against malicious attempts. As such a large number of services depend on the Internet and its increasingly important wireless access methods now and in the future, proper shielding is of the utmost importance. Along with the popularization of wireless communications systems such as Wi-Fi and cellular networks, the utilization of the services often takes place via wireless equipment such as smartphones and laptops supporting short and long range radio access technologies. Threats against these services and devices are increasing, one of the motivations of the attackers being the exploitation of user credentials and other secrets to achieve monetary benefits. There are also plenty of other reasons for criminals to attack wireless systems which thus require increasingly sophisticated protection methods by users, operators, service providers, equipment manufacturers, standardization bodies and other stakeholders.

Along with the overall development of IT and communications technologies, the environment has changed drastically over the years. In the 1980s, threats against mobile communications were merely related to the cloning of a user's telephone number to make free phone calls and eavesdropping on voice calls on the unprotected radio interface. From the experiences with the relatively poorly protected first-generation mobile networks, modern wireless communications systems have gradually taken into account security threats in a much more advanced way while the attacks are becoming more sophisticated and involve more diversified motivations such as deliberate destruction of the services and ransom-type threats. In addition to all these dangers against end-users, security breaches against the operators, service providers and other stakeholder are on the rise, too. In other words, we are entering a cyber-world, and the communications services are an elemental part of this new era.

The Internet has such an integral role in our daily life that the consequences of a major breakdown in its services would result in chaos. Proper shielding against malicious attempts requires a complete and updated cyber-security to protect the essential functions of societies such as bank institutes, energy distribution and telecommunications infrastructures. The trend related to the Internet of Things (IoT), with estimations of tens of billions of devices being taken into use within a short time period, means that the environment is becoming even more challenging due to the huge proportion of the cheaper IoT devices that may often lack their own protection mechanisms. These innocent-looking always-connected devices such as intelligent household appliances – if deployed and set up improperly – may expose doors deeper into the home network, its services and information containers, and open security holes even further into the business networks. This is one of the key areas in modern wireless security preparation.

As my good friend Alfredo so well summarized, the Internet can be compared to nuclear power; it is highly useful while under control, but as soon as security threats are present, it may lead to major disaster. Without doubt, proper protection is thus essential. This book presents the solutions and challenges of wireless security by summarizing typical, currently utilized services and solutions, and paints the picture for the future by presenting novelty solutions such as advanced mobile subscription management concepts. I hope you find the contents interesting and relevant in your work and studies and obtain an overview on both the established and yet-to-be-formed solutions of the field. In addition to this book, the contents are available in eBook format, and you can find additional information and updates from the topics at *www.tlt.fi*, which complement the overall picture of wireless security. As has been the case with my previous books published by Wiley, I would be glad to receive your valuable feedback about this *Wireless Communications Security* book directly via my email address: *jyrki.penttinen@hotmail.com*.

Jyrki T. J. Penttinen
Morristown, NJ, USA

Acknowledgements

It has been a highly interesting task to collect all this information about wireless security aspects into a single book. I reckon many of the presented solutions tend to develop extremely fast as the threats become increasingly sophisticated and innovative. The challenge is, of course, to maintain the relevancy of the written material. It is perhaps equally difficult for the stakeholders to ensure proper shielding of the wireless communications networks, devices, mobile apps and services along with all the advances in consumer and machine-to-machine domains – not forgetting the overall development of the Internet of Things (IoT), which is currently experiencing major interest. Even so, I believe that the foundations are worth describing in a book format, while the latest advances of each presented field can be checked via the identified key references and root sources of information.

An important part of this book, that is, describing the basics, is something I have been involved with throughout my career when I was working with mobile network operators as well as network and device vendors, while the rest of the contents complete the picture by presenting the most recent advances such as embedded SIM and respective subscription management which will be highly relevant in the near future for the most dynamic ways of utilizing consumers' mobile and companion devices as well as the ever growing amount of IoT equipment. I thank all my good colleagues I have had the privilege to work with and to exchange ideas related to mobile security. I want to especially mention the important role of Giesecke & Devrient in offering me the possibility to focus on the topic in my current position.

I warmly thank the Wiley team for the professional work and firm yet tender ways for ensuring the book project and schedules advanced according to the plans. Special thanks belong to Mark Hammond, Sandra Grayson, Tiina Wigley and Nithya Sechin, as well as Tessa Hanford, among all the others who helped me to make sure this book was finalized in good order.

I also want to express my warmest gratitude to the Finnish Association of Non-fiction Writers for their most welcomed support.

Finally, I thank Elva, Stephanie, Carolyne, Miguel, Katriina and Pertti for all their support.

Jyrki T. J. Penttinen
Morristown, NJ, USA

Abbreviations

3DES	Triple-Data Encryption Standard
3GPP	3rd Generation Partnership Program
6LoWPAN	IPv6 Low power Wireless Personal Area Network
AAA	Authentication, Authorization and Accounting
AAS	Active Antenna System
ACP	Access Control Policy
ADF	Application Dedicated File
ADMF	Administration Function
ADSL	Asymmetric Digital Subscriber Line
ADT	Android Developer Tool
AES	Advanced Encryption Standard
AF	Authentication Framework
AID	Application ID
AIDC	Automatic Identification and Data Capture
AIE	Air Interface Encryption
AK	Anonymity Key
AKA	Authentication and Key Agreement
ALC	Asynchronous Layered Coding
AMF	Authenticated Management Field
AMI	Advanced Metering Infrastructure
AMPS	Advanced Mobile Phone System
ANDSF	Access Network Discovery and Selection Function
ANSI	American National Standards Institute
AOTA	Advanced Over-the-Air
AP	Access Point
AP	Application Provider
APDU	Application Protocol Data Unit
API	Application Programming Interface
AR	Aggregation Router
ARIB	Association of Radio Industries and Businesses

AS	Access Stratum
AS	Authentication Server
ASIC	Application-Specific Integrated Circuit
ASME	Access Security Management Entity
ASN.1	Abstract Syntax Notation One
ATCA	Advanced Telecommunications Computing Architecture
ATR	Answer to Reset
ATSC	Advanced Television Systems Committee
AuC	Authentication Centre
AUTN	Authentication Token
AV	Authentication Vector
AVD	Android Virtual Device
BAN	Business/Building Area Network
BCBP	Bar Coded Boarding Pass
BCCH	Broadcast Control Channel
BE	Backend
BGA	Ball Grid Array
BIN	Bank Identification Number
BIP	Bearer-Independent Protocol
BLE	Bluetooth, Low-Energy
BM-SC	Broadcast – Multicast Service Centre
BSC	Base Station Controller
BSP	Biometric Service Provider
BSS	Billing System
BSS	Business Support System
BTS	Base Transceiver Station
C2	Command and Control
CA	Conditional Access
CA	Carrier Aggregation
CA	Certificate Authority
CA	Controlling Authority
CAT	Card Application Toolkit
CAT_TP	Card Application Toolkit Transport Protocol
CAVE	Cellular Authentication and Voice Encryption
CB	Cell Broadcast
CBEFF	Common Biometric Exchange Formats Framework
CC	Common Criteria
CC	Congestion Control
CCM	Card Content Management
CCMP	Counter-mode Cipher block chaining Message authentication code Protocol
CCSA	China Communications Standards Association
CDMA	Code Division Multiple Access
CEIR	Central EIR
CEPT	European Conference of Postal and Telecommunications Administrations
CFN	Connection Frame Number
CGN	Carrier Grade NAT

CHV	Chip Holder Verification
CI	Certificate Issuer
CK	Cipher Key
CL	Contactless
CLA	Class of Instruction
CLF	Contactless Frontend
CLK	Clock
CMAS	Commercial Mobile Alert System
CMP	Certificate Management Protocol
CN	Core Network
CoAP	Constrained Application Protocol
CoC	Content of Communication
CPU	Central Processing Unit
CS	Circuit Switched
CSFB	Circuit Switched Fallback
CSG	Closed Subscriber Group
CSS7	Common Signaling System
CVM	Cardholder Verification Method
DBF	Database File
DD	Digital Dividend
DDoS	Distributed Denial-of-Service
DE	Data Element
DES	Data Encryption Standard
DF	Dedicated File
DFN	Dual-Flat, No leads
DHCP	Dynamic Host Configuration Protocol
DL	Downlink
DM	Device Management
DM	Device Manufacturer
DMO	Direct Mode Operation
DNS	Domain Name System
DoS	Denial-of-Service
DPA	Data Protection Act
DPI	Deep Packet Inspection
DRM	Digital Rights Management
DS	Data Synchronization
DSS	Data Security Standard
DSSS	Direct Sequence Spread Spectrum
DTLS	Datagram Transport Layer Security
DTMB	Digital Terrestrial Multimedia Broadcast
DVB	Digital Video Broadcasting
EAL	Evaluation Assurance Level
EAN	Extended Area Network
EAP	Extensible Authentication Protocol
EAPoL	Extensible Authentication Protocol over Local Area Network
EAP-TTLS	Extensible Authentication Protocol-Tunneled Transport Layer Security

ECASD	eUICC Controlling Authority Secure Domain
eCAT	Encapsulated Card Application Toolkit
ECC	Elliptic Curve Cryptography
ECDSA	Elliptic Curve Digital Signature Algorithm
ECO	European Communications Office
EDGE	Enhanced Data Rates for Global Evolution
EEM	Ethernet Emulation Mode
EEPROM	Electrically Erasable Read-Only Memory
EF	Elementary File
EGAN	Enhanced Generic Access Network
EID	eUICC Identifier
EIR	Equipment Identity Register
E-MBS	Enhanced Multicast Broadcast Service
EMC	Electro-Magnetic Compatibility
EMF	Electro-Magnetic Field
EMI	Electro-Magnetic Interference
EMM	EPS Mobility Management
EMP	Electro-Magnetic Pulse
eNB	Evolved Node B
EPC	Enhanced Packet Core
EPC	Evolved Packet Core
EPS	Electric Power System
EPS	Enhanced Packet System
ERP	Enterprise Resource Planning
ERTMS	European Rail Traffic Management System
eSE	Embedded Security Element
eSIM	Embedded Subscriber Identity Module
ESN	Electronic Serial Number
ESP	Encapsulating Security Payload
ETSI	European Telecommunications Standards Institute
ETWS	Earthquake and Tsunami Warning System
eUICC	Embedded Universal Integrated Circuit Card
EUM	eUICC Manufacturer
E-UTRAN	Enhanced UTRAN
EV-DO	Evolution Data Only/Data Optimized
FAC	Final Approval Code
FAN	Field Area Network
FCC	Federal Communications Commission
FDD	Frequency Division Multiplex
FDT	File Delivery Table
FEC	Forward Error Correction
FF	Form Factor
FICORA	Finnish Communications Regulatory Authority
FID	File-ID
FIPS	Federal Information Processing Standards
FLUTE	File Transport over Unidirectional Transport

FM	Frequency Modulation
FPGA	Field Programmable Gate Array
GAA	Generic Authentication Architecture
GBA	Generic Bootstrapping Architecture
GCSE	Group Communication System Enabler
GEA	GPRS Encryption Algorithm
GERAN	GSM EDGE Radio Access Network
GGSN	Gateway GPRS Support Node
GMSK	Gaussian Minimum Shift Keying
GoS	Grade of Service
GP	GlobalPlatform
GPRS	General Packet Radio Service
GPS	Global Positioning System
GRX	GPRS Roaming Exchange
GSM	Global System for Mobile Communications
GSMA	GSM Association
GTP	GPRS Tunnelling Protocol
GUI	Graphical User Interface
HAN	Home Area Network
HCE	Host Card Emulation
HCI	Host Controller Interface
HE	Home Environment
HF	High Frequency
HFN	Hyperframe Number
HIPAA	Health Insurance Portability and Accountability Act
HLR	Home Location Register
HNB	Home Node B
HRPD	High Rate Packet Data
HSPA	High Speed Packet Access
HSS	Home Subscriber Server
HTTPS	HTTP Secure
HW	Hardware
I/O	Input/Output
I^2C	Inter-Integrated Circuit
IAN	Industrial Area Network
IANA	Internet Assigned Numbers Authority
IARI	IMS Application Reference ID
ICAO	International Civil Aviation Organization
ICC	Integrated Circuit Card
ICCID	ICC Identification Number
ICE	In Case of Emergency
ICE	Intercepting Control Element
ICIC	Inter Cell Interference Control
ICT	Information and Communication Technologies
IDE	Integrated Development Environment
IDEA	International Data Encryption Algorithm

ID-FF	Identity Federation Framework
IDM	Identity Management
IDS	Intrusion Detection System
ID-WSF	Identity Web Services Framework
IEC	International Electrotechnical Commission
IEEE	Institute of Electrical and Electronics Engineers
IETF	Internet Engineering Task Force
IF	Intermediate Frequency
IK	Integrity Key
IKE	Internet Key Exchange
IMEI	International Mobile Equipment Identity
IMEISV	IMEI Software Version
IMS	IP Multimedia Subsystem
IMSI	International Mobile Subscriber Identity
IOP	Interoperability Process
IoT	Internet of Things
IOT	Inter-Operability Testing
IP	Internet Protocol
IPS	Intrusion Prevention System
IPSec	IP Security
IR	Infrared
IRI	Intercept Related Information
ISD	Issuer Security Domain
ISDB-T	Terrestrial Integrated Services Digital Broadcasting
ISD-P	Issuer Security Domain Profile
ISD-R	Issuer Security Domain Root
ISIM	IMS SIM
ISO	International Organization for Standardization
ISOC	Internet Society
ITSEC	Information Technology Security Evaluation Criteria
ITU	International Telecommunications Union
IWLAN	Interworking Wireless Local Area Network
JBOH	JavaScript-Binding-Over-HTTP
JTC	Joint Technical Committee
K	User Key
KASME	Key for Access Security Management Entity
KDF	Key Derivation Function
LA	Location Area
LAN	Local Area Network
LBS	Location Based Service
LCT	Layered Coding Transport
LEA	Law Enforcement Agencies
LEAP	Lightweight Extensible Authentication Protocol
LEMF	Law Enforcement Monitoring Facilities
LF	Low Frequency
LI	Legal/Lawful Interception

LIF	Location Interoperability Forum
LIG	Legal Interception Gateway
LLCP	Logical Link Control Protocol
LOS	Line-of-Sight
LPPM	Location-Privacy Protection Mechanism
LTE	Long Term Evolution
LTE-M	LTE M2M
LTE-U	LTE Unlicensed
LUK	Limited Use Key
LWM2M	Lightweight Device Management of M2M
M2M	Machine-to-Machine
MAC	Medium Access Control
MAC	Message Authentication Code
MBMS	Multimedia Broadcast and Multicast Service
MC	Multi Carrier
MCC	Mobile Country Code
MCPTT	Mission Critical Push To Talk
ME	Mobile Equipment
ME ID	Mobile Equipment Identifier
MF	Master File
MFF2	Machine-to-Machine Form Factor 2
MGIF	Mobile Gaming Interoperability Forum
MIM	Machine Identity Module
MIMO	Multiple In Multiple Out
MITM	Man in the Middle
MM	Mobility Management
MME	Mobility Management Entity
MMS	Multimedia Messaging
MNC	Mobile Network Code
MNO	Mobile Network Operator
MPLS	Multiprotocol Label Switching
MPU	Multi Processing Unit
MRTD	Machine Readable Travel Document
MSC	Mobile services Switching Centre
MSISDN	Mobile Subscriber's ISDN number
MSP	Multiple Subscriber Profile
MST	Magnetic Secure Transmission
MT	Mobile Terminal
MTC	Machine-Type Communications
MVNO	Mobile Virtual Network Operator
MVP	Minimum Viable Product
MWIF	Mobile Wireless Internet Forum
NAA	Network Access Application
NACC	Network Assisted Call Control
NAF	Network Application Function
NAN	Neighborhood Area Network

NAS SMC	NAS Security Mode Command
NAS	Non-Access Stratum
NAT	Network Address Translation
NB	Node B
NCSC-FI	National Cyber Security Centre of Finland
NDEF	NFC Data Exchange Format
NDS	Network Domain Security
NE ID	Network Element Identifier
NFC	Near Field Communications
NGMN	Next Generation Mobile Network
NH	Next Hop
NHTSA	National Highway Transportation and Safety Administration
NIS	Network and Information Security
NIST	National Institute of Standards and Technology
NMS	Network Monitoring System
NMT	Nordic Mobile Telephony
NP	Network Provider
NPU	Numerical Processing Unit
NTP	Network Time Protocol
NWd	Normal World
OAM	Operations, Administration and Management
OBU	Onboard Unit
OCF	Open Card Framework
OCR	Optical Character Recognition
ODA	On-Demand Activation
ODM	Original Device Manufacturer
OEM	Original Equipment Manufacturer
OFDM	Orthogonal Frequency Division Multiplexing
OM	Order Management
OMA	Open Mobile Alliance
OP	Organizational Partner
OPM	OTA Provisioning Manager
OS	Operating System
OSPT	Open Standard for Public Transport (Alliance)
OTA	Over-the-Air
OTT	Over-the-Top
PAN	Personal Account Number
PAN	Personal Area Network
PC/SC	Personal Computer/Smart Card
PCC	Policy and Charging Control
PCI	Payment Card Industry
PCI-DSS	Payment Card Industry Data Security Standard
PDA	Personal Digital Assistant
PDCP	Packet Data Convergence Protocol
PDN	Packet Data Network
PDP	Packet Data Protocol

PDPC	Packet Data Convergence Protocol
PDS	Packet Data Services
PDU	Protocol/Packet Data Unit
PED	PIN-Entry Device
PGC	Project Coordination Group
P-GW	Proxy Gateway
PICC	Proximity ICC
PIN	Personal Identification Number
PITA	Portable Instrument for Trace Acquisition
PIV	Personal Identity Verification
PKI	Public Key Infrastructure
PLI	Physical Layer Identifier
PLMN	Public Land Mobile Network
PMR	Private Mobile Radio
PNAC	Port-based Network Access Control
POS	Point-of-Sales
PP	Protection Profile
PTM	Point-to-Multipoint
PTP	Point-to-Point
PTS	PIN Transaction Security
PTS	Protocol Type Selection
PUK	Personal Unblocking Key
PWS	Public Warning System
QoS	Quality of Service
QR	Quick Read
RA	Registration Authority
RAM	Random Access Memory
RAM	Remote Application Management
RAN	Radio Access Network
RANAP	RAN Application Protocol
RAND	Random Number
RAT	Radio Access Technology
RCS	Rich Communications Suite
REE	Rich Execution Environment
RES	Response
RF	Radio Frequency
RFID	Radio Frequency Identity
RFM	Remote File Management
RLC	Radio Link Control
RN	Relay Node
RNC	Radio Network Controller
RoI	Return on Investment
ROM	Read-Only Memory
RPM	Remote Patient Monitoring
RRC	Radio Resource Control
RRM	Radio Resource Management

RSP	Remote SIM Provisioning
RTC	Real Time Communications
RTD	Record Type Definition
RTT	Radio Transmission Technology
RUIM	Removable User Identity Module
SA	Security Association
SA	Services and System Aspects
SaaS	Software-as-a-Service
SAE	System Architecture Evolution
SAR	Specific Absorption Rate
SAS	Security Accreditation Scheme
SAT	SIM Application Toolkit
SATCOM	Satellite Communications
SBC	Session Border Controller
SC	Sub-Committee
SCD	Signature-Creation Data
SCP	Secure Channel Protocol
SCQL	Structured Card Query Language
SCTP	Stream Control Transmission Protocol
SCWS	Smart Card Web Server
SD	Secure Digital
SD	Security Domain
SDCCH	Stand Alone Dedicated Control Channel
SDK	Software Development Kit
SDS	Short Data Services
SE	Secure Element
SE	Service Enabler
SEG	Security Gateway
SEI	Secure Element Issuer
SES	Secure Element Supplier
SFPG	Security and Fraud Prevention Group
SG	Smart Grid
SGSN	Serving GPRS Support Node
S-GW	Serving Gateway
SIM	Subscriber Identity Module
SIP	Session Initiation Protocol
SiP	Silicon Provider
SM	Short Message
SMC	Security Mode Command
SM-DP	Subscription Manager, Data Preparation
SMG	Special Mobile Group
SMS	Short Message Service
SMSC	Short Message Service Centre
SM-SR	Subscription Manager, Secure Routing
SN ID	Serving Network's Identity
SN	Sequence Number

SN	Serving Network
SoC	System on Chip
SON	Self-Organizing Network
SP	Service Provider
SPI	Serial Peripheral Interface
SQN	Sequence Number
SRES	Signed Response
SRVCC	Single Radio Voice Call Continuity
SS	Service Subscriber
SSCD	Secure Signature-Creation Device
SSD	Shared Secret Data
SSDP	Simple Service Discovery Protocol
SSID	Service Set Identifier
SSL	Secure Sockets Layer
SSO	Single Sign On
SubMan	Subscription Management
SVLTE	Simultaneous Voice and LTE
SVN	Software Version Number
SW	Software
SWd	Secure World
SWP	Single Wire Protocol
TAC	Type Approval Code
TACS	Total Access Communications System
TC	Technical Committee
TCAP	Transaction Capabilities Application Part
TCP	Transmission Control Protocol
TDD	Time Division Multiplex
TDMA	Time Division Multiple Access
TE	Terminal Equipment
TEDS	TETRA Enhanced Data Service
TEE	Trusted Execution Environment
TETRA	Terrestrial Trunked Radio
TIA	Telecommunications Industry Association
TKIP	Temporal Key Integrity Protocol
TLS	Transport Layer Security
TMO	Trunked Mode Operation
TMSI	Temporary Mobile Subscriber Identity
TOE	Target of Evaluation
ToP	Timing over Packet
TPDU	Transmission Protocol Data Unit
TSC	Technical Sub-Committee
TSG	Technical Specification Group
TSIM	TETRA Subscriber Identity Module
TSM	Trusted Service Manager
TTA	Telecommunications Technology Association
TTC	Telecommunications Technology Committee

TTLS	Tunneled Transport Layer Security
TUAK	Temporary User Authentication Key
TZ	Trusted Zone
UART	Universal Asynchronous Receiver/Transmitter
UDP	User Data Protocol
UE	User Equipment
UHF	Ultra High Frequency
UICC	Universal Integrated Circuit Card
UIM	User Identity Module
UL	Uplink
UMTS	Universal Mobile Telecommunications System
UN	United Nations
UP	User Plane
URI	Uniform Resource Identifier
USAT	USIM Application Toolkit
USB	Universal Serial Bus
USIM	Universal Subscriber Identity Module
UTRAN	Universal Terrestrial Radio Access Network
UWB	Ultra-Wide Band
UX	User Experience
VLAN	Virtual Local Area Network
VLR	Visitor Location Register
VoIP	Voice over Internet Protocol
VoLTE	Voice over LTE
VPLMN	Visited PLMN
VPN	Virtual Private Network
WAN	Wide Area Network
WAP	Wireless Access Protocol
WCDMA	Wideband Code Division Multiplexing Access
WEP	Wired Equivalent Privacy
WG	Working Group
WIM	Wireless Identity Module
WISPr	Wireless Internet Service Provider roaming
WLAN	Wireless Local Area Network
WLCSP	Wafer-Level re-distribution Chip-Scale Packaging
WPA	Wi-Fi Protected Access
WPA2	Wi-Fi Protected Access, enhanced
WPS	Wi-Fi Protected Setup
WRC	World Radio Conference
WSN	Wireless Sensor Network
WWW	World Wide Web
XOR	Exclusive Or
XRES	Expected Response

1

Introduction

1.1 Introduction

Wireless Communications Security: Solutions for the Internet of Things presents key aspects of the mobile telecommunications field. The book includes essential background information of technologies that work as building blocks for the security of the current wireless systems and solutions. It also describes many novelty and expected future development options and discusses respective security aspects and protection methods.

This first chapter gives an overview to wireless security aspects by describing current and most probable future wireless security solutions, and discusses technological background, challenges and needs. The focus is on technical descriptions of existing systems and new trends like the evolved phase of Internet of Things (IoT). The book also gives an overview of existing and potential security threats, presents methods for protecting systems, operators and end-users, describes security systems attack types and the new dangers in the ever-evolving mobile communications networks and Internet which will include new ways of data transfer during the forthcoming years.

Chapter 1 presents overall advances in securing mobile and wireless communications, and sets the stage by summarizing the key standardization and statistics of the wireless communications environment. This chapter builds the base for understanding wireless network security principles, architectural design, deployment, installation, configuration, testing, certification and other security processes at high level while they are detailed later in the book. This chapter also discusses the special characteristics of the mobile device security, presents security architectures and gives advice to fulfil the regulatory policies and rules imposed. The reader also gets an overview about the pros and cons of different approaches for the level of security.

In general, this book gives the reader tools for understanding the possibilities and challenges of wireless communications, the main weight being on typical security vulnerabilities and practical examples of the problems and their solutions. The book thus functions as a practical guide to describe the evolvement of the wireless environment, and how to ensure the fluent continuum of the new functionalities yet minimize potential risks in the network security.

Wireless Communications Security: Solutions for the Internet of Things, First Edition. Jyrki T. J. Penttinen.
© 2017 John Wiley & Sons, Ltd. Published 2017 by John Wiley & Sons, Ltd.

1.2 Wireless Security

1.2.1 Background and Advances

The development of wireless communications, especially the security aspects of it, has been relatively stable compared to the overall issues in the public Internet via fixed access until early 2000. Nevertheless, along with the enhanced functionalities of smart devices, networks and applications, the number of malicious attacks has increased considerably. It can be estimated that security attacks, distribution of viruses and other illegal activities increase exponentially in a wireless environment along with the higher number of devices and users of novelty solutions. Not only are payment activities, person-to-person communications and social media types of utilization under constant threat, but furthermore one of the strongly increasing security risks is related to the Machine-to-Machine (M2M) communications which belong in the IoT realm. An example of a modern threat is malicious code in an Internet-connected self-driving car. In the worst case, this may lead to physically damaging the car's passengers.

There is a multitude of ideas to potentially change the role of the current Subscriber Identity Module (SIM), or Universal Integrated Circuit Card (UICC) which has traditionally been a solid base for the 3rd Generation Partnership Program (3GPP) mobile communications as it provides a highly protected hardware-based Secure Element (SE). Alternatives have been presented for modifying or for replacing the SIM/UICC concept with, e.g., cloud-based authentication, authorization and payment solutions. This evolution provides vast possibilities for easing the everyday life of end-users, operators, service providers and other stakeholders in the field, but it also opens unknown doors for security threats. The near future will show the preferred development paths, one of the logical possibilities being a hybrid solution that keeps essential data like keys within hardware-protected SEs such as SIM/UICC cards while, e.g., mobile payment would benefit from the flexibility of the cloud concept via dynamically changing tokens that have a limited lifetime.

In the near future, the penetration of autonomously operated devices without the need for human interactions will increase considerably, which results in much more active automatic communication, e.g., the delivery of telemetric information, diagnostics and healthcare data. The devices act as a base for value-added services for vast amounts of new solutions that are still largely under development or yet to be explored. Nevertheless, the increased share of such machines attached to networks may also open new security threats if the respective scenarios are not taken into account in early phases of the system, hardware (HW) and software (SW) development.

The field of new subscription management, along with the IoT concept, automatised communications and other new ways of transferring wireless data, will evolve very quickly. The updated information and respective security mechanisms are highly needed by the industry in order to understand better the possibilities and threats, and to develop ways to protect end-users and operators against novelty malicious attempts. Many of the solutions are still open and under standardization. This book thus clarifies the current environment and most probable development paths interpreted from the fresh messages of industry and standardization fields.

1.2.2 Statistics

In the mobile communications, wireless Local Area Networks (LANs) are perhaps the most vulnerable to security breaches. Wi-Fi security is often overlooked by both private individuals and companies. Major parts of wireless routers have been equipped in advance with default settings in order to offer fluent user experience for installation especially for non-technical people. Nevertheless, this good aim of the vendors leads to potential security holes for some wireless routers and access points in businesses and home offices due to poor or non-existing security. According to Ref. [21], around 25% of wireless router installations may be suffering from such security holes. From tests executed, Ref. [21] noted in 2011 that 61% of the studied cases (combined 2133 consumer and business networks) had a proper security set up either via Wi-Fi Protected Access (WPA) or Wi-Fi Protected Access, enhanced (WPA2). For the rest of the cases, 6% did not have security set up at all while 19% used low protection of Wired Equivalent Privacy (WEP), 11% used default credentials, and 3% used hidden Service Set Identifier (SSID) without encryption.

Ref. [26] presents recent statistics of Internet security breaches, and has concluded that the three most affected industries are public, information and financial services. Typical ways for illegal actions include the following:

- **Phishing**. Typically in the form of email, the aim is to convince users to change their passwords for banking services via legitimate-looking web pages. The investigations of Ref. [26] shows that phishing is nowadays more focused and continues being successful for criminals as 23% of users opened the phishing email, and 11% clicked the accompanying attachments.
- **Exploitation of vulnerabilities**. As an example, half of the common vulnerabilities and exposures during 2014 fell within the first two weeks which indicates the high need for addressing urgent breaches.
- **Mobile**. Ref. [26] has noted that Android is clearly the most exploited mobile platform. Not necessarily due to weak protection as such, but 96% of malware was focused on Android during 2014. As a result, more than 5 billion downloaded Android apps are vulnerable to remote attacks, e.g., via JavaScript-Binding-Over-HTTP (JBOH) which provides remote access to Android devices. Nevertheless, even if the mobile devices are vulnerable to breaches, after filtering the low-grade malware, the amount of compromised devices has been practically negligible. An average of only 0.03% of smartphones per week in the Verizon network during 2014 were infected with higher grade malicious code.
- **Malware**. Half of the participating companies discovered malware events during 35 or fewer days during the period of 2014. Malware is related to other categories like phishing which is the door for embedding malicious code to user's devices. Depending on the industry type, the amount of malware varies, so, e.g., financial institutes protect themselves more carefully against phishing emails which indicates a low malware proportion.
- **Payment card skimmers and Point-of-Sale (POS) intrusions**. This breach type has gained big headlines in recent years as there have been tens of millions of affected users per compromised retailer.
- **Crimeware**. The recent development indicates the increase of Denial-of-Service (DoS) attacks, with Command and Control (C2) continuing to defend its position in 2014.
- **Web app attacks**. Virtually all the attacks in this set, with 98% share, have been opportunistic in nature. Financial services and public entities are the most affected victims. Some methods related to this area are the use of stolen credentials, use of backdoor or C2, abuse of functionality, brute force and forced browsing.
- **Distributed Denial-of-Service (DDoS) attacks**. This breach type is heavily increasing. Furthermore, DDoS attacks are being prepared increasingly via malware. The attacks rely on improperly secured services like Network Time Protocol (NTP), Domain Name System

(DNS) and Simple Service Discovery Protocol (SSDP) which provide the possibility to spoof IP addresses.

- **Physical theft and insider misuse**. These are related to human factors; in general, this category belongs to the 'opportunity makes theft', which is very challenging to remove completely as long as the chain of trust relies on key personnel who might have the possibility and motivation to compromise or bypass security. Detecting potential misuse by insiders is thus an important role to prevent and reveal fraudulent attempts early enough. This detection can be related to deviation of the data transfer patterns, login attempts, time-based utilization and, in general, time spent in activities that may indicate dissatisfaction at the working place.
- **Cyber espionage**. According to Ref. [26], especially manufacturing, government and information services are noted to be typical targets of espionage. Furthermore, the most common way to open the door for espionage seems to be the opening of an email attachment or link.
- Any other errors that may open doors for external or internal misuse.

More detailed information about data breach statistics and impacts in overall IT and wireless environments can be found in Ref. [26].

1.2.3   Wireless Threats

1.2.3.1   General

Wireless communications systems provide a functional base for vast opportunities in the area of IoT including advanced multimedia and increasingly real-time virtual reality applications. Along with the creation and offering of novelty commercial solutions, there also exist completely new security threats that are the result of such a fast developing environment such that users and operators have not yet fully experienced the real impacts. Thus, there is a real need for constant efforts to identify the vulnerabilities and better protect any potential security holes. The following sections present some real-world examples of the possibilities and challenges of wireless communications, the weight being in the discussion of security vulnerabilities and their solutions.

Protection in the wireless environment largely follows the principles familiar from fixed networks. Nevertheless, the radio interface especially, which is the most important difference from the fixed systems, opens new challenges as the communications are possible to capture without physical 'wire-tapping' to the infrastructure. Knowledgeable hackers may thus try to unscramble the contents either in real time or by recording the traffic and attacking the contents offline without the victims' awareness. The respective protection level falls to the value of the contents – the basic question is how much end-users, network operators and service providers should invest in order to guarantee the minimum, typical or maximum security. As an example, the cloud storage for smart device photos would not need to be protected too strongly if a user uploads them to social media for public distribution. The scenery changes, though, if a user stores highly confidential contents that may seriously jeopardize privacy if publicly exposed. There are endless amounts of examples about such incidences and their consequences, including the stealing and distribution of personal photos of celebrities. Regardless of the highly unfortunate circumstances of these security breaches, they can also work as very useful lessons. Some of the easiest means to minimize the damage is to apply additional application-layer security by encrypting the contents via a separate password, and simply to reconsider the uploading of the most sensitive data to external data storages.

The selection of the security level, whether it is done by the end-user, network operator or service provider, can be optimized by balancing the cost of the protection and the fluency of the utilization. This easy user experience may be an important aspect because a highly secured service may require such complicated procedures to authenticate and protect the contents that it is not practical for the average user. One of the most reliable yet fluent ways is to utilize two-fold authentication, e.g., based on permanent user ID and password as well as a one-time code that is sent to the user via an alternative route such as mobile communications messaging. Along with increasing mobile device penetration, the majority of users already have some kind of mobile device, so one of the most logical bearers for such messaging authentication is based on the robust, widespread Short Message Service (SMS).

1.2.3.2 Wireless Environment

First-generation mobile communications systems, such as the Nordic Mobile Telephone (NMT), British Total Access Communications System (TACS) and American Advanced Mobile Phone System (AMPS), were analogue and based on Frequency Modulated (FM) radio channels for solely voice communications. The conversations of users could be intercepted by tuning a simple commercial-grade radio scanner to the utilized frequencies of the base station and mobile device as there was no contents protection mechanism applied against potential eavesdropping. Also, copying and reutilization of the device credentials such as the telephone number was possible via the non-protected radio interface and Common Signaling System (CSS7) messages. The analogue mobile communications networks have been obsolete for many years, but these early experiences about security breaches have been educational for developing more advanced systems.

Still widely in commercial use, the Global System for Mobile Communications (GSM) is the most popular second-generation mobile communications system that was standardized by applying proper shielding against the obvious security holes noted during the operations of analogue systems. Thanks also to digital transport technologies, protection of the system was easier than in preceding systems. Not only has the radio interface been protected by encrypting the signalling and communications but also procedures for authenticating and authorizing subscribers have provided additional mechanisms for preventing misuse of the systems. However, along with the ageing of the original technology, vulnerabilities of the protection mechanisms have been found. One of the concrete threats of the basic GSM system is that it is possible to set up a spoof Base Transceiver Station (BTS) to capture the call attempts in such a way that the non-Mobile Network Operator (MNO) base station does not need to utilize scrambled channels, since it acts as a mere relay station without the legitimate user's awareness. As the principle for making this happen is based on the replication of the GSM BTS protocol layers which are publicly available, the actual equipment may be constructed by emulating the minimum set of BTS functionalities used in a laptop and by utilizing a commercial Gaussian Minimum Shift Keying (GMSK) modulated transceiver and antenna system [1]. According to the European Telecommunications Standards Institute/3rd Generation Partnership Program (ETSI/3GPP) GSM specifications, the unsecure radio channel, which is not protected by any of the A5 algorithm variants, meaning that the A5/0 is in use, must be indicated to the user. In practice, this unsecure channel indicator may be displayed as a small symbol such as an open lock, which the end-user might not be able to relate to unprotected communications. In some cases, the symbol might be missing completely regardless of the standards requirements. The basic reason for including the support of unsecured communications into GSM handsets is due to the fact that some network operators do not activate the secure communications, and the handset devices need to be able to function in all of the networks while roaming.

This vulnerability was identified in the early stage of third-generation (3G) standardization, and thus the ETSI/3GPP Universal Mobile Telecommunications System (UMTS) included mutual authentication as one of the enhanced security items since its first release in 1999. The 3G mobile communications are relatively secure against such threats as spoof base stations, although there are other threats that apply to any mobile communications network. One of these is the end-to-end path from user equipment up to the MNO infrastructure which is secured up to the unscrambling equipment, but the rest of the path up to the answering subscriber in a fixed telephony network or up to the receiver's mobile network's scrambling equipment is typically unsecure. Furthermore, even if the MNO's internal network is assumed to be isolated, and focused wire-tapping is challenging due to the increased utilization of fibre optics, the internal transmission of the 2G and 3G communications may be based on unsecured radio links which may expose the possibility to intercept the communications by applying the respective protocol layer stacks for capturing the contents from the bit stream.

The security level is again further increased for the 3GPP Release 8 Long Term Evolution (LTE) and its enhanced phase as of Release 10, which is referred to as LTE-Advanced (LTE-A). The enhanced items include, e.g., new communications algorithms.

Unlike the mobile communications networks that have been traditionally well protected, wireless solutions like Wi-Fi and WiMAX do not contain such a large-scale infrastructure and are thus more vulnerable to security breaches. Despite the deployment of authorization passwords in Wi-Fi hotspots in home use, as well as hiding the ID of the access point and applying new encryption algorithms, wireless LANs tend to be vulnerable to malicious attacks. The consequence may be exposure of the user's communications and stored files, and the attacker might set up an illicit server for spam mailing or illegal contents storage without the user's knowledge.

1.2.3.3   Examples from the Real World

With the improving security of wireless networks, malicious attempts have been increasingly focusing on devices and applications. Not only smart devices but also IoT devices are fruitful targets due to their often under-developed security. The following list summarizes a small snapshot of some of the real-world cases published in 2014–15.

Wired reported that hackers can silently control Google Now and Siri from 16 feet away by using local connectivity of Radio Frequency (RF) to trigger voice commands on commercial phones that have such applications enabled and external headphones/microphone attached to the device. The threat is related to the headphones' cord which functions as an antenna, transporting the captured RF signal and confusing the phone's operating system, which assumes the signal to be the user's own audio commands via the microphone. This attack would serve to command Siri or Google Now to send texts and to force the phone to dial other mobile devices thus forming a simple eavesdropping device. According to Ref. [74], the commands can also be used to force the phone's browsers to enter malicious sites, to generate spam and phishing messages via email.

Interference Technology has reported on the low-cost Portable Instrument for Trace Acquisition (PITA) developed by Tel Aviv University and the Israeli research centre. It is a hacking device that can steal encryption keys over the air. It is based on the interpretation of the RF emission of computer processors to reveal encryption keys, and the method does not thus depend on standard communication methods like Wi-Fi or Bluetooth. The device is able to work up to 19 inches away from the processors, and may store data encrypted with RSA and ElGamal and decrypt it. Furthermore, the device can transmit the decrypted data over Wi-Fi to the attacker's computer [75].

Ref. [76] reports about remote baby-sitter devices which are possible to hack and then use to spy on people. Rapid7, a US-based company, revealed the magnitude of the risk in a number of commercial devices. Some of the compromised models include iBaby M3S. Upon connecting them to the Internet, the attacker may take over control and use them as hidden cameras and eavesdropping devices. Furthermore, via these devices, it is possible to utilize them as vectors to break through further to the users' home and business networks, which generates a risk for the private and business utilization of the connectivity. The issues related to these security holes include the possibility of externals being able to monitor the home via video and audio. If these devices are close to users, potentially confidential calls can thus be eavesdropped upon.

Ref. [77], together with reports from CNN and Ars Technica, informs about the danger of innocent-looking home and office devices like printers, which may expose security holes even without an Internet connection. Red Balloon Security demonstrated sending text wirelessly by modifying the functions of a printer at the Red Hat event in 2015. Typically, IoT security breaches are based on Internet holes, but less focus has been put on the RF leaking from the devices' components, which can be captured within short distances of the devices. Furthermore, this methodology may expose security holes in computers that are completely isolated from the public Internet, including the highest security environments such as nuclear power plants and banks. The commonly used term for such devices leaking information locally is 'zombie'. More details can be found via the demo presentation of Ref. [77] about data exfiltration using malware.

These examples indicate that not only are networks and devices under threat via typical connectivity technologies but also that many 'out-of-the-box' methods are being constantly invented. The challenging yet highly needed counter-measure is to assess the existing and potential security threats. One solution is that the service or device provider may try to deliberately hack its own systems. This approach is called 'white hat' hacking, as the intentions for finding security holes is done in cooperation with the hacking experts to find and protect the security holes. As an example, Ref. [78] discusses MasterCard's digital security lab for proving the security level of its payment environment. In this case, manual and automatized methods are applied in pre- and post-crime forensics. The aim of the lab is to figure out the ways of thieves trying to attack digital payment systems, such as old-style magnetic stripe credit cards, contactless chip bank cards, smartphone-based biometric systems and new device-based payment methods like those planned for wearables using biometrics, e.g., heartbeat pattern for authentication. Some methods for exploring the exposure and to break the payment technology encryption, passwords and Personal Identification Numbers (PINs) and their potential issues are based on electron beams, lasers and ionizing radiation. Furthermore, the lab also has the means to investigate physical traces of the DNA of the criminals on ATMs, cards and hacked PIN-entry machines. One example of such illegal intentions is the tampering of payment cards by providing a malicious Radio Frequency Identity (RFID) chip which could broadcast account and PIN details via an RF signal which could be received, e.g., near a Point-of-Sales (POS) terminal or within close proximity of an ATM. The magnetic stripe cards are still widely used, and expose an important risk for the easiness of copying the card (e.g., simply spraying iron fillings on the magnetic stripe which indicates visually the respective binary code that the stripe contains, including account number and other key data). According to Ref. [78], the lab has not yet seen cloned chip cards. As for more sophisticated physical hacking methods, the electrical charge across the Europay, MasterCard, Visa chip connections can be monitored via an electron microscope by observing respective visual flashes to reveal the binary messages that in turn may help hackers reverse engineer the cryptographic keys. To protect against this type of possibility, the EMV chip's connecting tracks can be buried or rerouted, or logic gate positions shuffled, to head off such attacks, as concluded in Ref. [78]. Yet another threat is power analysis, which refers to the monitoring of the power profile of the chip during a cryptographic operation which may give hints about the encryption methods of the chip, thus proper counter-measures have been developed for this case.

Not only the chip cards as such but also a PIN-Entry Device (PED) located at the POS may be vulnerable to tampering efforts such as adding a Secure Digital (SD) card and connectors inside the device so that an attacker may have access to the information the PED executed, including card numbers and associated PINs. Protection mechanisms against such efforts include perfecting the tamper resistance functions in the PED such as device lock and memory cleaning upon tampering efforts.

As Ref. [79] indicates, criminals can try to attack any remote location, including even jail doors by hacking the respective office automatic central control points which manage the heating, lights, air conditioning, water, alarms, web cameras, etc. According to the report, the National Cyber Security Centre of Finland (NCSC-FI) at the Finnish Communications Regulatory Authority (FICORA) has noted the presence of a surprisingly large amount of unprotected devices related to such automatic control systems. If an unauthorized hacker gets access to such a system, costly damages may result. Also, home control systems are equally vulnerable due to default passwords that users do not always change, even if this is one of the simplest ways to increase the protection level. A potential threat of entering such systems is that the criminals may get hold of important information about the hours when the inhabitants are not present, in order to plan the timing for a subsequent burglary.

The unprotected device refers especially to the environment with Internet connectivity for entering the respective system. Password protection does not necessarily guarantee proper safety because the devices may have known vulnerabilities – which may often be very easy to detail from Internet sources for attack intentions. Aalto University has investigated automatized control system vulnerabilities in Finland and found that in the majority of cases there are known vulnerabilities per device with easily tracked instructions on the Internet. Such devices are used in energy production, electricity companies and water services.

One lifestyle-changing innovation is the connected, self-driving car. It is easy to guess that this environment attracts hackers to try to access the car control systems. There is publicly available information about the surprisingly easy ways of hacking some of the current Internet-connected cars as Ref. [80] informs, including more advanced wireless hijacking of the control system even during driving.

These examples merely scratch the surface, but they prove the importance of enhanced protection techniques to ensure safety of home and business environments. One of the challenges, though, is that there are increasingly activities concerned with hacking the IP network infrastructure and consumer gear, including very old components like routers, bridges and consumer accessories like Wi-Fi routers which do not have such systematic SW upgrading procedures as is the case with up-to-date computers, laptops and smart devices.

The importance of protection mechanisms is understandably considerably higher in environments like control systems that are meant for public transportation or other functions involving human well-being. As an example, Ref. [81] discusses the British signalling system for train control which could potentially be hacked to cause a crash. The conclusion of this specific case is that ensuring adequate protection is of utmost importance, in particular in replacing the old signal lights with new computers – if done without proper assessment and a prevention plan – could leave the rail network exposed to cyber-attacks which in turn may lead to a major accident. The system in question, the European Rail Traffic Management System (ERTMS), dictates critical safety information including how fast the trains should go and how long they will take to stop, so potential hackers could theoretically cause trains to travel too quickly with dramatic consequences.

With all the new and existing potential security holes resulting from the growing numbers of Internet-connected devices, it is clear that M2M security requires very special attention. As Ref. [34] summarizes, there is an increasing amount of known (and unimaginable) opportunities in the IoT, some examples being remote home thermostat control, self-driving cars, factories communicating with the container terminal and digitized city infrastructure. At the same time, new business and service models are emerging as our lifestyle becomes more convenient and mobile as a result of the ongoing digital revolution. We are now part of the connected world, which is in general beneficial for all, but as the amount of shared data and information grows, the risks also increase – not only for stored and transferred information but also for networked environments which include the means for controlling safety-critical systems like physical access rights and chemical processes.

The difference between fixed and mobile device security is on the whole not so huge, the openly radiating air interface being the most important differentiator in the wireless environment. The trend seems to indicate that as protection mechanisms of the networks are improving, the interest of the hacking efforts against the infrastructure is lowering. At the same time, with the popularity of the smart devices with vast amounts of applications, security breaches are focusing increasingly on the application level, e.g., via embedded malicious code hidden in the apps or via viruses that alter the functionality of the device or open doors for further attacks. The smart devices, including advanced SW executable mobile phones and tablets with radio connectivity, are assumingly thus becoming the primary targets for hackers, which requires active protection from the end-users as well as the operators providing the connectivity, and the service providers offering the backend for the app communications.

1.2.4   M2M Environment

One of the basic benefits of IoT is to facilitate always-connected devices to automatically control and report without human interaction. A simple example of this M2M communications is a refrigerator alerting about food items that require replenishing. The IoT environment also includes human interactions with machines and systems, e.g., making it possible to turn lights on and off remotely. The environment may include awareness of various environmental and user-related items. For instance, when a user visits a supermarket, stored information about recently purchased items may trigger reminders on the mobile device as the user walks nearby, to suggest repurchasing the same products. The IoT environment also helps to optimize the logistics chain, presenting items that purchasers are able to carry home and highlighting heavier and less frequently purchased items which may be transported via alternative ways. This environment with everything connected (connected society) in an intelligent way (intelligent homes, offices and transportation) is actually a groundbreaking step in human history. It leverages the automatized Information and Communications Technology (ICT) society to the next level, optimizes the techno-economics and influences positively on the green values as energy consumption and transportation of goods and people is minimized via high level of awareness. This awareness is possible in real time due to collected data and post-processed information of IoT devices as they communicate with each other and systems. IoT is the next big thing to change our living and working environments.

IoT comprises this advanced environment, with a huge amount of networked devices, and objects and users enabling and benefiting from data. Nevertheless, IoT is still in a relatively early stage. The first concrete solutions that are starting to form IoT include smart devices, the cloud and sensors. The combination of various access technologies like RFID, wireless and cellular connectivity, as well as the evolved, miniature components and devices are essential enablers for advancing the connected IoT world. The sub-categories of IoT include industrial Internet and M2M communications, and smart consumer environments with devices and services like health devices and smart wristwatches which are easing mobile banking and many other daily functions.

The M2M environment is currently developing strongly with new, evolved technologies and services coming into commercial markets. It creates a great deal of challenges especially for the management of such a huge amount of always-connected device subscriptions and traffic, and for the security of the communications.

1.3   Standardization

The following sections summarize the standardization bodies relevant to wireless security, and lists the respective key standards.

1.3.1   The Open Mobile Alliance (OMA)

The OMA is a non-profit organization producing open specifications. The aim of the OMA is to create interoperable, end-to-end global services on any bearer network. The OMA was formed in 2002 by the key mobile operators, device and network suppliers, information technology companies and content and service providers. The OMA's specifications support fixed and mobile terminals, such as established cellular operator networks as well as emerging networks with M2M device communications. The OMA drives service enabler architectures and open enabler interfaces independently from the underlying wireless platforms, and has developed programs for testing the interoperability of new products [28].

In addition, the OMA has integrated the WAP Forum, Location Interoperability Forum (LIF), SyncML Initiative, Multimedia Messaging Interoperability Process (MMS-IOP),

Wireless Village, Mobile Gaming Interoperability Forum (MGIF), and the Mobile Wireless Internet Forum (MWIF) into the OMA for promoting end-to-end interoperability across different devices, geographies, service providers, operators and networks. The OMA drives the development of mobile service enablers such as Device Management (DM), M2M communications, Application Programming Interfaces (APIs) and Augmented Reality.

The Device Management Working Group of the OMA (DM WG) specifies protocols and mechanisms to achieve the management of mobile devices, services access and software on connected devices [29]. The OMA's suite of DM specifications includes 21 mobile service enablers and more than 60 management objects that provide ways to deploy new applications and services with low risk. There are an additional 21 management objects defined by other standards organizations and forums in cooperation with the OMA to minimize fragmentation. As an example, the OMA Diagnostics and Monitoring Management Object is used by 3GPP and WiMAX Forum, and other industry bodies have extended the OMA DM to the IP environment for use with remote sensors and in automotive scenarios. The aim of the OMA DM is to manage converged and multi-mode devices in technology-agnostic networks, including devices that do not have a SIM card which makes the OMA DM also suitable for M2M communications [30].

1.3.1.1 OMA Lightweight M2M 1.0

Network operators and enterprises are actively using device management in mobile communications consumer space. The current M2M DM environment relies partially on mobile devices, being typically proprietary as has been the consumers' DM technologies. Today, the OMA DM provides a more standardized way, although even in this case the handset providers normally implement proprietary mechanisms. The OMA's M2M Lightweight Device Management (LWM2M) standard is designed for this M2M market to reduce fragmentation.

The LWM2M stabilized in 2013. It is designed for mobile communications and M2M device environments for enhancing interoperability based on the Internet Engineering Task Force (IETF) standards. It is simple yet provides an efficient set of protocols, interfaces and payload formats. It includes pre-shared and public key methodologies, provisioning and bootstrapping. It is applicable to mobile systems, Wi-Fi and other IP-based devices and networks, and it is possible to be combined with other DM solutions.

The LWM2M defines a strong, holistic security solution via the Datagram Transport Layer Security (DTLS) v1.2 for Constrained Application Protocol (CoAP) communications. CoAP is a SW protocol designed for highly simplified electronic devices that communicate interactively over the Internet, and is especially useful for low-power sensors, switches and other remotely located components requiring supervision and controlling – in other words, it is suitable for IoT and M2M environments. More detailed information about CoAP can be found in Ref. [32].

The DTLS is similar to the Secure Sockets Layer (SSL) and Transport Layer Security (TLS) protocols providing the same integrity, authentication and confidentiality services, but instead of relying on the Transmission Control Protocol (TCP), the DTLS is transported via the User Data Protocol (UDP) which works for securing unreliable datagram traffic. It thus provides communications security for datagram protocols. The defined DTLS security modes of LWM2M are pre-shared key, raw public key and certificate mode. More information about the DTLS can be found in Ref. [33].

Table 1.1   OMA DM specifications as of December 2015

Enabler (document theme, category)	Release type, version
Device management, protocol	Enabler, 1.1.2, 1.2.1, 1.3*, 2.0*
Client side enabler API, protocol	Reference, 1.0
Device management smartcard, protocol	Enabler, 1.0
Client provisioning, protocol	Enabler, 1.1
M2M device classification, white paper	Reference, 1.0
Management object design guidelines, white paper	Reference, 1.0
Provisioning objects, device management application characteristics management object, white paper	Reference, 1.0.1
Browser, management object	Reference, 1.0
Connectivity, management object	Reference, 1.0
Device capability, management object	Enabler, 1.0
Delta record, management object	Enabler, 1.0
Diagnostics and monitoring, management object	Enabler, 1.0, 1.1, 1.2
Firmware update, management object	Enabler, 1.0.4
Gateway, management object	Enabler, 1.0, 1.1*
Lock and wipe, management object	Enabler, 1.0
Management policy, management object	Enabler, 1.0*
Scheduling, management object	Enabler, 1.0
SW component, management object	Enabler, 1.0, 1.1
SW and application control, management object	Enabler, 1.0*
Virtualization, management object	Enabler, 1.0*
OMA LWM2M, protocol	Enabler, 1.0*

*indicates draft or candidate

The LWM2M also includes bootstrapping methodologies that are designed for provisioning and key management via pre-configured bootstrapping (flash-based), and smartcard bootstrapping (SIM-based). The OMA LWM2M Version 1.0 was released in 2013.

1.3.1.2   OMA Standards

Table 1.1 summarizes the current and planned OMA DM specifications as indicated in Ref. [29,31,35].

1.3.2   The International Organization for Standardization (ISO)

The ISO together with the International Electrotechnical Commission (IEC) are worldwide standard-setting bodies for smartcards, among various other technologies related to electronics. They jointly have an important role in the standardization of the SIM card, which is the variant if smartcards are adapted into mobile communications systems. The SIM was introduced along with the 2G systems, firstly via the GSM, and it has been further developed in 3G systems which apply the 2G SIM and 3G Universal SIM (USIM) functionalities representing applications within the UICC.

The ISO provides an open process for participating stakeholders with the aim to facilitate the creation of voluntary standards. ISO 7816 defines Integrated Circuit Cards (ICCs), commonly called smartcards as defined in Refs. [37] through [53]. This standard defines contact cards, which means that the communication between the card and external devices like card readers happens via the electrical circuit contacts of the card. It also functions as a base for the contactless cards extended via the ISO 14443 standard, which defines the communications via the RF channel. The contactless card is based on Near Field Communications (NFC). Both ISO 7816 and ISO 14443 are provided by the American National Standards Institute (ANSI).

The key standards for smartcards are ISO/IEC 7816, ISO/IEC 14443, ISO/IEC 15693 and ISO/IEC 7501. The ISO/IEC 27000 series is also relevant to smartcards, describing information security management [90]. A complete list of the ISO/IEC JTC1/SC17 working groups and respective ISO standards can be found in Ref. [91].

ISO/IEC 7816 includes multiple parts from which Parts 1, 2 and 3 are related to contact cards and their essential aspects for interfaces, dimensions and protocols whereas Parts 4–6, 8, 9, 11, 13 and 15 include definitions for both contact and contactless card, e.g., for file and data element structures of the cards, API commands for the card use, application management, biometric verification, cryptographic functions and application naming. Part 7 defines a secure relational database for smartcards via Structured Card Query Language (SCQL) interfaces. Part 10 is related to applications of memory cards, including pre-paid telephone and vending machine cards.

The most important reference for the SIM/UICC card used in mobile communications systems is ISO 7816. Parts 1, 2 and 3 define the physical and communications characteristics as well as the application identifiers for embedded chips and data. This standard creates the base for mobile communications smartcards and is referenced in major part of the other standards. Moreover, ISO/IEC 7816 describes, among other definitions, the fundamental physical and logical aspects of the smartcard, voltage levels and file systems. Table 1.2 details the definitions of ISO/IEC 7816. ISO/IEC 7816 is jointly defined by the ISO and IEC, and edited by the Joint Technical Committee (JTC) 1 and Sub-Committee (SC) 17, Cards and Personal Identification [1] and adapted by ETSI, 3GPP and 3GPP2. More details about ISO/IEC 7816 sub-standards can be found in Chapter 4.

Table 1.2 ISO/IEC 7816 standard definitions

Standard	Description
7816-1	Physical characteristics
7816-2	Cards with contacts; dimensions and location of the contacts
7816-3	Cards with contacts; electrical interface and transmission protocols
7816-4	Organization, security and commands for interchange
7816-5	Registration of application providers
7816-6	Inter-industry data elements for interchange
7816-7	Inter-industry commands for SCQL
7816-8	Commands for security operations
7816-9	Commands for card management
7816-10	Electronic signals and answer to reset for synchronous cards
7816-11	Personal verification through biometric methods
7816-12	Cards with contacts; USB electrical interface and operating procedures
7816-13	Commands for application management in multi-application environment
7816-15	Cryptographic information application

ISO/IEC 14443 defines the interfaces of contactless smartcards that are used in NFC within about 10 cm from the respective reader. It includes the electrical and RF interfaces as well as the communications protocols. The cards operate on a 13.56 MHz frequency. This standard is the base for the contactless environment in access control, transit and financial applications as well as in electronic passports and in Federal Information Processing Standards (FIPS) 201 PIV cards.

ISO/IEC 15693 defines the so-called vicinity cards, including the base for the physical aspects, RF power, interface levels, collision management and protocols. The idea of vicinity cards is to operate from a maximum distance of 1 m from the reader. ISO/IEC 7501, in turn, has descriptions for machine-readable travel documents.

1.3.2.1 Other Relevant ISO/IEC Standards

The ISO/IEC has produced various other standards related to ICCs and payment cards. More information of the overall ISO/IEC standards can be found in Ref. [54].

1.3.3 The International Telecommunications Union (ITU)

The ITU is a worldwide organization that takes care of the global requirements of telecommunications. The ITU belongs to the United Nations (UN) and is specialized for Information and Communication Technologies (ICT). The tasks of the ITU include the allocation of global radio spectrum and satellite orbits, and development of technical standards. In addition, the ITU paves the way to enhance globally the access to ICTs to underserved communities. The ITU facilitates people's right to communicate via the developed communications [3]. The ITU is divided into three sectors from which the ITU-T concentrates on the standardization of the telecommunications area whilewhile the ITU-R standardizes the radio area, and the ITU-D focuses on the development of the telecommunications area. The recommendations of the ITU-T can be found in Ref. [4] and the set of recommendations of the ITU-R is located in Ref. [5]. The most relevant documentation of the ITU for the security aspects of telecommunications is found in the ITU-T series X, which describes data networks, open system communications and security.

1.3.4 The European Telecommunications Standards Institute (ETSI)

ETSI produces globally applicable standards for ICT including fixed, mobile, radio, broadcast, internet, aeronautical and other areas. ETSI created the first GSM and UMTS standards. The GSM standardization work was executed in various groups under the Special Mobile Group (SMG) until the 3GPP took over the major part of GSM and UMTS development in 1999. Nevertheless, an important part of the security aspects of mobile communications remains under ETSI [62]. This can be clearly seen from the continuum of the 13-series standards in ETSI. In practice, the security definitions of, e.g., UICC for GSM, 3G and advanced LTE systems are in fact found as a combined set of ETSI and 3GPP specifications.

As for the SIM/UICC, some of the most important ETSI standards are ETSI 102 221 (UICC definition) and ETSI 102 241 (UICC), which are based on the original GSM standards for the SIM as defined in TS 11.11 and TS 11.14.

The European Union recognizes ETSI as an official European standards organization, and one of the aims of ETSI is to provide access to European markets. ETSI clusters represent ICT standardization activities which are divided into Security, Home and Office, Better Living with ICT, Content Delivery, Fixed Networks, Wireless Systems, Transportation, Connecting Things, Interoperability, and Public Safety. ETSI executes standardization work under all of these clusters in various Technical Committees (TCs) and Working Groups (WGs).

As an example of SIM development, ETSI TS 103 384 defines the embedded UICC. It is aligned with the SIMalliance interoperable profile specification v1.0 and GSMA eSIM Technical Specification for M2M v2.0. There is also new development in the field such as the AppleSIM, and open items for handling the SIM-related data beyond SMS as in the case of IP-only remote management via the Bearer-Independent Protocol (BIP), which has been an active discussion topic at 3GPP SA2. The development of the embedded SIM solutions by ETSI and other parties is highly relevant for both consumer and M2M environments along with IoT and new types of end-user devices like wearables. This topic thus falls into the area of interoperable subscription management which is discussed in more detail later in this book.

1.3.5 The Institute of Electrical and Electronics Engineers (IEEE)

The IEEE provides a range of publications and standards that make the exchange of technical knowledge and information possible among technology professionals. One of the most important areas of IEEE standards is the IEEE 802 series, which defines a set of wired and wireless networks, including the Wireless Local Area Network (WLAN). IEEE publications and standards can be accessed via Refs. [6,7].

The IEEE is involved in standardization activities related to network and information security as well as in anti-malware technologies. Areas include encryption, fixed and removable storage, hard copy devices, and applications of these technologies in smart grids [8]. The IEEE Computer Society has a technical committee focused on computer security and privacy, with respective publications. IEEE also has an Industry Connections Security Group for tackling the malware environment. Some of the relevant IEEE standards in encryption domain are presented in Table 1.3.

Furthermore, the IEEE produces a multitude of standards for fixed and removable storage, security for hardcopy devices and Network and Information Security (NIS) for smart grids.

Table 1.3 Some of the most important IEEE standards related to encryption

Standard	Description
IEEE 1363-2000/ 1363a-2004	IEEE Standard Specifications for Public Key Cryptography
IEEE 1363.1-2008	IEEE Standard Specification for Public Key Cryptographic Techniques Based on Hard Problems over Lattices
IEEE 1363.2-2008	IEEE Standard Specification for Password Based Public Key Cryptographic Techniques
IEEE P1363.3	Draft Standard for Identity Based Cryptographic Techniques using Pairings

1.3.6   The Internet Engineering Task Force (IETF)

The IETF is an international and open community of network operators, vendors, designers and researchers. The task of the IETF is to be actively involved in the evolution of Internet architecture. The outcome of the work of the IETF is documented as recommendations, which are numbered as RFC *n*, where *n* is an integer number referring to a specific area of the definition. The list of documents can be found in [10], and the specific documents can be searched in [9,12].

The way of work of the IETF is based on email lists and exchanging comments online. This method greatly differs from the more formal approach of other international standardization organizations. The IETF also organizes meetings, though only three times per year. In addition to the actual standardization, the IETF performs tasks like assignment of parameter values for Internet protocols, which is taken care of in a centralized way by the Internet Assigned Numbers Authority (IANA), chartered by the Internet Society (ISOC). For more detailed information of the work and structure of IETF, Ref. [11] presents useful guidelines.

Specifically related to working group information sharing of Internet security, the IETF has a dedicated web page named the IETF Security Area [13].

1.3.7   The 3rd Generation Partnership Project (3GPP)

The 3GPP was established by several standardization bodies in 1998. Its original aim was to develop and enhance the 3G mobile communications systems. The current cooperating parties within 3GPP are the Association of Radio Industries and Businesses (ARIB), ATIS (Alliance for Telecommunications Industry Solutions), CCSA (China Communications Standards Association), ETSI, TSDSI (Telecommunications Standards Development Society of India), Telecommunications Technology Association (TTA) and Telecommunications Technology Committee (TTC). In addition to the set up of the 3GPP, there is also a group that takes care of the American variant – the 3GPP2.

Since the first introduction of the GSM-based SIM card, the concept has been further developed. Along with the handover of the mobile communications standardization work from the ETSI SMG group to the 3GPP, with the exception of few security-related items that still are taken care of by ETSI, the mobile communications part of the SIM card is included to an extended version of the original, UICC, which contains mobile communications functionalities per radio network type treated as separate applications. Not only the technical aspects have been enhanced, but also the SIM card's frame materials have been evolved, and there are e.g. eco-friendlier cards in the market nowadays [36].

Currently, the 3GPP is formed by a Project Coordination Group (PGC) and a total of five sub-groups which are: Services and System Aspects (SA), Radio Access Network (RAN), Core Network (CN), Terminals (T) and GSM EDGE Radio Access Network (GERAN). The original GSM standardization work of ETSI was taken over by the 3GPP GERAN almost completely in July 2000, excluding some of the security-related topics. The security algorithm specification work is not public. The User Equipment (UE), SIM and USIM specifications are listed in the original 11-series, and in the 34-series in later phases. Some relevant 3GPP specifications for the mobile communications security aspects can be found in 3GPP TS 42.009, V4.1.0, 23 June 2009 (security aspects). The security aspects of 3GPP are taken care of by SA working group 3.

The complete set of 3GPP technical specifications and recommendations as well as study reports can be found in Ref. [18]. For the LTE phase, Refs. [19,20,22] present overall security aspects. Furthermore, some of the most relevant 3GPP LTE security specifications are listed in Tables 1.4 and 1.5.

To summarize, a few of the highest level requirements of the 3GPP specifications, most importantly, the continued usage of current USIM needs to be ensured. The earlier USIM still needs to be able to access the EPS network of the LTE/LTE-A to ensure that previous USIM variants are also functional in future 3GPP networks. Furthermore, the level of the security should be at least equal or better compared to the UMTS.

In principle, the 3GPP specifications are the root information sources for the GSM, UMTS and LTE security aspects. Ref. [20] presents general aspects of the 3GPP security, and more detailed security aspects of the 3GPP systems are discussed in Chapter 2 in this book. The following sections detail further some of the key aspects of the SIM/UICC specifications defined by 3GPP.

The GSM triggered the introduction of the SIM for storing subscription related data. In the beginning, as defined in the first phase ETSI standard TS 11.11, the SIM was merely a physical

Table 1.4 Some of the key 3GPP security specifications

Topic	Specification
LTE security principles	TS 21.133, Security Threats and Requirements
	TS 33.120, Security Principles and Objectives
	TS 33.401, System Architecture Evolution (SAE); Security Architecture
	TS 33.402, System Architecture Evolution (SAE); Security Aspects of Non-3GPP
Security architecture	TS 22.022, Personalization of Mobile Equipment
	TS 33.102, Security Architecture
	TS 33.103, Integration Guidelines
Algorithms	TS 33.105, Cryptographic Algorithm Requirements
	Specifications for 3GPP Confidentiality and Integrity Algorithms: 1) f8 and f9; 2) KASUMI; 3) Implementers' Test Data; 4) Design Conformance Test Data
Lawful interception	TS 33.106, Lawful Interception Requirements
	TS 33.107, Lawful Interception Architecture and Functions
	TS 33.108, Handover Interface for Lawful Interception
Key derivation	TS 33.220, GAA: Generic Bootstrapping Architecture (GBA)
Backhaul security	TS 33.310, Network Domain Security (NDS); Authentication Framework (AF)
Relay Node security	TS 33.816, Feasibility Study on LTE Relay Node Security (also 33.401)
Home (e) Node B security	TS 33.320, Home (evolved) Node B Security
Technical reports of 3GPP	TR 33.901, Criteria for Cryptographic Algorithm Design Process
	TR 33.902, Analysis, 3G Authentication Protocol
	TR 33.908, Report, Design, Specification and Evaluation of 3GPP Standard Confidentiality and Integrity Algorithms

Table 1.5   The complete list of 3GPP security-related 33-series documents

Document	Description
TS 33.102	3G Security; Security Architecture
TS 33.103	3G Security; Integration Guidelines
TS 33.105	3G Security; Cryptographic Algorithm Requirements
TS 33.106	3G security; Lawful Interception Requirements
TS 33.107	3G security; Lawful Interception Architecture and Functions
TS 33.108	3G security; Handover Interface for Lawful Interception
TS 33.110	Key Establishment between a Universal Integrated Circuit Card (UICC) and a Terminal
TS 33.116	Security Assurance Specification for the MME Network Product Class
TS 33.117	Catalogue of General Security Assurance Requirements
TS 33.120	Security Objectives and Principles
TS 33.141	Presence Service; Security
TS 33.187	Security Aspects of Machine-Type Communications (MTC) and Other Mobile Data Applications Communications Enhancements
TS 33.200	3G Security; Network Domain Security (NDS); Mobile Application Part (MAP) Application Layer Security
TS 33.203	3G Security; Access Security for IP-Based Services
TS 33.204	3G Security; Network Domain Security (NDS); Transaction Capabilities Application Part (TCAP) User Security
TS 33.210	3G security; Network Domain Security (NDS); IP Network Layer Security
TS 33.220	Generic Authentication Architecture (GAA); Generic Bootstrapping Architecture (GBA)
TS 33.221	Generic Authentication Architecture (GAA); Support for Subscriber Certificates
TS 33.222	Generic Authentication Architecture (GAA); Access to Network Application Functions Using Hypertext Transfer Protocol over Transport Layer Security (HTTPS)
TS 33.223	Generic Authentication Architecture (GAA); Generic Bootstrapping Architecture (GBA) Push Function
TS 33.224	Generic Authentication Architecture (GAA); Generic Bootstrapping Architecture (GBA) Push Layer
TS 33.234	3G Security; Wireless Local Area Network (WLAN) Interworking Security
TS 33.246	3G Security; Security of Multimedia Broadcast/Multicast Service (MBMS)
TS 33.259	Key Establishment between a UICC Hosting Device and a Remote Device
TS 33.269	Public Warning System (PWS) Security Architecture
TS 33.303	Proximity-Based Services (ProSe); Security Aspects
TS 33.310	Network Domain Security (NDS); Authentication Framework (AF)
TS 33.320	Security of Home Node B (HNB) / Home evolved Node B (HeNB)
TS 33.328	IP Multimedia Subsystem (IMS) Media Plane Security
TS 33.401	3GPP System Architecture Evolution (SAE); Security Architecture
TS 33.402	3GPP System Architecture Evolution (SAE); Security Aspects of Non-3GPP Accesses
TR 33.769	Feasibility Study on Security Aspects of Machine-Type Communications Enhancements to Facilitate Communications with Packet Data Networks and Applications
TR 33.803	Coexistence between TISPAN and 3GPP Authentication Schemes
TR 33.804	Single Sign On (SSO) Application Security for Common IP Multimedia Subsystem (IMS) based on Session Initiation Protocol (SIP) Digest
TR 33.805	Study on Security Assurance Methodology for 3GPP Network Products

Table 1.5 (Continued)

Document	Description
TR 33.806	Pilot Development of Security Assurance Specification for MME Network Product Class for 3GPP Network Product Classes
TR 33.810	3G Security; Network Domain Security/Authentication Framework (NDS/AF); Feasibility Study to Support NDS/IP Evolution
TR 33.812	Feasibility Study on the Security Aspects of Remote Provisioning and Change of Subscription for Machine to Machine (M2M) Equipment
TR 33.816	Feasibility Study on LTE Relay Node Security
TR 33.817	Feasibility Study on (Universal) Subscriber Interface Module (U)SIM Security Reuse by Peripheral Devices on Local Interfaces
TR 33.820	Security of Home Node B (HNB)/Home evolved Node B (HeNB)
TR 33.821	Rationale and Track of Security Decisions in Long Term Evolution (LTE) RAN/3GPP System Architecture Evolution (SAE)
TR 33.822	Security Aspects for Inter-Access Mobility between Non-3GPP and 3GPP Access Network
TR 33.823	Security for Usage of Generic Bootstrapping Architecture (GBA) with a User Equipment (UE) Browser
TR 33.826	Study on Lawful Interception Service Evolution
TR 33.828	IP Multimedia Subsystem (IMS) Media Plane Security
TR 33.829	Extended IP Multimedia Subsystem (IMS) Media Plane Security Features
TR 33.830	Feasibility Study on IMS Firewall Traversal
TR 33.831	Study on Security on Spoofed Call Detection and Prevention (Stage 2)
TR 33.832	Study on IMS Enhanced Spoofed Call Prevention and Detection
TR 33.833	Study on Security Issues to Support Proximity Services
TR 33.838	Study on Protection against Unsolicited Communication for IMS (PUCI)
TR 33.844	Security Study on IP Multimedia Subsystem (IMS) based Peer-To-Peer Content Distribution Services (Stage 2)
TR 33.849	Study on Subscriber Privacy Impact in 3GPP
TR 33.859	Study on the Introduction of Key Hierarchy in Universal Terrestrial Radio Access Network (UTRAN)
TR 33.860	Study on Security Aspects of Cellular Systems with Support for Ultra-Low Complexity and Low Throughput Internet of Things
TS 33.863	Study on Battery Efficient Security for Very Low Throughput Machine Type Communication Devices
TR 33.865	Security Aspects of WLAN Network Selection for 3GPP Terminals
TR 33.868	Study on Security Aspects of Machine-Type Communications (MTC) and Other Mobile Data Applications Communications Enhancements
TR 33.871	Study on Security for Web Real Time Communications (WebRTC) IP Multimedia Subsystem (IMS) Client Access to IMS
TR 33.872	Security Enhancements To Support WebRTC Interworking
TR 33.879	Study on Security Enhancements for Mission Critical Push To Talk (MCPTT) over LTE
TR 33.888	Study on Security Issues to Support Group Communication System Enablers (GCSE) for LTE
TR 33.889	Feasibility Study on Security Aspects of Machine-Type Communications Enhancements to Facilitate Communications with Packet Data Networks and Applications

(Continued)

Table 1.5   (Continued)

Document	Description
TR 33.895	Study on Security Aspects of Integration of Single Sign-On (SSO) Frameworks with 3GPP Operator-Controlled Resources and Mechanisms
TS 33.897	Study on Isolated E-UTRAN Operation for Public Safety; Security Aspects
TR 33.901	Criteria for Cryptographic Algorithm Design Process
TR 33.902	Formal Analysis of the 3G Authentication Protocol
TR 33.905	Recommendations for Trusted Open Platforms
TR 33.908	3G Security; General Report on the Design, Specification and Evaluation of 3GPP Standard Confidentiality and Integrity Algorithms
TR 33.909	3G Security; Report on the Design and Evaluation of the MILENAGE Algorithm Set; Deliverable 5: An Example Algorithm for the 3GPP Authentication and Key Generation Functions
TR 33.916	Security Assurance Scheme for 3GPP Network Products for 3GPP Network Product Classes
TR 33.918	Generic Authentication Architecture (GAA); Early Implementation of Hypertext Transfer Protocol over Transport Layer Security (HTTPS) Connection between a Universal Integrated Circuit Card (UICC) and a Network Application Function (NAF)
TR 33.919	3G Security; Generic Authentication Architecture (GAA); System Description
TR 33.920	SIM Card Based Generic Bootstrapping Architecture (GBA); Early Implementation Feature
TR 33.924	Identity Management and 3GPP Security Interworking; Identity Management and Generic Authentication Architecture (GAA) Interworking
TR 33.937	Study of Mechanisms for Protection against Unsolicited Communication for IMS (PUCI)
TR 33.969	Study on Security Aspects of Public Warning System (PWS)
TR 33.978	Security Aspects of Early IP Multimedia Subsystem (IMS)
TR 33.980	Liberty Alliance and 3GPP Security Interworking; Interworking of Liberty Alliance Identity Federation Framework (ID-FF), Identity Web Services Framework (ID-WSF) and Generic Authentication Architecture (GAA)
TR 33.995	Study on Security Aspects of Integration of Single Sign-On (SSO) Frameworks with 3GPP Operator-Controlled Resources and Mechanisms

card that represented the GSM application. Along with the 3G system development, the SIM was enhanced and took the term of UICC. It refers to a physical card that contains logical functionality as described in 3GPP TS 31.101. Furthermore, the USIM refers to the 3G application residing on the UICC as defined in 3GPP TS 31.102. In addition to the 3G-USIM application, the UICC is capable of storing various other applications, among which the SIM, or 2G-SIM, is used for GSM, and the IMS SIM (ISIM) is used for the IP Multimedia Subsystem (IMS) of the LTE/LTE-A. With the further development of the 3GPP system, the UICC still continues to be the trusted element for the LTE/SAE user domain, including newly evolved applications and security. In fact, the importance of the role of the UICC is increasing as of the Release 8 features.

The following sections summarize some of the key aspects in the development path of the SIM/UICC up to the 3GPP Release 12.

1.3.7.1 Release 8

The USIM in the LTE phase is defined in 3GPP TS 31.102 as is the case for the 3GPP 3G. Along with Release 8, the earlier USIM features continue to support LTE. Furthermore, there are now new features that better take into consideration the non-3GPP radio access, Mobility Management (MM) and emergency cases. Unlike previous releases, Release 8 dictates that the USIM authenticates and secures access to the Evolved Packet Core (EPC) for non-3GPP access systems as a result of the USIM support for advanced network selection mechanisms. The additional radio access networks are Code Division Multiple Access (CDMA) 2000 and High Rate Packet Data (HRPD) by enabling USIM Public Land Mobile Network (PLMN) lists for the selection of roaming between CDMA, UMTS and LTE.

As the secure element based on the USIM is highly protected, it is further enhanced for storing the LTE/ System Architecture Evolution (SAE) MM parameters. It also stores important data, such as location and EPS security context instead of the user equipment. Furthermore, Release 8 enhances personal safety aspects by allowing the USIM to store a user's In Case of Emergency (ICE) information, like allergies, blood type and emergency contact information that the emergency personnel can retrieve even if the user is not able to answer or establish a call. The 3GPP has also included a USIM storage for eCall parameters. As soon as it is activated, the eCall is able to establish a voice call (either manually or by relying on the vehicle's sensors) with emergency personnel services and sends key information like the user's location and vehicle identification data in order to speed up the emergency service's action time.

1.3.7.1.1 Release 8 Toolkit features
The 3GPP specification TS 31.111 defines improvements for the Toolkit features of Release 8. The Release 8 Toolkit supports NFC as the contactless interface is integrated with the UICC which allows the UICC applications to proactively trigger contactless interfaces. The Toolkit also supports the triggering of OMA-DS (Data Synchronization) and OMA-DM (Device Management) sessions for easier device support and data synchronization. The OMA-DS protocol (SyncML DS) is a data-type and transport-agnostic protocol for any transport, such as Bluetooth and any mobile radio access, and can synchronize any data type such as contacts, calendar, files and Java credit card records [92].

Furthermore, the Toolkit takes into account devices with limited capabilities like M2M devices and data cards which might not contain a normal user interface such as a screen and keypad. The Toolkit event informs the UICC application about network rejection, e.g., in the case of failure in a registration attempt. Operators may use this functionality for issue solving, e.g., radio outage areas. Finally, the Toolkit supports UICC proactive commands for transferring radio signal strength measurement reports from eNodeB.

1.3.7.1.2 Contact manager
Release 8 defines a multimedia phone book for USIM as stated in 3GPP TS 31.220, based on OMA-DS and its corresponding JavaCard API (3GPP TS 31.221).

1.3.7.1.3 Remote management
Along with the enhanced ability to create IP sessions, Release 8 defines evolved remote management as described in 3GPP TS 31.115 and TS 31.116. It provides capability to utilize remote application and file management over a Card Application Toolkit Transport Protocol

(CAT_TP) link based on a BIP session between the UE and USIM. Also, the related algorithms have been renewed for updating UICC from the Data Encryption Standard (DES) to the Advanced Encryption Standard (AES) algorithm which provides additional flexibility and security.

1.3.7.1.4 Application management with third parties
Release 8 provides means for confidential application management of UICC for third parties, thus providing possibility for hosting confidential third-party applications. This is beneficial for managing and ensuring feasible business models in Mobile Virtual Network Operator (MVNO) and mobile payment environments, e.g., via memory rental of UICC in such a way that this memory and respective contents can be managed remotely and securely by the third party.

1.3.7.1.5 Secure channel
Release 8 provides a trusted and secure communications channel for Universal Serial Bus (USB) and ISO interfaces between UE and UICC, or between applications that are located on UICC and UE. For the overall Secure Channel Protocols (SCPs), please refer to section 4.11.1.

1.3.7.1.6 Other aspects
Among other topics, the 3GPP CS6 group has worked on the following key items that have an impact on UICC.

The home operator and the user are allowed to prioritize certain Radio Access Technologies (RATs) in the PLMN selection via the *support of multi-system PLMN selection*. The added networks are E-UTRAN, CDMA2000-1xRTT and CDMA2000 HRPD. The user may thus select certain operators in the list that are favoured over others. This RAT prioritization functionality can apply to the RATs supported and available in the Home PLMN (HPLMN) as well in the Visited PLMN (VPLMN). The *storage of Electric Power System (EPS) MM parameters on the USIM* provides the possibility to optimize the initial network selection when on an LTE network by provisioning of the EPS MM (EMM) parameters on the USIM. *The provisioning for Home (e)NodeB* refers to the possibility for the UE to maintain a list of allowed Closed Subscriber Group (CSG) identities, and it will be possible to store them in the USIM in the future. The *support of LTE in the EF (Elementary File) Operator PLMN List* is an item that adds the possibility to assign an operator name tag for LTE-based networks to the operator's PLMN list. The *support of EPS in the USIM Application Toolkit (USAT); extension of Call Control* is an item that provides extension of the procedures and commands for the call control when an EPS Packet Data Network (PDN) connection is requested.

1.3.7.2 Release 9

3GPP Release 9 introduces enhancements for M2M and Home eNodeB (femtocell) environments of LTE which have also been taken into account in the UICC development. As the femtocell deployment scenario is highly relevant in paving the way for LTE-Advanced (ITU-compliant 4G), the importance of the respective USIM is increasing for managing provisioning information to the femtocell. Being a highly reliable security element, access to femtocells is an ideal option when relying on the USIM that is based on the CSG list controlled by the operator or user.

The femtocell is in practice a consumer's plug-in element, meaning that it can be deployed by users without any operator control. The challenge of the femtocell is that its location cannot be predicted beforehand by the respective MNO and thus the respective radio network optimization is challenging. The location discovery of femtocells could be provided via enhanced toolkit commands, which are under consideration at 3GPP. Operators could thus use this information for possibly adjusting network services within that area.

The upcoming releases will develop and capitalize on the IP layer for UICC Remote Application Management (RAM) over HTTP or HTTPS. The network can also send a push message to the UICC to initiate a communication using the TCP protocol. Also the M2M dedicated form factor for the UICC is one of the development topics for extending the requirements beyond the currently defined temperatures and mechanics. Yet another discussion topic is the full IP UICC integration to the IP-based UE along with services over Ethernet Emulation Mode (EEM) and USB, as well as the UICC capability to register on multicast-based services.

The main changes of 3GPP CS6 work in Release 9 are related to CSG management. The following lists key changes. The UE shall contain both independent lists: (a) a list of allowed CSG identities controlled by both the operator and the user; and (b) an operator-controlled list of allowed CSG identities. It shall be possible to store both lists on the USIM, and the allowed CSG list on the USIM may be inhibited. It shall be possible to store the CSG type in textual and/or graphical format which has been a Release 8 unfulfilled requirement. It shall also be possible to provide UICC applications with a list of CSG identities available for selection, and to inform UICC applications when selecting and leaving a CSG cell.

Other CS6 key items that impact the UICC in Release 9 are the following. *Introduction of operator controlled CSG list for H(e)NodeB* is a new operator-controlled CSG list. *Correction of controlled CSG list for H(e)NodeB* presents correction of access conditions, names and indicators. *Introduction of an indicator for inhibition of the allowed CSG list* refers to the item based on home operator preferences, and the use of the allowed CSG list on the USIM that may be inhibited.

Furthermore, the following Card Application Toolkit (CAT) commands have been extended. *Terminal Profile* refers to the support of CSG cell discovery and cell selection event. *Provide Local Information* refers to the support of the CSG list and corresponding H(e)NB names of surrounding CSG cells. *Terminal Response* refers to the inclusion of the CSG ID list identifier. There is also a definition produced for the *CSG cell selection* event, and new 'q' class is added to state the support of *Physical Layer Identifier (PLI)* in CSG cell discovery and *event download* in CSG cell selection.

1.3.7.3   Release 10

1.3.7.3.1   *Key enhancements*
The following summarizes the key highlights of Release 10 enhancements for the UICC. (a) CAT over AT commands and encapsulated CAT (eCAT); (b) UICC access to IMS: IMS Application Reference ID (IARI) of UICC-based applications; Session Initiation Protocol (SIP) session management on the handset; BIP for IMS; (c) USIM for Home(e)NBs (femtocells) is optional for authentication of the hosting party; (d) USIM on Relay Nodes (RNs) has two new applications, i.e., USIM-INI and USIM-RN, and Binding of RN to USIM via Secure Channel (TS 102 484).

Furthermore, the most important 3GPP specifications that impact UICC in Release 10 are TS 31.101 (Physical and Logical Characteristics of the UICC-Terminal Interface), TS 31.102 (Characteristics of the USIM Application), TS 31.103 (Characteristics of the ISIM Application), TS 31.111 (USAT), and TS 31.130 (USIM API for Java Card).

1.3.7.3.2 Main features in Release 10

The RN will have two flavours of USIM, which are USIM-INI and USIM-RN. The USIM-INI is used for the link establishment (initialization) whereas the USIM-RN is used for authenticating with the Core Network (CN) and offering RN services. These have an impact on the 3GPP TS 31.101 and TS 31.102. The *Smartcard Web Server (SCWS)* will have an introduction of an SCWS launch functionality. A dedicated EF for the Mobile Equipment (ME) is defined to check which icon and text should be indicated in the phone's main menu in order to access the SCWS on the UICC. The item called *Communication Control by USIM for IMS* is similar to the Call Control but it is meant for the IMS Services. There is also an addition of *USAT facility control* which refers to the USAT commands over the AT modem commands support. The *CSG lists display Control Management by USIM* refers to the possibility to store and manage the CSG lists by USIM with user interaction. The *UICC access to IMS* means an IARI list and SIP Push to UICC Apps [93]. Finally, the *NAS Parameters storage update and corrections* refers to the storage rate of security context and non-access stratum configuration storage.

1.3.7.3.3 Other Release 10 features

The following lists other relevant Release 10 features impacting SIM/UICC. *The SIP Push*, i.e., the UICC access to IMS refers to the item that only applies if class 'e' (BIP) and 't' (UICC Access to IMS) are supported by the ME. In this solution, the IMS access is managed by the ME in the following cases: (a) PDN context establishment; (b) registration and data flow; (c) USIM and ISIM use is possible. Furthermore, the USIM or ISIM include a parameter called EF_UICCIARI, which refers to an EF that contains a list of IMS Application Reference Identifiers associated with active applications installed on the UICC that are included in the SIP Register.

Other items resulting in a modified flow chart are the *discovery of the UICC's IARI and IMS registration*, and *notification of incoming IMS data*. Furthermore, the OTA HTTPS and SIP Push are defined. In this solution, the SIP Push can be used to trigger the HTTPS client in the UICC, and IMS SIP dialogue can be used for indicating the AdminAgent in the UICC the information for the HTTPS access to the OTA server.

1.3.7.4 Release 11

Release 11 includes the following enhancements. The *4FF* (Fourth Form Factor, i.e., nano-SIM) which has been driven by Apple and first used in the iPhone 5. The *secure channel for CAT* has been defined to protect all CAT communication; this item has been inspired by a special-use case for the MVNO environment. The *secure channel for eCAT* has been defined to protect encapsulated CAT exchanged with an eCAT client which could be the Trusted Execution Environment (TEE) in the ME or an endpoint outside the ME; the goal is to enable a trusted user nterface based on CAT. The *adoption of key establishment scenario 3* by GlobalPlatform is an item to establish initial keysets in a service provider's SD; this provides AES-level security based on elliptic curve cryptography. The *Forced Refresh* is an item meant for cases where the terminal keeps a data connection always on; it is also required for the profile management of the embedded UICC (eUICC). Finally, the *API for HTTPS* has been defined to allow an applet to use HTTPS communication.

1.3.7.5   Release 12

As for the 3GPP Release 12 highlights, the Temporary User Authentication Key (TUAK) is a second standardized authentication algorithm next to Milenage. It is based on the Keccak Hash algorithm, and is the winner of the SHA-3 contest. It is specified as of Release 12 in the 3GPP TS 35.231, with the aim of using it in the eUICC.

Another key topic of Release 12 standardization is the retrieval of the DNS server addresses from the network. This item is similar to the Dynamic Host Configuration Protocol (DHCP) on a personal computer which allows applications based on domain names only. All resolutions are thus provided by the network which means that the reconfiguration has no impact on applets.

1.3.8   The 3rd Generation Partnership Project 2 (3GPP2)

The 3GPP2 is a 3G telecommunications specifications project meant for the North American and Asian markets. It works in cooperation with 3GPP, and its aim is to develop global specifications for ANSI/TIA/EIA-41 -based 3G cellular radio telecommunications evolution and global specifications for the Radio Transmission Technologies (RTTs) supported by ANSI/TIA/EIA-41.

In addition to the communications with the 3GPP, the 3GPP2 is a collaborative effort between five officially recognized standardization development organizations, or Organizational Partners (OPs): Japanese Association of Radio Industries and Businesses (ARIB), TTC, Chinese China Communications Standards Association (CCSA), North American Telecommunications Industry Association (TIA) and Korean Telecommunications Technology Association (TTA). The Technical Specification Groups (TSGs) of 3GPP2 are TSG-A (Access Network Interfaces), TSG-C (CDMA2000), TSG-S (Services and Systems Aspects), and TSG-X (Core Networks). As is the case for freely downloadable specifications and reports of 3GPP, each 3GPP2 TSG has a dedicated section on the website which can be accessed for the downloading of specifications and reports. More information about the 3GPP2 can be found in Ref. [89].

1.3.9   The GlobalPlatform

The GlobalPlatform is an industry organization which consists of several committees for the standardization of subscriber cards and systems [55]. The GlobalPlatform is an internationally recognized non-profit association with the aim to establish, maintain and promote adoption of interoperable infrastructure standards for smartcards, devices and systems. The focus of the GlobalPlatform is to simplify and speed up the development, deployment and management of interoperable security applications and solutions. The GlobalPlatform standards establish mechanisms and policies for enabling secure communications, and many banks have adopted them at a global level for cryptographic data loading based on JavaCard.

One of the focus areas of the GlobalPlatform is secure deployment and management of multiple applications on secure chip technology by using standardized infrastructure, providing service providers means to develop digital services and deploy them uniformly across different devices and channels. This interoperable approach in security and privacy parameters enables dynamic combinations of secure and non-secure services from multiple providers on the same device. The GlobalPlatform is thus an international industry standard for trusted end-to-end secure deployment and management solutions, aiming for global adoption across finance, mobile and telecommunications, government, premium content, automotive, healthcare, retail and transit sectors via interoperability and scalability of application deployment and management through its secure chip technology open compliance program. the GlobalPlatform has over 120 members working on the alignment with existing and emerging market requirements. The GlobalPlatform's interoperability solves service provider issues in ensuring compatibility with different security architectures and APIs which are meant for secure access services offered by devices. The aim of the GlobalPlatform is thus to eliminate compatibility and scalability issues via standardized infrastructure and APIs for the management of applications on secure chips compatible with connected devices.

A concrete example of the recent work by the GlobalPlatform is the development of the interoperable Subscription Management (SubMan) standard. Some other key areas of GlobalPlatform's work are the following.

First, service providers' services rely on a backend server which is under their control. This solution ensures the respective end-point's security level. The GlobalPlatform further gives possibility for service providers to establish a second trusted and secure endpoint, which is a secure chip of the end user's device. This second trusted endpoint provides the service provider and the end-user with end-to-end security.

Secondly, the GlobalPlatform defines two secure components based on the SE and TEE. In addition to protecting service providers and consumers from external hackers, these secure components prevent competing service providers or the consumer from accessing sensitive application information. Each service provider may load secret keys into the SE to protect its own applications.

Thirdly, the GlobalPlatform messaging technology standardizes messaging to ensure the correct utilization of data and formats to load and provision a service into a secure component.

1.3.10   The SIMalliance

The SIMalliance is a global, non-profit industry association with the aim to simplify SE implementation and thus to drive the creation, deployment and management of secure mobile services. The SIMalliance also promotes the useful role of the SE in delivering secure mobile applications and services across all devices with an access to wireless networks. It also identifies and addresses SE-related technical issues, clarifies and recommends existing technical standards relevant to SE implementation, and promotes an open SE ecosystem to facilitate and accelerate delivery of secure mobile applications globally [56]. As is the case with the GlobalPlatform and GSMA, the SIMalliance is also involved in the development of interoperable subscription management solutions.

1.3.11   The Smartcard Alliance

The Smartcard Alliance is a multi-industry association focusing on the information sharing, adoption, use and application of smartcard technology. The Smartcard Alliance coordinates projects such as education programmes, market research, advocacy, industry relations and open forums and eases the networking and innovation of its members and cooperating parties. The Smartcard Alliance acts thus as a centralized industry interface for smartcard technology, and follows the impact and value of smartcards in the United States and Latin America. The worldwide coverage of members include representatives from financial, government, enterprise, transportation, mobile telecommunications, healthcare and retail industries, and therefore has views from issuers and adopters of smartcard technology to understand the implementation of smartcard based systems for secure payments, identification, access and mobile communications [73].

1.3.12   The GSM Association (GSMA)

The GSMA has a long history, going back to the early days of GSM development. Its roots originate from the GSM Memorandum of Understanding (MoU), which was the basis for the agreement of the GSM standards in a form of an association representing the participating operators [58].

Currently, the GSMA represents the interests of mobile operators worldwide. There are about 800 operators and 250 companies involved, including handset and device manufacturers, software companies, equipment providers, Internet companies and other related industry sectors [57]. Some of the concrete key areas of the GSMA are the following.

The first item is the spectrum for mobile broadband. The GSMA is involved in spectrum initiatives to ease the mobile broadband deployments. The GSMA investigates the importance of future availability of affordable, ubiquitous, high-speed mobile broadband services. It can be estimated that the industry is likely to require around 1600–1800 MHz of spectrum to ensure widespread access to mobile broadband services. Also, the data traffic may increase by nearly 10 times by 2019 according to Ref. [70], which is a report authored by GSMA Intelligence, a source of global mobile operator data, analysis and forecasts.

Secondly, the GSMA is involved with public policy. The GSMA contributes, e.g., to the *Mobile Policy Handbook* and policy case studies and is active in areas of capacity building in mobile sector regulation, privacy in the mobile environment, mYouth, energy efficiency, roaming, health, connected society, mobile commerce and economy such as payment, retail and transport, as well as utilities and disaster response.

Thirdly, the GSMA is involved with Network 2020 which is paving the way for the forthcoming 5G, including evolved items such as Voice over LTE (VoLTE), Rich Communications, HD Voice and IP Interconnect.

The fourth key item is Connected Living, which covers a variety of topics such as automotive, health, transport, utilities and trackers. In addition, there are also numerous other activities such as globally recognized events.

Apart from the above-mentioned items, the GSMA actively contributes to subscription management development, and is also involved with the embedded SIM development for the M2M environment. As an example, GSMA eSIM (M2M) Technical Specification v2.0, together with the contribution of the SIMalliance interoperable profile specification v1.0, resulted in the GSMA eSIM (M2M) TS v3.0 in 2015, combined with the ideas of SCP03t-version for the secure channel.

1.3.13   The National Institute of Standards and Technology (NIST)

The US National Institute of Standards and Technology (NIST), together with various industries, develops a voluntary, non-regulatory cyber-security framework to address critical infrastructure and new challenges especially as a result of the IoT and M2M. The NIST Framework was initiated officially in 2014. The NIST Framework development has involved a Presidential Executive Order, the mobile industry and the telecommunications sector. The focus of the framework is thus on the security and privacy for the evolution of M2M and IoT [23,24].

1.3.14   The National Highway Transportation and Safety Administration (NHTSA)

The focus of the NHTSA is on improving safety and mobility on US roadways. One of its topics is wirelessly connected vehicle technology which involves vehicles such as cars and trains, and facilitates the communication of safety and mobility information to each other. The ultimate aim of the initiative is to help save human lives, prevent injuries, ease traffic congestion and improve the environment. The NHTSA is thus remarkably engaged with vehicle cyber security [25].

1.3.15   Other Standardization and Industry Forums

1.3.15.1   The European Conference of Postal and Telecommunications Administrations (CEPT)

CEPT was initiated in 1959. It soon expanded as new members joined, and currently has 48 members, covering nearly all of Europe. Having first represented monopoly-holding postal and telecommunications administrations, CEPT continues to cooperate on commercial, operational, regulatory and technical standardization issues [14]. CEPT activities are coordinated by the permanent European Communications Office (ECO).

   CEPT publishes documents related to requirements for approval, approval authorities, certification bodies and testing laboratories for telecommunications terminal services for attachment to public telecommunications networks or access to public telecommunications services.

1.3.15.2   The Accredited Standards Committee on Telecommunications (T1)

T1 develops telecommunications standards, definitions and technical reports for the needs of the United States. T1 has six Technical Sub-Committees (TSCs), which are administered by the T1 Advisory Group (T1AG). As an example, GSM development is handled by T1P1 (Wireless/Mobile Services and Systems), which is further divided into five sub-groups: T1P1.1 for International Wireless/Mobile Standards Coordination, T1P1.2 for Personal Communications Service Descriptions and Network Architectures, T1P1.3 for Personal Advanced Communications Systems (PACS), T1P1.5 for PCS 1,900, and T1P1.6 for CDMA/TDMA.

1.3.15.3   The American National Standards Institute (ANSI)

The aim of ANSI is to strengthen the position of the United States in the global economy. It also takes care of the important safety and health issues of consumers, as well as the protection of the environment [15]. One of its tasks is to promote the use of the US standards. It also facilitates the adoption of international standards for the US environment if these are seen to be suitable for the needs of the user community. ANSI is the sole representative of the United States at the ISO and the IEC.

ANSI offers an online library with open documents, as well as restricted documents for members only [16].

1.3.15.4   The Association of Radio Industries and Businesses (ARIB)

ARIB is the Japanese standardization body, which has strongly developed 3G definitions of mobile communications, and interacts in cooperation with other bodies developing further stages of mobile communications.

1.3.15.5   The Telecommunications Technology Committee (TTC)

The TTC is another Japanese standardization body, which has been actively developing mobile communications systems. The TTC Standard Summary can be found in Ref. [17].

1.3.16   The EMV Company (EMVCo)

Europay, MasterCard and Visa have formed the EMV Company (EMVCo) and produced the *Integrated Circuit Card Specifications for Payment Systems*, based on the ISO/IEC 7816 standard. The aim of this initiation is to ease the card and system implementation of a stored value system [63].

1.3.17   The Personal Computer/Smartcard (PC/SC)

The PC/SC is an international specification for cards and readers, and is applicable to contact cards. The further evolution of v2.0 also introduces a PIN pad for the card communications. Various Operations System (OS) companies support the PC/SC, and it is currently a common middleware interface for PC logon applications.

1.3.18   The Health Insurance Portability and Accountability Act (HIPAA)

HIPAA is an umbrella for national US standards for the implementation of electronic health transaction systems in a secure way. Some examples of the respective transaction functions include patient claims, enrolment, eligibility, payment and benefit coordination. HIPAA sets requirements for smartcards in this environment to ensure data security and patient privacy.

1.3.19   The Common Criteria (CC)

The CC refers to an international security evaluation framework. The basic idea of the CC is to provide reliable IT product evaluation for the security capabilities. These products include the hardware of the secure integrated cards as well as the software of smartcard operating systems and applications. The CC is thus used for an independent assessment with the result indicating the ability of the product to comply with security standards. As the CC is globally recognized and established, it is especially useful for customers requiring very high security, such as governments that increasingly want the CC certification as a part of their security solution investments. Furthermore, the CC allows the vendors to tackle security solutions based on concrete needs more efficiently while offering a broad set of products. The CC is thus an international standard for computer security certification, including evaluation of information technology products and protection profiles [59].

1.3.20 The Evaluation Assurance Level (EAL)

An important security-related topic for UICC security is compliance with the CC. It refers to standards denoting the EAL from 1 through 7 as summarized in Table 1.6. The SIM/UICC complies with the level 4 EAL which makes it one of the most protected solutions in mobile communications.

The EAL indicates a numeric value that is based on the completion of a CC evaluation, the higher values meaning more enhanced assurance levels. It should be noted that the value of EAL does not explicitly indicate the absolute security level as such but it states at which level the system has been tested to meet specific assurance requirements.

Table 1.6 The EAL classes of CC

Level	Item	Description
EAL 1	Functionality tested	Indicates some confidence in correct operation although the threats are not considered as serious. Evidence that the Target of Evaluation (TOE) functions consistently with the respective documentation, providing useful protection
EAL 2	Structure tested	Applicable in environments where developers users require low or moderate level of independently assured security, e.g., legacy systems
EAL 3	Methodically tested and checked	Applicable in environments where developers/users require a moderate level of independently assured security, with thorough investigation of TOE
EAL 4	Methodically designed, tested and reviewed	Highest level at which it is assumed to be economically feasible to retrofit to an existing product line. Applicable in environments where developers/users require a moderate or high level of independently assured security. Many operating systems belong to this category
EAL 5	Semi-formally designed and tested	Provides developers with maximum assurance from security engineering. Numerous smartcard devices have been evaluated at this level
EAL 6	Semi-formally verified design and tested	Applicable to the development of security TOEs for application in high-risk situations, with additional costs involved but justified by the value of the protected assets
EAL 7	Formally verified design and tested	Applicable to the development of security TOEs for application in extremely high risk situations, with additional costs justified by the value of the protected assets

1.3.21   The Federal Information Processing Standards (FIPS)

FIPS are a set of standards developed by the Computer Security Division within the National Institute of Standards and Technology (NIST). FIPS are designed to protect federal assets, including computer and telecommunications systems. FIPS 140 (1–3) and FIPS 201 apply to smartcard technology and pertain to digital signature standards, advanced encryption standards and security requirements for cryptographic modules.

FIPS 140 (1–3) contains security requirements related to the secure design and implementation of a cryptographic module. The items include cryptographic module specification, cryptographic module ports and interfaces, roles, services and authentication, finite state model, physical security, operational environment, cryptographic key management, Electromagnetic Interference (EMI), Electromagnetic Compatibility (EMC), self-tests, design assurance and mitigation of other attacks.

FIPS 201 specification includes aspects for the multifunction card utilization in US government identity management systems [61].

1.3.22   Biometric Standards

The importance of biometrics as a part of the secure ID environment is increasing steadily. Some examples of the development include the wider utilization of fingerprint and eye iris scanning for identifying legitimate users. Thus, various current secure ID system implementations rely on biometrics as well as smartcards in order to provide increased levels of security and privacy. The following sections summarize some of the respective standards.

1.3.22.1   ANSI-INCITS 358-2002

ANSI-INCITS 358-2002 contains the BioAPI Specification, equivalent to ISO/IEC 19784-1. The BioAPI defines a generic biometric authentication model that fits to all biometric technologies. Among other definitions, it includes the enrolment, verification and identification, and database interface. The latter offers means for Biometric Service Providers (BSPs) to manage the involved device. The BioAPI framework has been ported to various operating systems.

1.3.22.2   ANSI-INCITS 398

ANSI-INCITS 398 defines the Common Biometric Exchange Formats Framework (CBEFF), which is equivalent to ISO/IEC 19785-1. The CBEFF defines data elements that support biometric technologies and exchange data. It provides interoperability for applications based on biometrics and vendor-independent systems.

1.3.22.3   Other ANSI-INCITS and ISO Standards

The following list summarizes other ANSI-INCITS biometric data format interchange standards that specify the data record interchange format for storing, recording and transmitting biometric sample information within a CBEFF-defined environment [60].

- ANSI-INCITS 377-2004: Finger pattern data interchange format.
- ANSI-INCITS 378-2004: Finger minutiae format for data interchange.
- ANSI-INCITS 379-2004: Iris interchange format.
- ANSI-INCITS 381-2004: Finger image-based data interchange format.
- ANSI-INCITS 385-2004: Face recognition format for data interchange.
- ANSI-INCITS 395-2005: Signature/sign image-based interchange format.
- ANSI-INCITS 396-2004: Hand geometry interchange format.

1.3.22.4 ISO/IEC 19794

ISO/IEC 19794 contains biometric data interchange formats including various parts for, e.g., framework, finger minutiae data, finger pattern spectral data, finger image data, face image data, iris image data, signature/sign time series data, finger pattern skeletal data and vascular image data.

1.3.23   Other Related Entities

The International Civil Aviation Organization (ICAO) provides guidance on the standardization and specifications for Machine Readable Travel Documents (MRTDs) such as passports and visas. ICAO has also published specifications for electronic passports based on contactless smartcards.

MiFare is a technology involving ticketing applications, which has a major part of the total contactless transit ticketing card market. Other variants in this domain include, e.g., Cipurse [69].

The Payment Card Industry (PCI) Security Standards include the Data Security Standard (PCI DSS), Payment Application Data Security Standard (PA-DSS) and PIN Transaction Security (PTS) [65].

The NFC Forum is an NFC industry association. It promotes specification and use of NFC short-range wireless interaction in consumer electronics, mobile devices and PCs [64].

The other documents that are related in one or another way with wireless security include Information Technology Security Evaluation Criteria (ITSEC) security standards [67] which have been largely replaced by the CC, as well as industry initiatives such as MULTOS [66] and the Open Card Framework (OCF) [68].

For the IoT environment, one of the relevant parties is the IoT Security Foundation [82].

1.4   Wireless Security Principles

1.4.1   General

The security and safety in wireless environment do not differ too much from wired set ups. The current smart devices are actually kinds of powerful miniature computers, so basically the executable software may have the very same vulnerabilities that we have seen in the personal computer environment. The official app store concept of testing and pre-authorization aims to minimize the potential threats with malicious code hidden in the applications, but as indicated in various instances with all the operating systems, this is not bullet-proof.

As indicated in Ref. [27], typical attacks from outside the networks are not the only issue, there also exist concrete new forms of threats related to consumer devices via embedded malware. According to this reference, there have been incidents of malware-infected consumer hardware and software products, and even tainted peripheral devices. Some examples of pre-loaded malware environments include products like USB sticks, microchips, cameras, battery chargers, digital photo frames, webcams, printers, cell phones, motherboards, system boards and hard drives.

1.4.2   Regulation

The security assurance for wireless networks as well as for the fixed environment is largely related to the protection of user devices and network elements. Some of the integral tasks of the assurance include the protection of the networks, devices and applications. A systematic and controlled way to execute the protection via security architecture design already in an early phase (standardization, research and prototyping) is highly recommended, as well as in all further phases like network planning, deployment, installation, and configuration. As a useful tool for all this is the security process which applies to the testing, certification and other security phases.

These tasks are done by the involving parties such as device manufacturers and mobile network operators based on the available tools and processes. The operators may and typically will create specific requirements that the equipment manufacturers must comply with. Additional help in this can be provided by the regulation with the respective policies.

One example of the fulfilment of regulatory policies and rules imposed is NERC CIP 5 which refers to CIP V5 Transition Program in the United States. It is designed to protect the bulk power system against cyber-security compromises that without proper protection could lead to misbehaviour or instability. This critical infrastructure protection cyber-security standard set represents the progress in mitigating cyber risks to the bulk power system which, in fact, can be an Achilles' heel for the wireless communication networks as for the remote base station powering.

1.4.3   Security Architectures

There are various wireless systems' security architectures outlined throughout this book. The security can be related to the internal protection mechanisms of the networks such as authentication, authorization and encryption of the communication, as well as the supporting functions such as subscription management which requires secure data transfer in a subscription's profile activation and – as supported by the very latest technologies – more advanced data transfer such as the complete SIM/UICC card OS download in a protected way.

1.4.4   Algorithms and Security Principles

Crypto techniques have improved considerably over time, originating from such ancient methods for hidden writing (steganography) documented by Johannes Trithemius in 1499 (and published in 1606), leading gradually to the most modern methods for actually encrypting the contents. In modern mobile communications, it is logically challenging to apply steganography even if the basic principle is that the encrypted text is typically easy to spot and hard to make plain, whereas the text hidden by steganography is hard to find but relatively easy to interpret once spotted. If not exactly in the mobile communications, modern steganographic techniques are used, e.g., for manipulating pixels of visual contents such as images by applying very small changes to the values that define the colour of the pixel. As an example, a 24-bit image pixel is based on information on the grade of red, green and blue colours. A small deviation to each pixel value modifies the colour but human eye cannot distinguish the difference from the original. The unscrambling of the contents from such an image requires the original reference image to reveal the additional information hidden in the pixel value altering. These techniques are typically called least significant bit insertions for transporting secret messages.

As for standardized mobile communications systems, the steganographic methods are not practical for the encryption of the communications, but they may be used by end-users on the application level. Instead, the algorithms that are applied to the modern mobile communications environment consist of both non-public and public variants of encryption. In general, it is thought that the publicly exposed algorithms provide the most efficient protection as they have been under the hardest and widest testing, and any possible remaining flaws have been encountered more efficiently simply due to the large size of the test community compared to the very limited resources of non-public variants. As an example, the original GSM A5/1 algorithm for encrypting the radio interface communications was kept confidential, which was thought to protect against attacks, but, at the same time, it limited the number of experts for proofing the adequate security level. This non-public approach may be the reason for the eventual exposure of the weaknesses of the algorithm to attacks, including the easier resolution of the contents by knowing the bit-by-bit form of the encrypted initialization message. Nevertheless, regardless of the public or non-public approach, the security of the algorithms needs to be revised every now and then as code-breaking techniques evolve and computer processing power increases.

The basic categorization of the current algorithms can be divided into symmetric and asymmetric variants. The symmetric approach has been applied in the securing of messages for the last two thousand years or so, including modern mobile communications systems like GSM and UMTS. One example of symmetric encryption is the AES algorithm. The drawback of symmetric encryption is that the same key is used for encrypting and decrypting the message, and thus the key delivery in a secure manner may cause vulnerabilities in this method. Code makers or code breakers have been, one at the time, leading the advances in this era.

The asymmetric approach refers to the security system that uses different keys for encrypting and decrypting the messages. One example of asymmetric ciphering techniques is RSA. In practice, the current asymmetric encoding solutions rely on the modular functions that are based on prime numbers and a pair of public and private keys [2]. The mathematical principles of prime numbers provide the possibility to make the encryption in such a way that the reverse analysis for finding the used prime numbers and for eavesdropping is not feasible in practice for sufficiently big numbers, whereas the legitimate receiving party can decrypt the message by knowing the pair of his/her own public key (which the sending party utilized as a part of the encryption) and own private key (which others are not aware of). Thus, once the sending party encrypts the message, they cannot decrypt it any more as it only opens in a feasible manner with the private key paired with the public key.

Table 1.7 lists and compares some of the most relevant ciphering techniques used in mobile and wireless communications today.

Table 1.7 Comparison of ciphering techniques relevant for mobile communications

Algorithm	Principle	Examples
AES	Block cipher, with block size of 128 bits and key length of 128, 192 and 256 bits. Symmetric algorithm. Includes key addition, byte substitution and diffusion layers	Provides efficient SW implementation. Well suited for 8-bit processors such as smartcards, but not efficient on 32- or 64-bit machines. Optimized via look-up tables (T-Box). AES forms part of numerous open standards like IPSec and TLS, and is a mandatory encryption algorithm for US government applications. More information: [83,84]
DES	Block cipher encrypting blocks of 64 bits with a key of 56 bits. Symmetric cipher (same key for encryption and decryption). Iterative algorithm with 16 rounds and respective sub-keys derived from main key	Very efficient in HW such as Field Programmable Gate Array (FPGA) and Application-Specific Integrated Circuit (ASIC). Suitable in very small devices with reduced space, such as RFID tags and low-cost smartcards (high-volume public transportation payment tickets). Vulnerable to differential and linear crypto-analysis attacks excluding the S-box. More information: Ref. [88]
Triple-DES (3DES)	Symmetric cipher and alternative to DES, consisting of three subsequent DES encryptions with three different keys, respectively	Resistant to brute-force attacks and analytical attacks. The advantage of 3DES is the single DES encryption if the keys are the same which benefits the support of legacy systems
RSA	Asymmetric cipher. Encryption and decryption based on integer ring and modular computations of bit strings via public and private keys	Most widely used asymmetric cryptographic scheme. Typically used for small pieces of data such as key transport, and digital signatures, e.g., for digital certificates on the Internet. Much slower than AES so RSA does not replace symmetric ciphers such as AES
Elliptic curve/ discrete logarithm schemes	Elliptic curves are used, e.g., for encryption, digital signatures and pseudo-random generators	Elliptic curve is one of the latest additions for mobile communications security solutions. More information can be found in Ref. [85], and information about related Brainpool can be found in Ref. [86]
Hash functions	Hash functions compute a digest of a message which is a short, fixed-length bit string. Does not have a key.	Hash function can be used as a fingerprint of the message and is thus suitable for digital signature schemes and message authentication. Can be used for storing password hashes or key derivation [87]
MILENAGE	A new algorithm taken into use in mobile communications, as defined in 3GPP TS 35.205, TS 35.206, TS 35.207, TS 35.208 3G security specifications	3GPP TR 35.909, Release 11 presents an example algorithm set for 3GPP authentication and key generation functions f1, f1*, f2, f3, f4, f5 and f5*, including algorithm specification, implementers' test data, design conformance test data as well as results of the design and evaluation
TUAK	A new algorithm as described in 3GPP TS 35.231, TS 35.232 and TS 35.233 (Release 12).	The specifications include the algorithm description, implementers' test data and design conformance test data

More background information about crypto techniques in general and in the mobile communications environment can be found, e.g., in references [2,20–23,27,71,72].

1.5 Focus and Contents of the Book

This book summarizes key aspects of the wireless security field, and is aimed to be especially useful for network operators and mobile/IoT device manufacturers, as well as for companies and organizations that are involved in mobile communications security like smartcard providers, embedded security element and 'traditional' SIM card manufacturers, and service providers. This book also clarifies the most important current and future solutions in a practical way for telecommunications students in order to map the theories with industry trends. The primary audience of the book is the wireless security industry, but the book has been aimed at all other interested parties, including personnel of regulators, the gaming industry, defence forces and the academic environment, to explain the overall picture of recent and expected future security solutions.

This book also aims to work as a useful guideline for personnel involved with standardization groups and alliances like the GlobalPlatform, GSMA, SIMalliance, MiFare, Cipurse, 3GPP, 3GPP2 and ETSI.

The first two chapters of the book form the **introductory module** containing a description of the wireless security environment and IoT. They describe the essential basics of the mobile and wireless systems relevant for understanding security aspects such as user authentication, authorization and securing the radio interface, as well as the focus and role of standardization bodies, the development of systems for consumers as well as M2M communications. The most important security algorithms are presented, with further references to the specialized literature for those wanting to study the principles in more detail.

The **second module** presents detailed secure solutions in the wireless environment, including contact and contactless smartcards, secure elements and evolved systems that are useful for securing communications apart from the hardware-based solutions such as cloud payment. Also the current and expected development of subscription management is described based on the most updated information from the involving standardization bodies and industry forums. The second module outlines the overall development of IoT with respective waves of the M2M solutions and mobile connectivity, and discusses the connected society concept as well as other industry forums, alliance and international standard body initiations. Novelty and expected future solutions are discussed, including wearable devices, household appliances, industry solutions and self-driving cars. One of the vast IoT device bases is related to utilities, so the contributing role and technologies of them are discussed with the ways utilities are relying on wireless technologies, including their role in the electric domain, mobility and smart grid applications.

An important part of module 2 is dedicated to smartcards, so reasons are presented on why smartcards are still a useful anchor for security in the era of the IoT. Along with the development of smartcard technologies and parallel solutions, this module also discusses the modifications needed for smartcards to support the IoT, and presents available or future alternatives. Technical descriptions of contact cards and contactless cards are given, including standards, current solutions, form factors, electrical and mechanical characteristics, use cases for, e.g., wireless payment and access systems, NFC and other wireless techniques. The payment and access environment is described by giving examples such as the EMVCo concept and other banking systems, e-commerce, transport and access systems.

The second module also describes wireless security platforms and functionality by presenting justifications of why each specific security mechanism is relevant. The secure element of both HW and SW-based solutions is described, and secure protocols are also discussed. Based on smartcard principles, this module also details the telecoms environment's SIM/UICC card and embedded SIM/UICC. The SIM-based Over-the-Air (OTA) techniques are presented for initiation of the subscription, subscription life-time management, as well as remote file management and application management. This leads to a more complete environment of subscription management for consumer subscriptions and M2M devices, which is detailed based on the latest knowledge from the industry and standardization fields.

Furthermore, alternative secure solutions for SIM are described, like TEE, cloud and Host Card Emulation (HCE), including the functionality of tokenization. Life-cycle management is also outlined, and the benefits and challenges of subscription management are discussed. The evolution of device types and costs for selection of techno-economically optimal subscription management are also considered, along with potential issues and their solutions for subscription management.

The **third module** details typical security threats in the wireless environment, and explains how to monitor and enhance protection against malicious attacks. This module also outlines future aspects of mobile security. More specifically, this module presents concerns of wireless security mechanisms, potential security holes including the role of human errors and flaws in the network, user equipment, applications, communications, signalling and production. Some attack types are discussed such as eavesdropping, overloading and RF-attacks. Along with the increasing importance of the IoT environment, module 3 outlines impacts of wireless security on the utilities domains and applications. Feasible protection techniques are thus discussed as well as monitoring techniques such as deep packet inspection, virus protection and legal interception.

The future section of the third module discusses the forthcoming wireless environment by summarizing trends, threats and solutions for the security mechanisms in order to avoid performance degradation of wireless technologies along with extensive data transmission. There is also a description of the evolution of sensors networks and their security. Finally, the third module introduces mobile communications systems of 5G and beyond, which is still a set of items under preparation for standardization, as well as the security challenges of future wireless technologies.

Figure 1.1 presents the main level contents of this book to ease the navigation between the modules. The modules and chapters are independent from each other so they can be read through in any preferred order. Nevertheless, for studying the area from scratch, it is recommended to get familiar with the topic gradually in the presented order as this makes it easier to understand the later contents.

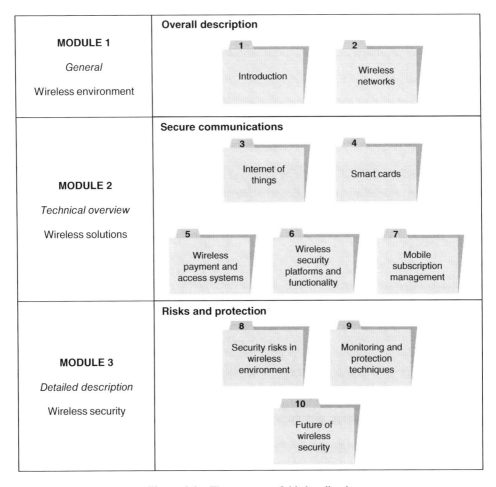

Figure 1.1   The contents of this handbook

References

[1] *Wired.* Hacker spoofs cell phone tower to intercept calls, October 2010. http://www.wired.com/2010/07/
    intercepting-cell-phone-calls/ (accessed 13 December 2014).

[2] Simon Singh. *The Code Book: The Evolution of Secrecy from Mary Queen of Scots to Quantum Cryptography.*
    Anchor Books, New York, 1999.

[3] ITU, 2015. http://www.itu.int/en/about/Pages/overview.aspx (accessed 4 July 2015).

[4] ITU-T Recommendations, 2015. http://www.itu.int/rec/T-REC/e (accessed 4 July 2014).

[5] ITU-R Recommendations, 2015. http://www.itu.int/pub/R-REC (accessed 4 July 2014).

[6] IEEE Publications, 2015. http://www.ieee.org/publications_standards/index.html (accessed 4 July 2015).

[7] IEEE Standards, 2015. http://standards.ieee.org/about/get/index.html (accessed 4 July 2015).

[8] IEEE Standards Association. IEEE standards activities in the network and information security (NIS) space. 19
    June 2013. 4 p.

[9] IETF RFC search page, 2015. http://www.rfc-editor.org/rfcsearch.html (accessed 4 July 2015).

[10] IETDF, Official Internet Protocol Standards, 2015. http://www.rfc-editor.org/rfcxx00.html (accessed 4 July
    2015).

[11] IETF RFC 1677. *The Tao of IETF – A Novice's Guide to the Internet Engineering Task Force,* 2015.

[12] IETF RFC List, 2015. http://www.ietf.org/rfc.html (accessed 4 July 2015).

[13] IETF Security Area, 2015. https://tools.ietf.org/area/sec/trac/wiki (accessed 4 July 2015).

[14] CEPT, 2015. http://www.cept.org (accessed 4 July 2014).

[15] ANSI, 2015. http://www.ansi.org/about_ansi/overview/overview.aspx?menuid=1 (accessed 4 July 2014).

[16] ANSI Standards (Restricted Area), 2015. http://www.ansi.org/library/overview.aspx?menuid=11 (accessed 4 July 2014).

[17] TTC, 2015. http://www.ttc.org.jp/cgi/summarydb/index.html (accessed 4 July 2015).

[18] 3GPP, 2015. www.3gpp.org (accessed 4 July 2015).

[19] 3GPP, Security Aspect, 30 May 2011. ftp://www.3gpp.org/Information/presentations/presentations_2011/2011_05_Bangalore/DZBangalore290511.pdf (accessed 4 July 2015).

[20] Anand R. Prasad. 3GPP SAE/LTE Security. NIKSUN WWSMC, 26 July 2011.

[21] Bogdan Botezatu. 25 percent of wireless networks are highly vulnerable to hacking attacks, Wi-Fi security survey reveals, 11 October 2011. http://www.hotforsecurity.com/blog/25-percent-of-wireless-networks-are-highly-vulnerable-to-hacking-attacks-wi-fi-security-survey-reveals-1174.html (accessed 4 July 2015).

[22] Michael Walker, chairman of 3GPP SA3 WG (Security). On the security of 3GPP networks. Eurocrypt 2000.

[23] CTIA, Mobile Cybersecurity and the Internet of Things; Empowering M2M Communication.

[24] Executive Order 13636 Improving Critical Infrastructure Cybersecurity, 12 February 2013. http://www.whitehouse.gov/the-press-office/2013/02/12/executive-order-improving-critical-infrastructurecybersecurity (accessed 6 July 2015).

[25] National Highway Traffic Safety Administration, Preliminary Statement of Policy Concerning Automated Vehicles, 30 May 2013. http://www.nhtsa.gov/staticfiles/rulemaking/pdf/Automated_Vehicles_Policy.pdf (accessed 6 July 2015).

[26] 2015 Data Breach Investigations Report. Verizon, 2015.

[27] S. Bosworth, M.E. Kabay and E. Whyne. *Computer Security Handbook*. Sixth edition, Volume 1. John Wiley & Sons, Inc., Hoboken, NJ, 2014.

[28] Open Mobile Alliance, 12 October 2015. http://openmobilealliance.org/ (accessed 12 October 2015).

[29] OMA Work Programs, 12 October 2015. http://openmobilealliance.org/about-oma/work-program/device-management/ (accessed 12 October 2015).

[30] OMA DM Development, 29 February 2012. http://technical.openmobilealliance.org/comms/documents/OMA_DM_1.4Billion_PR_Final.pdf (accessed 12 October 2015).

[31] OMA Releases, 12 October 2015. http://technical.openmobilealliance.org/Technical/technical-information/release-program/current-releases/gssm-v1-0 (accessed 12 October 2015).

[32] CoAP. IETF RFC 7252, June 2014. https://tools.ietf.org/html/rfc7252 (accessed 13 October 2015).

[33] DTLS version 1.2. IETF RFC 6347, January 2012. https://tools.ietf.org/html/rfc6347 (accessed 13 October 2015).

[34] Infineon, IoT overview, 1 November 2015. http://www.infineon.com/iot-security-ebrochure/en/index.html (accessed 1 November 2015).

[35] Open Mobile Alliance Release Program document package, 1 November 2015. http://technical.openmobilealliance.org/Technical/Release_Program/docs/GSSM/V1_0-20111220-A/OMA-ERP-GSSM-V1_0-20111220-A.zip (accessed 1 November 2015).

[36] The green life of a SIM card. Giesecke & Devrient, Smart! Telecommunications, 2/2009.

[37] ISO/IEC 7816-1:2011. Identification cards; Integrated circuit cards, Part 1: Cards with contacts; Physical characteristics. http://www.iso.org/iso/home/store/catalogue_ics/catalogue_detail_ics.htm?csnumber=54089 (accessed 1 November 2015).

[38] ISO/IEC 7816-2:2007. Identification cards; Integrated circuit cards, Part 2: Cards with contacts; Dimensions and location of the contacts. www.iso.org (accessed 30 December 2015).

[39] ISO/IEC 7816-3:2006. Identification cards; Integrated circuit cards, Part 3: Cards with contacts; Electrical interface and transmission protocols. www.iso.org (accessed 30 December 2015).

[40] ISO/IEC 7816-4:2013. Identification cards; Integrated circuit cards, Part 4: Organization, security and commands for interchange. www.iso.org (accessed 30 December 2015).

[41] ISO/IEC 7816-5:2004. Identification cards; Integrated circuit cards, Part 5: Registration of application providers. www.iso.org (accessed 30 December 2015).

[42] ISO/IEC 7816-6:2004. Identification cards; Integrated circuit cards, Part 6: Interindustry data elements for interchange. www.iso.org (accessed 30 December 2015).

[43] ISO/IEC 7816-7:1999. Identification cards; Integrated circuit(s) cards with contacts, Part 7: Interindustry commands for Structured Card Query Language (SCQL). www.iso.org (accessed 30 December 2015).

[44] ISO/IEC 7816-8:2004. Identification cards; Integrated circuit cards, Part 8: Commands for security operations. www.iso.org (accessed 30 December 2015).

[45] ISO/IEC 7816-9:2004. Identification cards; Integrated circuit cards, Part 9: Commands for card management. www.iso.org (accessed 30 December 2015).

[46] ISO/IEC 7816-10:1999. Identification cards; Integrated circuit(s) cards with contacts, Part 10: Electronic signals and answer to reset for synchronous cards. www.iso.org (accessed 30 December 2015).

[47] ISO/IEC 7816-11:2004. Identification cards; Integrated circuit cards, Part 11: Personal verification through biometric methods. www.iso.org (accessed 30 December 2015).

[48] ISO/IEC 7816-12:2005. Identification cards; Integrated circuit cards, Part 12: Cards with contacts; USB electrical interface and operating procedures. www.iso.org (accessed 30 December 2015).

[49] ISO/IEC 7816-13:2007. Identification cards; Integrated circuit cards, Part 13: Commands for application management in a multi-application environment. www.iso.org (accessed 30 December 2015).

[50] ISO/IEC 7816-15:2004. Identification cards; Integrated circuit cards, Part 15: Cryptographic information application. www.iso.org (accessed 30 December 2015).

[51] ISO/IEC 7816-15:2004/Cor 1:2004. www.iso.org (accessed 30 December 2015).

[52] ISO/IEC 7816-15:2004/Amd 1:2007. Examples of the use of the cryptographic information application. www.iso.org (accessed 30 December 2015).

[53] ISO/IEC 7816-15:2004/Amd 2:2008. Error corrections and extensions for multi-application environments. www.iso.org (accessed 30 December 2015).

[54] ISO/IEC Standards Summary, 1 November 2015. http://www.iso.org/iso/home/store/catalogue_ics/catalogue_ics_browse.htm?ICS1=35&ICS2=240&ICS3=15& (accessed 1 November 2015).

[55] GlobalPlatform, 1 November 2015. https://www.globalplatform.org/ (accessed 1 November 2015).

[56] SIMalliance, 1 November 2015. http://simalliance.org/ (accessed 1 November 2015).

[57] GSMA, 1 November 2015. http://www.gsma.com/aboutus/ (accessed 1 November 2015).

[58] GSMA history, 9 November 2015. http://www.gsma.com/aboutus/history (accessed 9 November 2015).

[59] Common Criteria, 9 November 2015. http://www.commoncriteriaportal.org/ (accessed 9 November 2015).

[60] Robert Yen. Overview of ANSI-INCITS biometric standards on data interchange format. Biometrics, U.S. Department of Defense, 19 January, 2005.

[61] Federal Information Processing Standards (FIPS), 11 November 2015. https://csrc.nist.gov (accessed 11 November 2015).

[62] European Telecommunications Standards Institute, 11 November 2015. http://www.etsi.org/ (accessed 11 November 2015).

[63] EMVCo, 11 November 2015. http://www.emvco.com/ (accessed 11 November 2015).

[64] NFC Forum, 11 November 2015. http://www.nfc-forum.org (accessed 11 November 2015).

[65] PCI Security Standards, 11 November 2015. https://www.pcisecuritystandards.org (accessed 11 November 2015).

[66] MULTOS, 11 November 2015. https://www.multos.com/ (accessed 11 November 2015).

[67] Information Technology Security Evaluation Criteria (ITSEC). Department of Trade and Industry, London, June 1991.

[68] The Open Card Framework, 11 November 2015. http://www.openscdp.org/ocf/ (accessed 11 November 2015).

[69] CIPURSE V2. Integrating CIPURSE V2 into an existing Automated Fare Collection system. OSPT Alliance, 2014.

[70] GSMA, The Mobile Economy Report 2015.

[71] Crypto tutorial. https://www.cs.auckland.ac.nz/~pgut001/tutorial/ (accessed 11 November 2015).

[72] SSL encryption evaluation. http://www.peerlyst.com/blog-post/a-technical-rant-about-the-different-e-s-in-ssl-tls (accessed 11 November 2015).

[73] Smartcard Alliance, 22 November 2015. http://www.smartcardalliance.org/alliance/ (accessed 22 November 2015).

[74] *Wired*. Siri hack, October 2015. http://www.wired.com/2015/10/this-radio-trick-silently-hacks-siri-from-16-feet-away/?mbid=social_twitter (accessed 22 November 2015).

[75] *Interference Technology*. OTA hacking device. http://www.interferencetechnology.com/unstoppable-new-hacking-device-steals-encryption-keys-out-of-the-air/ (accessed 22 November 2015).

[76] *Helsingin sanomat*. Baby monitoring device security threats, 7 September 2015. www.hs.fi/m/kotimaa/a1441508730024?ref=hs-mob-prio45-1 (accessed 22 November 2015).

[77] *Helsingin sanomat*. Home device security holes, 7 August 2015. http://www.hs.fi/m/ulkomaat/a1438918533873 (accessed 22 November 2015).

[78] BBC. Security breach lab, 20 July 2015. http://www.bbc.com/future/story/20150720-the-hidden-lab-where-bankcards-are-hacked (accessed 22 November 2015).

[79] *Helsingin sanomat*. Remote access breaches. http://www.hs.fi/m/kotimaa/a1435630410758 (accessed 22 November 2015).

[80] *Helsingin sanomat*. Security breach revealed by under aged. http://www.hs.fi/m/autot/a1425284972822 (accessed 22 November 2015).

[81] BBC. Rail signal upgrade could be hacked to cause crashes. http://m.bbc.com/news/technology-32402481 (accessed 22 November 2015).

[82] IoT Security Foundation. https://iotsecurityfoundation.org/ (accessed 22 November 2015).

[83] J. Daemen and V. Rijmen. *The design of Rijndael*. Springer, 2002.

[84] J. Daemen and V. Rijmen. AES proposal: Rijndael. First Advanced Encryption Standard (AES) Conference, Ventura, CA, USA, 1998.

[85] I. F. Blake, G. Seroussi and N. P. Smart. *Elliptic Curves in Cryptography*. Cambridge University Press, New York, 1999.

[86] ECC Brainpool. ECC Brainpool Standard Curves and Curve Generation. 2005. http://www.ecc-brainpool.org/ecc-standard.htm (accessed 12 December 2015).

[87] J. L. Carter and M. N. Wegman. New hash functions and their use in authentication and set equality. *Journal of Computer and System Sciences*, 22(3):265–277, 1981.

[88] W. Diffie and M. E. Hellman. Exhaustive cryptanalysis of the NBS Data Encryption Standard. *COMPUTER*, 10(6):74–84, 1977.

[89] 3GPP2 introduction, 2 January 2016. http://www.3gpp2.org/Public_html/Misc/AboutHome.cfm (accessed 2 January 2015).

[90] ISO 27000 Information Security Management. http://www.iso.org/iso/home/standards/management-standards/iso27001.htm (accessed 9 January 2016).

[91] ISO/IEC JTC1/SC17 Standing Document 3 – SC17 Work Programme including all published standards and target date summary for all work items under development, 23 August 2006. http://wg8.de/wg8n1255_17n3074_SC17_SD3_Work_Programme.pdf (accessed 10 January 2016).

[92] OMA. A Primer to SyncML/OMA DS. Approved 31 Mar 2008. 27 p. http://technical.openmobilealliance.org/Technical/release_program/docs/SyncML_Primer/V1_0-20080331-A/OMA-WP-SyncML_Primer-20080331-A.pdf (accessed 10 January 2016).

[93] 3GPP TR 31.828 V10.0.0 (2011-04). UICC access to IMS (Release 10). 25 p.

2

Security of Wireless Systems

2.1 Overview

This chapter describes the security architecture of the most relevant modern wireless and mobile systems, with the main weight on 3GPP and IEEE 802 networks. The security aspects of satellite systems, special systems represented by TETRA and broadcast networks are discussed, as well as local wireless connectivity. Practical examples are presented that identify potential security threats and also methods for protecting confidentiality of the mobile communications. The examples include user actions like calls and messaging, as well as operator-related functions like subscription management. This chapter also gives an overview of security aspects in the interoperability and interworking of the networks, including 3GPP networks, non-3GPP networks and related Wi-Fi Offload.

2.1.1 Overall Security Considerations in the Mobile Environment

The data services of current mobile communications networks are increasingly based on packet connectivity. This has greatly enhanced the efficiency of the cellular connections, providing superior user experiences compared to the preceding solutions that were based on circuit-switched techniques. At the same time, as the current mobile services are comparable to the principles of the Internet, so have similar types of security threats increased. It can be generalized that the security breaches familiar from the public Internet are in many ways relevant in mobile communications, and thus require similar protection mechanisms. This applies to both 'traditional' feature mobile phone equipment and advanced smart devices. Furthermore, not only is data transmission under threat but also the services related to voice communications may be vulnerable.

Some examples of the security threats of commercial mobile communications may be related to the economical exploitation of the user credentials (some concrete cases being the stealing of credit card information or other type of company secrets) or 'virtual vandalism'

Wireless Communications Security: Solutions for the Internet of Things, First Edition. Jyrki T. J. Penttinen.
© 2017 John Wiley & Sons, Ltd. Published 2017 by John Wiley & Sons, Ltd.

(as can be the case, e.g., in the interfering of calls). Apart from the threats related to end-user and business communications, there also exist potential threats in special environments which may jeopardize the safety of, e.g., governments and defence forces. An example is an attacker who may want to seek strategic level information from the mobile communications, either in real time or by post-processing the stored data that could be obtained via the publicly exposed but secured channels, or by trying to enter the internal network infrastructure. There may also be intentions to paralyze mobile networks and related services in order to block completely the legitimate communications.

Whatever the threat, its impact needs to be minimized via preparation from the known and expected threats and by setting up respective protection mechanisms prior to the moment when the concrete attacks take place. This is a joint effort which can ideally be done in cooperation between operators, equipment manufacturers and end-users.

As the mobile communications networks develop along with the new generations, the protection mechanisms also improve. As is the case with any other form of telecommunications, the latest mobile system generations may contain weaknesses that are exposed as a result of the overall development of electronics and novelty HW and SW attack types. Not only are the networks and user equipment under threats but today's environment, especially services and applications that are used on top of the protocol layers, brings along previously unknown challenges. One example is the increasingly utilized cloud storage that can be accessed via mobile communications networks. Typically deployed as a separate component from the mobile network infrastructure, the contents of the cloud server may be vulnerable against unauthorized access regardless of the security level of the mobile systems. Equally, the increasingly popular smart devices equipped with applications expose security vulnerabilities familiar from the public Internet, i.e., the malicious attacker can aim to access the user data, copy, modify and destroy the contents and manage the devices via viruses.

2.1.2 Developing Security Threats

One of the highly concrete security threats of today and in the near future is related to the IoT, which also includes M2M communications. A multitude of devices that are connected to Internet either wirelessly or via a wired manner, like printers and video surveillance cameras, audio systems or even light bulbs, may open unknown back doors right into the otherwise well-protected environment. One such security threat is a simple default password of the device manufacturer that the user does not change. This is especially problematic for devices that are able to jeopardize the personal health of the user, such as Internet-connected self-driving cars [8].

Whether it is about servers, laptops or simple IoT devices, the ways for ensuring adequate shielding against the most obvious attack attempts include changing the default passwords before utilization of the equipment, as well as monitoring the traffic flows of systems, updating virus protection, maintaining the latest and protected firewall settings, using common sense in the installing of new applications, and applying any other reasonable shielding principles familiar from the public Internet. For highly confidential communications, a feasible way to ensure protection is to use a point-to-point scrambling solution in the application layer. Furthermore, when storing confidential information outside of the local systems, like into clouds, a user managed data-scrambling solution provides extra protection.

As for the fixed and mobile network operators and service providers, monitoring of the fraudulent attempts has been a daily routine for a long time, and typically there is also at least some sort of basic virus protection mechanism applied. The obviously suspicious behaviour, such as intentions to modify user profiles as a part of hijacked end-user equipment for scam emailing, can be tracked from the network side but that alone cannot prevent security breaches without the active role of the end-user.

The importance of the IoT, as well as the increasing amount of IoT security threats, can be interpreted from the statistics presented by Verizon in Refs. [9,12]. According to the data, the year-to-year growth of global IoT devices can be estimated to present a steady 28% figure during 2015–2020. According to this source, the total number of IoT devices may be around 5.4 billion by 2020 while it was about 1.4 billion in 2014.

Regardless of the technology and level of the radio interface protection, the path of the mobile system's radio network represents similar security principles as is the case for fixed telephony networks. The telecommunications networks are by default closed environments and thus protected from external attacks. The modern networks are based on digital transmission and fibre optics which means that it is challenging to access them by external parties for unauthorized eavesdropping. Nevertheless, if the unauthorized party does have access to the communications, e.g., via an unprotected radio link of the transport network, there is not much to be done for detecting or preventing such attempts, apart from the deployment of additional encryption devices for those links.

Not only is eavesdropping a threat but also the intentional – or unintentional – damaging of the network may be a severe issue. As a result, part of the communications network service may break down, including emergency call attempts. Public news stories have reported occasionally about service level degradation of fixed or mobile communications networks, which can result in regional service blackout. This can happen when strategically important fibre optics cables are damaged, typically due to construction works in the area. The proper redundancy with clearly physically separated cabling of the transport network is thus one of the important preventive protection methods of the infrastructure provider to minimize the negative impacts of such incidences.

Back in the 1990s, the mobile networks were isolated from the public data networks, and only circuit-switched data methods were used for accessing early Internet-type of services. The General Packet Radio Service (GPRS) that was deployed at the beginning of the 2000s opened packet-switched channels into the Internet which, in turn, exposed a completely new world of potential security threats for the end-users as well as for the operators. One example of such an issue is the attempt to overload the internal network by repeatedly sending Packet Data Protocol (PDP) context activation requests from the external network to either real or imaginary GPRS users. Such attempts can trigger a big amount of signalling by the Home Location Register (HLR) and Visitor Location Register (VLR) of the GSM network as part of the procedure for resolving the location of the receiving party. These messages can overload the signalling resources of the GPRS, and as the GSM user registers are common for both data and voice users, the voice traffic of the GSM network may also be impacted. This type of relatively simple DoS attack can be prevented in the border of the GPRS network at the GPRS Gateway Support Node (GGSN) element by using additional traffic analysis and blocking procedures. Thus, the early deployment of GPRS has resulted in the introduction of such firewall-type of protection mechanisms familiar from the public Internet. In fact, operators tend to avoid the utilization of the option to activate the PDP context via the network due to

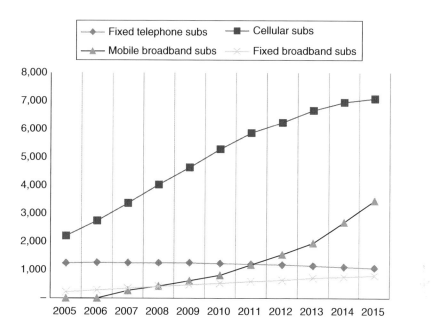

Figure 2.1   The statistics of data consumption of mobile laptop and smartphone users

the aforementioned reason, so the user device typically makes the opening – which can also happen after a push message in a form of a short message from the network side.

Since then the utilization of the packet data has increased exponentially as a result of considerably higher data rates and lower latency of the mobile communications networks, which makes the mobile data communications comparable or even more appealing for Internet traffic compared to fixed Internet lines. As a result, many more mobile applications have been developed for all kinds of purposes. One of the main drivers for future data utilization is related to the growth of smartphone penetration. As an example, Informa estimated that during 2010, 65% of global mobile data traffic was generated by the 13% of mobile subscribers who used smartphones, with the average traffic per user of 85 MB per month. As can be seen in the ITU statistics presented in Figure 2.1, the mobile broadband traffic has been growing exponentially, as interpreted from Ref. [60,36].

In order to cope with the increasing mobile data consumption, the LTE and LTE-A networks will provide highly needed capacity and higher data rates for end-users during the forthcoming years. The LTE/LTE-A networks are being actively deployed at the moment. As the protection level of these networks can be assumed to be superior compared to previous generations, new attack types will most probably focus on the application level instead of the network infrastructure.

2.1.3   RF Interferences and Safety

In the practical LTE/LTE-A radio network deployment, the Electro-Magnetic Field (EMF), Electro-Magnetic Compatibility (EMC) and Specific Absorption Rate (SAR) need to be taken into account. They refer to the radiation properties of radio transmitters, the shielding and safety regulation and bio-effects, respectively. They thus belong in the area of wireless safety. The EMF domain includes scientific and general debate about RF radiation and its interaction with living biological tissue. EMF generally covers RF ranging from 10 MHz to 300 GHz, as well as Intermediate Frequency (IF) from 300 Hz to 10 MHz, Low Frequency (LF) up to 300 Hz, and static fields. For the LTE and LTE-A deployments, the focus is on RF as the range for the potential practical frequency bands varies from a few hundred MHz up to a few GHz.

For various reasons (expenses of land, lack of site or tower sharing concept due to the competitive environment, etc.) it is not uncommon for operators to deploy two or more antenna towers relatively close to each other as the topology in those locations may be the most suitable for radio network planning. The potential exceeding of recommended RF radiation exposure should be taken into account in these situations when field personnel is working in towers, especially when antenna height matches the working height [40].

It is a universally acknowledged fact that negative health effects occur above a certain level of RF exposure in terms of SAR to living tissue, caused by the rise in tissue temperature. Even after various decades of mobile communications networks and devices available for public use, the question remains if there are other effects which may cause aforementioned thermal effects below this critical level, or if there are any other negative health effects occurring during the longer term. Furthermore, are there cumulative effects?

More detailed radiation aspects are outside the scope of this book. Further information about radiation safety calculations and considerations can be found, e.g., in Refs. [39–42]. Ref. [43] has investigated the interaction between the radiation of LTE Multiple In Multiple Out (MIMO) antennas in a mobile handset and the user's body. It concluded that in order to meet the regulations and MNO requirements, the SAR and body loss of the radio link budget need to be optimized to achieve an acceptable level. Ref. [43] presents guidelines and clarifies the general behaviour of the variation of SAR and body loss with different sets of parameters.

2.2   Effects of Broadband Mobile Data

2.2.1   Background

The growth of mobile communications network utilization, measured by any criteria, has been fast since the introduction of the first generation systems. As the integrated data services are improving and the penetration as well as the number of smart devices are growing, there is a revolutionary demand for more capacity and quality for data. Ref. [27] has identified that the gap between the available spectrum and network capacity together with the required mobile data is widening at a global level. The source estimates that the mobile data traffic is dominated by video growth and will possibly increase 20 times in volume between 2012 and 2017. This, in turn, results in high traffic and signalling in wireless networks.

The growth is so spectacular that merely providing more network capacity by increasing the number of cells and spectrum might not be sufficient to cope with the demand. Even if the deployment of LTE and LTE-A is an essential step in this evolution with new operators and licenses arriving into the market, only advanced functionality can assure that the networks are capable of handling the increased traffic and signalling. The concept of small cells has a key role in this evolution to balance the traffic between highly localized spots and the macro layer. This will lead to heterogeneous networks that include the layered approach and support various radio interface types, frequency bands and bandwidths. According to Ref. [27], over 80% of operators claim that small cells will be the first or second most important factor in meeting their capacity objectives between 2012 and 2017.

2.2.2   The Role of Networks

To cope with the challenges of planning and deployment, and to secure capacity in the future, prior optimization, accurate forecasts and sophisticated means to estimate the effects of the radio network planning options are essential. They directly impact the return of investments and profitability of the business. The items that require special attention are the optimal site-hunting strategies by taking into account the whole network (inter-cell distances and performance achieved via different locations), the core network quality and capacity (so that it does not create a bottle-neck along with the high data rates of radio interface) and efficient performance monitoring and fault management (to assure the close to real-time response to correct the non-optimal parameters). Also the interfaces between the small cells and the rest of the radio network need to be minimized. This could happen via enhanced methods offered by 3GPP standardization, and especially via careful network optimization that should take the new home evolved Node B (eNB) installations into account in a dynamic way – which would justify the deployment of the Self-Organizing/Self-Optimizing Network (SON) concept. For the core network planning, increasing amount of options are available [28], so the respective task for the operator is to select the most techno-economically feasible one in each environment.

For the planning purposes of LTE/LTE-A networks, it is good to remember that the peak data rate is not the most important criterion from the point of view of the user's experience. The peak data rate indicates the theoretical throughput while the relevant figure for subscriber experience as well as for network capacity is the average sector throughput [29]. It indicates the amount of bandwidth that can realistically be delivered within a sector, taking into account the practical limitations of the environment. The throughput achieved via aggregation can further be considered to calculate the realistic maximum amount of concurrent subscribers that can be served per sector. It is especially this average sector throughput rate that can be used as a base for estimating the cost of deployment and operations.

The deployment of LTE and LTE-A networks is more challenging than ever before due to the need for precise planning and respective strategies for the optimal site locations in align with the home eNB concept. Some special items to be considered in the deployment are the needs for enhanced planning tools that are capable of taking into account the small cell concept, in align with assisting SON, and the techno-economically optimal solutions for the core network. It is also important to assure the small cell installation is done by minimizing interferences, which may need additional tools and guidelines from the operators to the customers. Nevertheless, as a result of the small cell deployment, the operators together with the customers can assure the offered data rates are supporting the demand on time. Not only is the capacity increase as such an important aspect for the operators but also the major goals for any operator in global markets are the reduction of the cost related to content delivery, increasing data rates and reduction of power consumption.

A more concrete way to handle increased data delivery is the enhancement in the radio link budget, e.g., by utilizing remote radio heads (to lower the antenna feeder attenuation) and smaller radio cells (which provides more capacity). Also, data traffic offloading has been identified as feasible, and is used in practice. The respective development is ongoing in order to assure efficient offloading use cases, policy and control functions as well as seamless user authentication, authorization and accounting – not forgetting the security aspects related to the real-time switching between the radio access technologies. In addition to offloading data traffic from the LTE/LTE-A networks (and other mobile communications networks) to Wi-Fi hotspots, there are also possibilities to use other networks for offloading such as femtocells.

LTE and LTE-A networks offer better spectral efficiency compared to previous generations. The possibility to further expand the data delivery via, e.g., carrier aggregation and higher order MIMO via supported standards, network equipment and user devices, will increase the data rates. LTE/LTE-A networks further provide the means for data increasing via, e.g., Active Antenna Systems (AASs).

A next step in the evolution of the small cell concept of LTE/LTE-A networks is the convergence with cellular networks and Wi-Fi, which forms a heterogeneous network. This phase consists of a considerable amount of small cells which are a combination of radio network elements consisting of both Wi-Fi and the mobile communications radio interface. They may also be separate elements of Wi-Fi access points and base stations that form single macro cells. The coordination of the traffic between different access networks happens via the dynamic capacity pool in the available RF bands. Logically, the operations and management system takes care of the traffic flow coordination in this case.

Ref. [50], among various others, emphasizes the importance of the Wi-Fi Offload, which is expected to grow considerably in the near future. The forecasts indicate that in 2017, close to 80% of operators will have multiband 3G and 4G networks with typically Wi-Fi Offload included. The related management challenges are the adaptation to different topologies, including heterogeneous networks (LTE-A being the driving force), the integration of Wi-Fi with feasible functionalities, and the optimization of the legacy systems while they are still serving as part of the network infrastructure.

There are also special opportunities related to the increased capacity offering. Ref. [30] has investigated business cases for the use of TV white spaces for mobile broadband services. As a conclusion, the source has identified that actors that are deploying new networks, costs of the radio equipment and transmission solutions are higher compared to incumbent operators. The already established operators, in turn, can deploy mobile broadband access services via, e.g., the LTE network, by adding new radio equipment to existing sites. This indicates the challenge of high costs for deploying new sites, even if the cost for cognitive radio equipment should be at the same level as LTE equipment. Ref. [30] further concludes that for the case that cognitive radio is substantially more expensive the business case is even worse. Nevertheless, the use of TV white space, when applicable, is worth considering in the selection of potential deployment scenarios.

In addition to the radio interface, the backhaul deployment also requires attention along with the LTE/LTE-A deployment. The main aim is to assure that the core is not creating bottlenecks as the radio interface offers much higher average and peak throughput values compared to the previous 3GPP releases. Source [31] has concluded that in the early phase of LTE, the backhaul capacity needs are being overstated in a similar fashion as happened in the 3G expectations in the early 2000s. Nevertheless, this will changes as the LTE technology establishes itself. Ref. [31] further claims that LTE probably has an accelerated cycle, although the industry might have initial frustrations. The planned LTE data capacity per site is fixed and the maximum resources are shared between the users. The maximum available capacity thus depends on the commercial Node B (NB) models and available channel bandwidth. The addition of capacity requires more LTE channel bandwidth, which might not be possible based on the LTE licensing. Furthermore, the adding of macro cell sites increases the operating expenditure of the networks. Thus, Ref. [31] has concluded that smaller picocells with more cost-effective backhaul options are a feasible compromise for LTE capacity offering. The source further claims that operators should consider this solution instead of aiming for theoretical maximum coverage of macro cells, especially in high traffic density urban areas. This claim is based on the calculation showing that the capacity limit of 150 Mb/s for a three-sector site with 10 MHz bandwidth available is below some LTE backhaul deployments. Nevertheless, this assumption takes into account radio propagation limits. Source [31] also notes that LTE backhaul needs to fulfil requirements of an evolving mobile network like high availability, low latency, low packet loss, Quality of Service (QoS), direct site-to-site connectivity and high network capacity. The source has concluded that ring or mesh topologies with multiple paths from access to the core fit better to assure the requirements are fulfilled compared to long chains of single-path links. The additional benefit of a ring or mesh topology is that it assures that new network nodes can be deployed without affecting the already deployed network for capacity distribution and connectivity relationships.

Many user actions, like Internet browsing, online gaming, video streaming, content uploading in social media and downloading email attachments require high LTE/LTE-A bandwidth and low latency in order to assure adequate user experience. Some LTE capabilities for assuring the sufficiently high-quality experience are the short idle-to-active state transition time, low latency and adequate QoS [32]. Also the new IoT devices may require various types of QoS, some located in very remote areas and sending only sporadically low-data rate contents while others may require higher capacity in real time. One important aspect in the IoT device behaviour is the 'keep-alive' signalling which may become a bottleneck in very densely populated cells if not managed properly.

The reduced time for idle-to-active stage transition is possible in LTE/LTE-A via the flat IP architecture. The simple solution with a single RAN element (eNB) and reduced amount of core elements, i.e., Mobility Management Entity (MME) for signalling and Serving Gateway (S-GW)/Proxy Gateway (P-GW) for the user plane, assures that the time for accessing the radio and core network resources is fast. Furthermore, the LTE connection states are reduced to two states compared to the previous four-state approach in High Speed Packet Access (HSPA).

The benefit of a low latency value is increased data throughput as well as enhanced user experience. As a comparison, HSPA networks may have delays of two seconds or longer for setting up the initial connection, followed by 75–150 ms roundtrip latency values. LTE has typical delays of 50 ms for the initial connection time, followed typically by 12–15 ms roundtrip latency values. As LTE has low latency and high average sector throughput compared to previous systems, it is especially suitable for demanding service delivery like Voice over Internet Protocol (VoIP) calls, video-on-demand services and online gaming while IoT/M2M connections within the same LTE/LTE-A networks, or within separate 2G/3G networks, do not require by default such demanding latency values.

Thirdly, QoS refers to the network's ability to manage service priorities for different types of applications, subscribers and sessions, yet assuring the expected performance. It should be noted that in 3GPP networks prior to LTE, the QoS classification was not available. In comparison, LTE provides the possibility to apply QoS for supporting variable levels of

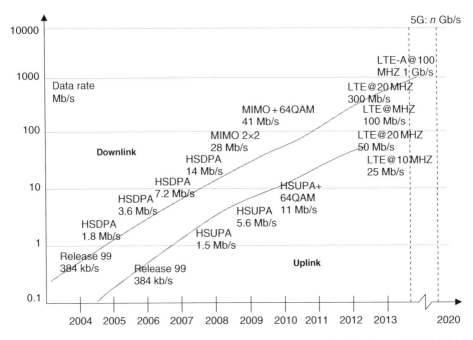

Figure 2.2   The general trends of 3G and 4G data rates. The planned 5G will offer considerably higher speeds

service for applications that are latency and/or bit-rate sensitive. The QoS of LTE has a class-based model for differentiating between services and varying the levels of service quality.

LTE/LTE-A networks thus address these current and future challenges for both consumer environments and for the IoT era. Figure 2.2 depicts the overall mobile communications evolution and the LTE development towards the fully equipped LTE-Advanced, and finally towards the 5G phase as defined by ITU-R.

2.2.3   The Role of Apps

Since the early introduction of the Wireless Access Protocol (WAP) and respective browsers, mobile communications has evolved towards the increasing utilization of the Internet. HTML browsers on smartphones have provided further a logical access to the web, making WAP as such unnecessary. Also the provision of email access has been important in the high utilization of wireless access, and it can be generalized that the overall business environment has greatly benefited from email. Integrated browsers and email clients are nowadays included in smartphones and in many basic phones by default. The next step towards an even more interoperable era is the deployment of HTML5 and respective applications. In this way, the utilization of the network is optimized for a high-quality user experience.

The provision of apps has also had a high impact on wireless communications. They enrich enormously user experiences and are highly popular due to the generally low or zero cost. This development has made the ecosystem for each smart device platforms an essential element – the availability or lack of popular apps may even mark the difference between successful device

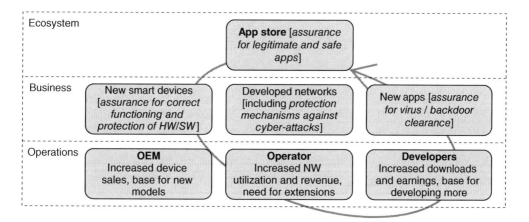

Figure 2.3   The app ecosystem depends on the available technologies and services

sales, so Original Equipment Manufacturers (OEMs), in addition to the hardware, need to guarantee the existence of this ecosystem. As a part of this ecosystem, the distribution channel, or application stores, functions as a logical means for the delivery of the apps. Furthermore, the creation of the ecosystem has initiated completely new business models where app developers together with operators and OEMs benefit from the app utilization. The business models are typically based on advertisements for the free-of-charge apps, low-cost price for the payable apps and sharing of the revenue between the participating stakeholders [33].

Figure 2.3 summarizes this business environment. If any element is missing from the complete picture, or the element is not performing fully, such as insufficiently complete offering of apps per operating system, the business does not achieve optimal results which decreases the sales of the devices and further lowers the interest on developing new apps for that specific Operating System (OS). It is thus a matter of cooperation between all the stakeholders in order to ramp up the sufficiently profitable balance for all the elements of the ecosystem which eventually starts feeding itself automatically. Based on the short history of smart device markets, the app ecosystem is one of the most significant forces influencing the market share of user devices.

After the apps have been developed and the offering of many kinds of private and professional solutions have taken off, including games, utility solutions, integrated navigation and mapping solutions, banking and stock market solutions, the next significant step will be the evolved voice and video solutions. LTE/LTE-A as such facilitates this development as the circuit-switched domain is completely missing from the standards, making the system work purely in the packet-switched domain. Thus, after the transition phase of voice calls handled via, e.g., Circuit Switched Fallback (CSFB), the more integrated solutions are taking voice and video services to next level.

Fully capable Voice over LTE (VoLTE) and Rich Communications Suite (RCS) will make the earlier mobile communications networks gradually unnecessary, although they will still serve as a base for operators for many years to come. It would logically not make sense to ramp down the already deployed networks at a fast schedule as the network utilization per generation is evolving gradually. As an example, the recent lowering of the GSM utilization means that there is huge amount of GSM-capable handsets in the field, including GSM-only equipment especially in an M2M environment. The transition strategies for the 2G, 3G and LTE/LTE-A networks with frequency re-farming optimization is thus one of the most important tasks for operators at the moment.

The evolution also brings along the offering of HTML5 technologies, with apps that are able to manage better data persistence, native code execution and multitasking. This evolvement will be seen increasingly in websites and browser-based applications, which will pave the way for converging browser-based applications and native applications. At least in theory, the role of operating systems and respective separated app ecosystems will thus lower in status. Also the always-on connectivity will continue to evolve, with multitasking applications capable of keeping the connections alive becoming much more efficient. This, in turn, brings further challenges for the operators because of the probably exponentially increasing amount of signalling, e.g., due to keep-alive or heart-beat signalling which forces the apps to maintain the reserved bearer connections active. This may mean dramatic re-dimensioning and increasing of signalling capacity – or actions from operators to guideline app developers not to waste the valuable capacity from the networks if other or less signalling can be used.

Observing the new mobile ecosystem as presented in Figure 2.3, the obvious security-related question arises on how well the authenticity of the applications can be ensured. The official app store per OS representative and responsible aims to guarantee that the SW is free of malicious code via sufficiently thorough testing, evaluation and certification programs. Nevertheless, especially with the tight competitor environment and 'rush' for new app releases, there have been cases that revealed inappropriate SW embedded into the apps, or unnecessarily wide rights to access users' data. Furthermore, the delivery of the apps via any other than the official ways always exposes a risk for embedded viruses, adware and other malware. If the installation of such non-official apps is prevented by default by the device vendors, there are 'temptations' for end-users to root the device, which opens potential security holes, so one of the easiest ways to ensure the protection level is that the end-users hold off modifying their devices and utilize only trusted sources for SW downloading.

2.2.4 UE Application Development

The LTE/LTE-A does not change as such the principles of the applications or app development that end-users experience via terminals. Nevertheless, the LTE/LTE-A does provide a base for more advanced apps due to the network data rate enhancements, lower delays and response times. Some possible new, higher quality apps might be related to time and jitter critical environments such as real-time gaming, which requires heavy data transmission and high processor power which can be provided via the most advanced LTE/LTE-A user devices.

The app development requires a Software Development Kit (SDK), which depends on the OS. The mobile OS is designed for smartphones, tablets, PDAs and other mobile devices. The typical mobile OS has elements familiar from the PC OS combined with other features optimized for touchscreen, mobile communications technologies, local connectivity like NFC, USB, Bluetooth and Wi-Fi, navigation systems like GPS, additional integrated elements like cameras and sensors, and functionalities supporting, e.g., audio and video playback. Smartphones that contain mobile communications capabilities are typically based on two mobile OSs which are divided into the main software platform for the user interactions and a lower level OS managing the needed functions for the radio communications and hardware modules. Thus, the apps need to be developed for each operating system separately as they also manage the proprietary native functions of the devices.

The best way to initiate the mobile app development is to select the proper OS and to familiarize with the respective SDK functionality. This happens most fluently by creating initial, simple apps by reviewing examples of available apps in order to understand the philosophy of the code, libraries, APIs and manners to control the blocks of the smart device, such as GPS, sensors and input/output of characters. As is typical in any other computer language, the best way to understand the possibilities and limitations of the code is to investigate existing codes.

The following presents the main principles of the most popular SDK environments, i.e., Android that has the major part of the markets measured by the number of the devices and applications, followed by iOS. The rest of the OSs represent the minority in 2015 but can be an equally interesting base for new app developers. According to the statistics of Ref. [52] Android represented 85% share of global smartphone shipments in the second quarter of 2014 while Apple iOS, Microsoft, Blackberry and others had 11.9%, 2.7%, 0.6% and 0.2%, respectively.

2.2.4.1 Android

For app development in the Android environment, which is supported by a vast variety of device manufacturers, Ref. [51] contains the needed SDK and instructions for developers. For a new Android developer, the most straightforward way to proceed is to download the Android Developer Tool (ADT) bundle. It contains Android SDK components and a version of the Eclipse Integrated Development Environment (IDE) with built-in ADT which is basically sufficient to initiate the actual programming. The components are: Eclipse and ADT plugin, Android SDK tools, Android platform tools, the latest Android platform and the latest Android system image for the emulator.

In the SW development, one of the tasks is to decide which Android version to select. The latest versions are by default backwards compatible with the earlier versions. Figure 2.4 summarizes the steps needed for the publication of an Android app, from the initial phase to its publicity.

2.2.4.2 iOS

The current SW development environment for the mobile environment of Apple is based on Xcode 5. It assists developers to create iOS apps. It includes automatic functions such as configuration of the apps in order to take advantage of the updated Apple services. It also manages images based on an asset catalogue, and assists in designing interfaces for iOS 7 and OS X. Furthermore, Xcode 5 is capable of analyzing the code, monitoring performance and testing apps. Ref. [53] includes the needed SW for the mobile environment of Apple and instructions for the app development.

2.2.4.3 Other Systems

BlackBerry is an open platform that provides a variety of development languages and runtimes according to the level of the skills of the developer. BlackBerry 10 offers the possibility to build integrated apps via C/C++/Qt, JavaScript, CCS, HTML, ActionScript, AIR and Java

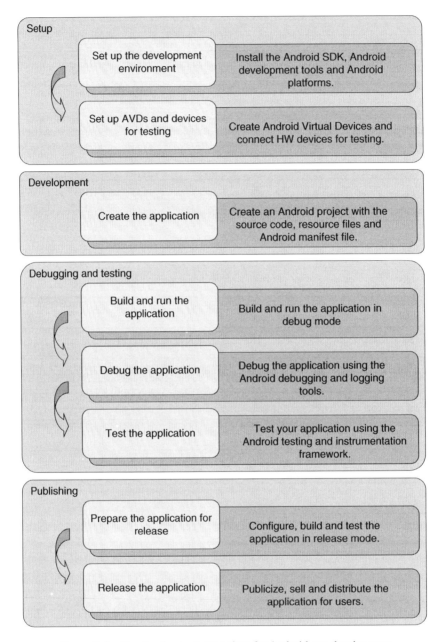

Figure 2.4   The development procedure for Android app development

Android. The development can be done via native BlackBerry 10 SDK, HTML5 WebWorks, Runtime for Android apps, PlayBook and BlackBerry OS. More information about app development in the BlackBerry environment can be found in Ref. [54].

For app developers interested in the Windows mobile environment, a logical strategy would be to select the most recent OS version to ensure an up-to-date consumer base. The evolution version of Windows 10 contains further native functionalities that not all the previous HW variants are capable of supporting. Ref. [55] presents the SW and instructions for the Windows Phone app development.

In addition to the above-mentioned variants, other OSs relevant to smart devices, tablets and other mobile devices include Firefox OS (Mozilla Foundation), Mer (Linux Foundation), Sailfish (Sailfish Alliance and Jolla), Tizen (Linux Foundation, Tizen Association, Samsung and Intel) and Ubuntu Touch (Canonical Ltd., Ubuntu community). These are typically based on free and open-source license while the iOS (Apple), Windows Phone (Microsoft) and BlackBerry OS (Blackberry Ltd.) are proprietary solutions.

2.2.5   Developers

The app ecosystem includes an essential element, the SDK, for each mobile application platform. The SDK provides access to the tools and set of Application Programming Interfaces (APIs) that are required for developing applications. Furthermore, the application framework facilitates the reuse of development components which results in ever richer and attractive apps in a fast schedule. The application framework also provides the means to take advantage of the HW of the user device, including supported sensors and the possibility to access location information of the user.

Distribution of the apps happens via the mobile application marketplace. The typical procedure is that the respective provider verifies the correct functioning, sufficiently high quality and safety of the apps, and places the certified apps to be downloaded by the users. The marketplace also functions as a forum where users can provide comments and evaluations. The marketplace typically is operated by OEMs, service providers, mobile application platform vendors and/or other parties.

The amount of apps that are based on intensive signalling between the UE and network are increasing. As identified in Ref. [34], chat-type of mobile apps may constantly poll servers for updates and for providing always-on immediacy which is thought to increase user experience. This background signalling results in devices to constantly connect to the mobile communications network which, in turn, causes connection attempts with numerous signalling messages. In many cases, this background signalling is merely additional load from the operator's point of view due to the fact that no updates are available.

The level of the background signalling load and thus the impact on the network capacity utilization depends heavily on the application profile. Video streaming requires wide bandwidth yet only a low amount of signalling. On the other hand, the messaging type of apps and social media utilization function normally with minimum bandwidth requirements but triggers at the same time a big amount of related signalling, which in turn may cause challenges for the operators along with the increased amount of users. The challenge is getting more critical because operators need to optimize constantly the limited resources, which means that these background signalling generating apps need to be optimized too, in order to minimize the impact on the QoS level experienced by other users.

The operators had strong control over the devices prior to the introduction of smart devices. The new technical solutions, together with a highly changed business environment where apps are the critical element for assuring competitiveness, mean that this control is lowering. The smart device markets are growing fast in all the regions worldwide. According to Ref. [35],

30% of all handsets sold in 2011 were smartphones, compared to approximately a 20% share of the handset market in 2010. About 10% of the global user base had subscribed via smartphones in 2011, and the Figure 2.1 indicates that the growth of smart device utilization has been significant.

As the hardware develops and the memory storage increases, users are able to download and utilize very large amounts of apps. The sufficient resources for the downloading and app signalling must thus be assured. As a solution for the possible non-optimal messaging of some apps, there may be a need for guidelines for the app developers to minimize the signalling. It is thus a matter of balancing the bit transfer in such a way that the user experiences are sufficiently good, the app business is sustainable, and the operator's network resources are not exploited.

The critical elements for the over-signalling are the Radio Network Controller (RNC) in 3G, and the Mobility Management Entity (MME) in LTE/LTE-A. In the worst case, heavy background signalling can cause temporal overload which has effects on the network resource availability and QoS for all the users. The optimization of the background signalling is thus an increasingly important task of the network operators, which requires cooperation between the operators, app developers and app businesses.

2.2.6   The Role of the SIM/UICC

An important innovation from the GSM era for current and future mobile systems is the SIM which still defends its position, providing one of the best protection mechanisms in the telecommunications networks. SIM, and its variant for the 3G era, the UICC with respective internal applications for different mobile networks, is a smartcard fulfilling the role of an SE. It is based on ISO/IEC 7816 standard with extensions for the wireless payment and access solutions via NFC via the ISO 14443 standard.

The SIM provides high security as a basis for the communications. In addition to the protection, it provides the radio network communications, it is a tamper-resistant physical HW element which is able to maintain the data protected. One of the security threats may be related to the intention to physically replicate the card information by making an exact copy of it, although even this does not reveal the cryptographic data in plain format. Furthermore, if such a copied card is utilized in the network (and thus a copy of the Mobile Subscriber's ISDN Number (MSISDN) is used), the operator's security breach analysis would reveal the suspicious communications and prevent the calls.

The role of the embedded Security Element (eSE) is currently changing the ways of how subscription management is handled. Some examples are autonomic telemetric or home utility equipment which can be managed remotely to switch the serving operator or subscriber's profile. As another example, the automotive industry is currently including mobile subscriptions into their vehicles. The fast development of connected cars is causing issues which require novelty protection mechanisms, the HW-based SE being one of the strongest components in the overall shielding [6]. The remote management of such subscriptions is especially beneficial to optimize the logistics as the cars can be transported to various countries from the production facilities, and the local subscription can be loaded, activated, changed and deleted during the lifetime of the vehicle depending on the preferences of the current and future owners.

One of the current trends in the consumer markets is the increased utilization of wearables like smart watches and even computers that are attached to clothes [13]. As they may be considerably small, they benefit greatly from the smallest Form Factors that the permanently installed eSE can provide compared to the physically removable SIM/UICC cards. As these devices are obviously equally vulnerable to security attacks, it is highly important to ensure the proper shielding of the remote management of the subscriptions. At present, standardization of interoperable and dynamic subscription management is under work, and is explained in more detail later in this book.

The SIM/UICC, either as a removable or as an embedded variant, is a suitable base for the combined functionalities that include telecommunications (authentication, authorization and secure radio interface) and other services like banking and physical access. Furthermore, it can be used as a secure data storage for the user's information. Along with the new methods under work and deployment, like HCE that may be based on one-time tokens and cloud storage, the combination of related services with physical SIM/UICC could provide a good level of security.

2.2.7   Challenges of Legislation

Protection against fraud as well as ways for sanctioning the malicious acts related to the early mobile communications networks was challenging as the legislation typically became outdated compared to the new aspects of such fast developing technologies. As an example, in the United States the Telecommunications Act of 1996 was the first major overhaul of telecommunications laws. The main goal of the new law is to liberate the communications business by allowing fair competition in any market.

One of the first type of fraud – in addition to the by then straightforward eavesdropping of unprotected radio channels – was the cloning of mobile equipment credentials and thus the telephone number of analogue systems. This resulted in the need for updating the legislation by criminalizing the use, possession, manufacture and sale of cloning hardware and software. As the digital mobile communications systems were already much better prepared against such frauds, based on the lessons learned, this types of fraud has practically disappeared.

A cloned mobile device refers to equipment that is reprogrammed and signals to the network the copied or modified HW ID. Nevertheless, the cloning of the hardware is in practice only relevant for 1G networks like Advanced Mobile Phone System (AMPS) and Total Access Communications System (TACS) that are now obsolete, and the still existing 'basic' Code Division Multiple Access (CDMA) devices with the subscription ID embedded into the HW. As has been the case with GSM and other enhanced 3GPP systems, the subscription is now stored in the removable SIM card, which makes the cloning of the device irrelevant. To provide an extra layer of security against mobile equipment cloning, CDMA device HW have a unique Electronic Serial Number (ESN) and telephone number (MIN) while the 3GPP devices supporting GSM, Universal Mobile Telecommunications System/HSPA (UMTS/HSPA) and LTE are equipped with an International Mobile Equipment Identity (IMEI) code.

Not only has the legislation been updated but also the preventing technologies have taken major steps in order to protect users and operators. On the other hand, the development of the mobile communications towards IP technologies, the same type of threats familiar from the public Internet are becoming more relevant in mobile communications. The Long Term

Evolution/System Architecture Evolution (LTE/SAE) network is based completely on IP which means that the same threats are valid there as in any other packet network. As for the security processes, the main aim of the LTE/SAE operator is to reduce the opportunities for the misuse of the network.

2.2.8   Updating Standards

The original ETSI specifications, GSM 02.09 and GSM 03.20, created the basis for GSM security solutions. Since the early days of the 3GPP 3G system, security has been identified as an essential part. From the first Release 99 specifications, there have been numerous completely new specifications produced by the SA3 Working Group, including the main definitions found in TS 33.102 (3G Security Architecture). The 3GPP has also produced advanced specifications for security, taking into account IP principles as the mobile networks are developing towards IP Multimedia Subsystem (IMS) and all-IP concepts.

The 3GPP SA3 Working Group has created new specifications for the LTE/SAE protection under TS 33.401 (Security Architecture of the SAE) and TS 33.402 (Security of SAE with Non-3GPP Access). The LTE system provides confidentiality and integrity protection for the signalling between the LTE-UE and MME. The confidentiality protection refers to the ciphering of the signalling messages. The integrity protection, in turn, assures that the signalling message contents are not altered during transmission.

The LTE radio interface traffic is secured by using the Packet Data Convergence Protocol (PDCP). In the control plane, the PDCP provides both encryption and integrity protection for the Radio Resource Control (RRC) signalling messages that are delivered within the PDCP packet payload. At the user plane, the PDCP performs encryption of the user data without integrity protection. It should be noted that the protection of the internal LTE/SAE interfaces like *S1* is left as optional.

2.2.9   3GPP System Evolution

Figure 2.5 presents the main elements and interfaces of the 3GPP GSM EDGE Radio Access Network (GERAN) (GSM and GPRS), Universal Terrestrial Radio Access Network (UTRAN), Core Network (CN), Enhanced UTRAN (E-UTRAN) (LTE radio network) and Enhanced Packet Core (EPC) of the LTE.

Being digital, the first phase of ETSI-defined GSM networks included the basic security like authorization and radio interface encoding. As another difference between earlier, 1G analogue systems, the concept of the smartcard in the form of a SIM was introduced to provide fluent user information portability between devices. Along with the further development of the 3G system, this base has remained compatible with GSM network architecture. Since the first GSM networks, the 3GPP systems have been developing. After the introduction of 3G UMTS and Release 8/9 LTE, operators have largely deployed also Release 10 and more advanced LTE networks in a global level. The Rel. 10 and beyond introduces completely new concepts like the home eNB that considerably change the network planning from highly controlled networks with known base station locations towards highly fragmented and unpredictable configurations as the end-users are able to easily purchase home base stations and freely deploy and switch them on and off.

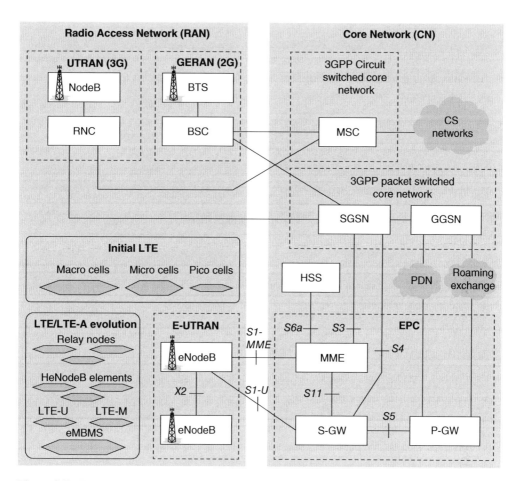

Figure 2.5   The main elements of 3GPP networks. The evolution of LTE brings new elements for, e.g., eMBMS, as well as cell extensions like relay nodes and Home eNB elements, while LTE also extends to unlicensed bands (LTE-U) and is optimized for IoT/M2M environment (LTE-M)

Release 10 also brings new aspects in the wireless security as the home eNBs are exposed to potentially malicious intentions. Thus, the network topology converts as a mix of different types of base stations. In this book, the term HetNet refers to the environment with many different eNB sizes within the same Radio Access Technology (RAT), while the term 'Heterogenous Network' refers to the wider concept which consists of different base stations over various RATs, such as a combination of LTE and Wi-Fi with data offloading properties.

2.3   GSM

When the first phase of GSM was under specification back in the late 1980s, the security aspects were quite different from the current world. Many of the modern threats we face today had not been experienced by the operators or users, such as mobile phone viruses and DoS attack attempts. At that time, the GSM system needed to have simply as good as or better security level than that of fixed networks. Since both GSM and fixed telephony networks were isolated from the external world, it was straightforward to comply with this requirement. The essential additional security aspects of GSM that ensured the radio interface protection at that time are the subscriber authentication and authorization, radio interface encryption for both signalling and communications, and the utilization of temporary identity during the communications.

2.3.1   The SIM

The SIM of GSM is a contact-based smartcard relying on the definitions of ISO/IEC 7816. SIM cards come in several sizes which are referred to as Form Factors (FFs). The first form factor 1FF refers to the credit card sized frame that was planned to be used in the first phase deployments in the early 1990s. In practice, the form factor 2FF with a size of approximately 2.5×1.5 cm initiated the commercial GSM era. After that, the further reduced cards are 3FF (micro card) and 4FF (nano card). Chapter 4 presents a more detailed description of the physical aspect of the SIM and its evolved variants.

The GSM SIM contains permanent records of the subscriber's identity (Ki key) and operator-specific authentication algorithms A3 and A8. In contrast, the standard-defined A5 algorithm is stored in the mobile equipment hardware. There are currently several versions of the radio interface encryption algorithm defined and available:

- A5/0, which refers to the unprotected connection without encryption algorithm.
- A5/1, which has offered good protection in the initial phase of GSM but has been compromised as a result of processor development and enhanced attack techniques. It is possible to find the Ki via brute force attacks combined by rainbow tables which significantly reduces the time for exploration [18].
- A5/2, which has been less protected since its deployment. The communications can be intercepted in real time by authorities [14].
- A5/3, which is a further developed variant but which may suffer from modern attack techniques [15].
- A5/4, which is the most recent updated algorithm currently providing the highest GSM security level.

The A5 algorithm is stored in the mobile terminal for the GPRS service, too. In that case, there is an adjusted A5 algorithm in use, the GPRS Encryption Algorithm (GEA), so the same SIM card functions in order to use both GSM and GPRS services.

In addition to the permanently stored information by the MNO, the user can store dynamic data such as phone numbers connected with alphanumeric information and short messages on the SIM card, according to the memory limitations of the card. The subscriber can, in principle, use any GSM terminal with the SIM card. Thus a subscriber to a European GSM 900 or GSM 1,800 network can use his/her own SIM card and a suitable mobile phone to use GSM 850 or GSM 1,900 networks in the United States, as long as the operator in question has signed a roaming agreement with the subscriber's home network operator. In practice, the GSM base band chip contains by default these quadra-bands and often also other GSM band extensions so the same GSM device HW are typically capable for global roaming.

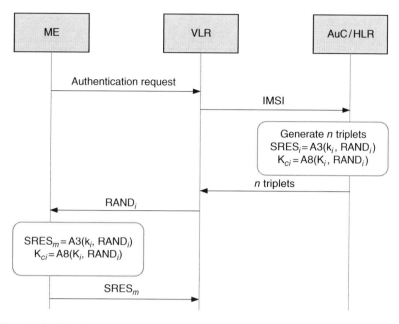

Figure 2.6   The signalling chart for the delivery of triplets from the AuC/HLR to VLR

The benefits of the SIM card are, e.g., the safety functions, interoperability with in principle any GSM phone and network, and the flexible adding and removing of services. As the SIM card is based on a regular smartcard, it is possible to combine functions unrelated to the GSM network such as an NFC-based bankcard app, making it a multifunctional card for a multitude of different services requiring high levels of security.

The SIM card of the GSM system is one example of contact smartcards; other examples include prepaid calling cards, bankcards and access control cards. Such smartcards have been specified in ISO series 7816 standards for the contact cards, and in ISO series 14443 for the contactless environment, which in mobile communications systems is done via NFC. More detailed descriptions of these SIM variants can be found in Chapters 4 and 5.

The user can activate and specify a Personal Identity Number (PIN) code, which the terminal requests when the device is powered on. An exception is the European emergency number 112, US emergency number 911, and other national variants, which are possible to dial without the PIN code. By default, it is possible to call emergency numbers even without a SIM card. The PIN code consists of a set of digits. If the PIN code function has been activated on the SIM card and an incorrect PIN code is entered three times, the SIM card starts to request a Personal Unblocking Key (PUK) code. If an incorrect PUK code is entered 10 times, the SIM card is blocked permanently.

The same principle of the physical card is utilized in 3G UMTS/LTE and 4G LTE-Advanced devices in the form of the more advanced UICC and its respective telecom apps, and the security functions have also been further developed as can be depicted from Figure 2.6.

2.3.2   Authentication and Authorization

The GSM system may check the user's permission to utilize the network as a part of the call set up, location area update and in the termination phase of the call. This happens with the help of the Authentication Centre (AuC). Even though the AuC is a separate logical functionality, it is typically integrated physically into the HLR because they do rely on the joint communication link.

Each subscriber is given a subscriber-specific key Ki which is stored in the user's SIM as well as in the AuC. This happens in the provision phase when the new subscription is created, and the subscriber profile is activated in the HLR of their home network. Figure 2.7 presents the security procedure in the call set up phase.

The AuC calculates in advance a security parameter triplet, or a set of these. The triplet contains parameter values for random number (RAND), SRES (Signed Response) and Kc (Temporal Key). This triplet is transferred and stored into the VLR to which the user is currently subscribed as shown in Figure 2.8. It should be noted that the user-specific Ki key is never transferred or exposed from the AuC or SIM.

As shown in Figure 2.9, the authentication happens in such a way that the AuC generates a new value of RAND. The value range of RAND is $0.2^{128}-1$. This value is sent as a part of the initial signalling via the Stand Alone Dedicated Control Channel (SDCCH) to the SIM.

In the next step, based on the received RAND and the Ki key and A3 algorithm stored in the SIM, the SIM calculates the SRES. Based on the same information, the AuC has also calculated the SRES in advanced. Now, the user device sends the SRES value to the corresponding VLR via SDCCH, and the VLR compares the SRES values it has received from the SIM and AuC. If the SRES values are different, the call attempt is terminated. The SRES may be incorrect if the user has an incorrect A3 or Ki key. The network reasons thus that the call attempt is not authorized.

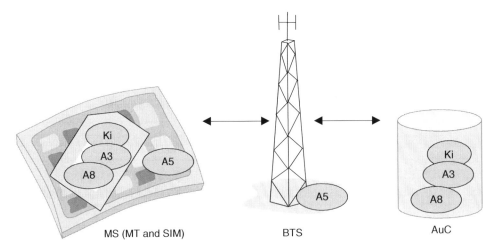

Figure 2.7   The subscriber-specific Ki, as well as the A3 and A8 algorithms are stored in the SIM and the AuC for the authentication, authorization and session key creation. The A5 algorithm is stored, in turn, in the HW of the Mobile Terminal (MT) and in the Base Transceiver Station (BTS) equipment for protecting the radio interface

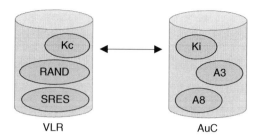

Figure 2.8   By utilizing Ki, A3 and A8, the AuC calculates the triplet, i.e., values for the Kc, RAND and SRES. The triplet is stored in the VLR

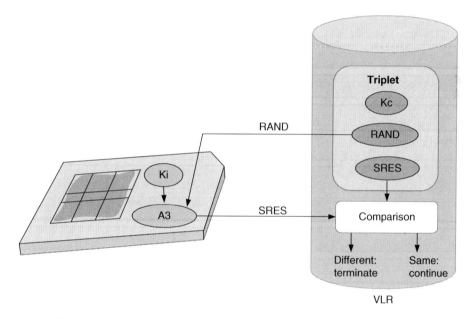

Figure 2.9   The authentication and authorization is done by A3, RAND and Ki

The A3 algorithm is operator-specific and resides in the SIM and AuC. It is recommended to use a sufficiently secure and complicated algorithm in order to assure that the Ki key cannot be calculated via RAND and SRES values. The A3 may be public, but typically the intention is to keep it classified. The length of the RAND is 128 bits and the SRES always consists of 32 bits. The length of the Ki key is operator-specific.

2.3.3   Encryption of the Radio Interface

As soon as the network has concluded that the user is authorized, i.e., the SRES is noted correct, the signalling and all the communications will be encrypted. This happens in two phases. First, a temporal key is generated, and secondly the key is utilized for the actual radio interface encryption.

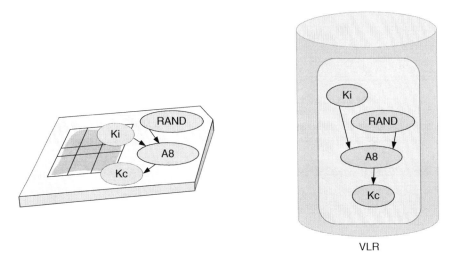

Figure 2.10   Kc is calculated with the A8 algorithm, based on Ki stored permanently within SIM, and RAND produced in the AuC/VLR

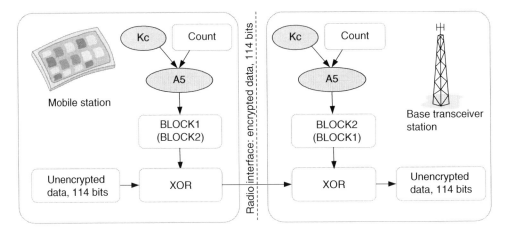

Figure 2.11   The encryption of the GSM radio interface takes place via the A5 algorithm

Both the SIM and AuC calculate the connection-specific key Kc by utilizing the same RAND value as described previously, and by applying the A8 algorithm that is stored in the SIM and AuC. Figure 2.10 clarifies the procedure. The length of the Kc is always 64 bits. If the key is shorter, zero bits are padded. The Kc is calculated separately for each connection to provide additional protection. The A8 algorithm is operator-specific, and the same principles for its recommended complexity apply as for the A3. As the A3 and A8 are both operator-specific, they can also be combined. It is of high importance to assure the combined algorithm A3/8 is not too simple and that the Kc can be protected.

As soon as the Kc is calculated, the actual encryption of the radio interface takes place as shown in Figure 2.11. From this point onwards, all the communications between the UE and

BTS is scrambled, including the signalling such as location area updates and handovers, voice and data calls as well as short messages and multimedia messaging.

For the radio interface encryption, the same connection-specific Kc and A5 algorithm are used. The algorithm takes as an input the Kc and the superframe number COUNT of the GSM radio interface timeslot structure, and the result is fed into a logical Exclusive Or (XOR) operation together with the single burst information of 114 bits.

The superframe cycle of the GSM system is about 3.5 hours, and the length of the COUNT is 22 bits. The block that A5 produces is called BLOCK1 or BLOCK2. BLOCK1 is utilized for the encryption and BLOCK2 is utilized for the decryption as presented in Figure 2.11.

Because the COUNT gets different value for each block, the scrambling of each burst's 114 information bit are different from previous ones. If the connection takes longer than the repetition cycle of the superframe, the COUNT values start repeating accordingly. The values for Kc and RAND are calculated from scratch each time a new connection is established.

The efficiency of the A5 algorithm does not depend on the non-publicity as each operator has a description of it. In any case, the GSM Association has decided to keep it confidential, so it has not been officially published. According to the GSM security specifications, there can be seven different A5 algorithms. At the moment, four algorithms have been produced, A5/1, A5/2, A5/3 and A5/4. The original A5/1 is designed as a strong one whereas the A5/2 is lighter, the latter being easier to decrypt, e.g., by governments. For the case where no encryption is used, A5/0 is utilized. The most recent and currently best protected A5/4 algorithm for 128 bit cases is defined as of Release 9.

At the start of the connection, the UE informs the network about the desired version of the A5 algorithm. If the AuC does not have the common algorithm, and the network does not support non-encrypted calls, the call attempt is unsuccessful. If the network supports non-encrypted connections, the call is established without encryption. In this case, the user will notice respective information on the display of the device to indicate that the communications is not encrypted. If common algorithms are present, the network decides which of the versions will be utilized for that connection.

2.3.4   Encryption of IMSI

The International Mobile Subscriber Identity (IMSI) is always a priority identifier within the GSM network, but operators typically want to avoid exposing it in the radio interface signalling. Instead, the Temporary Mobile Subscriber Identity (TMSI) may replace the IMSI. This gives additional means for the user identity protection. Nevertheless, the support of TMSI is not mandatory, so each operator decides the strategy for its use. If it is supported, the UE gets a new value each time it registers to a new Location Area (LA). The TMSI is sent over the radio interface encrypted via SDCCH.

2.3.5   Other GSM Security Aspects

2.3.5.1   Radio Parameters and Topology

Additional protection for eavesdropping is achieved via slow frequency hopping which has an interval of a single burst. Another aspect that increases the challenges in systematic user-specific eavesdropping or for getting location information is the relatively small cell sizes in urban areas and in the moving environment. Furthermore, the cells are typically sectorized, and power control as well as discontinuous transmission cause extra challenges for focused monitoring.

2.3.5.2 Equipment Identity Register

Each mobile phone (GSM and the devices of later generations) are tagged with an identity number, which is also stored in the Equipment Identity Register (EIR). The identity number has nothing to do with the subscriber number, and it indicates only the HW equipment. The EIR signals with the Mobile services Switching Centre (MSC), so that the network can, if necessary, check the equipment information. Information on faulty or stolen phones or mobile phones under surveillance can be stored in the EIR. It is possible to prevent the use of a mobile phone via the EIR which can store the IMEI codes.

The EIR consists of white, grey and black lists. The white list contains all equipment types that have been type approved and allowed to be used in the networks. The grey list contains equipment which needs be tracked or followed up, such as equipment which has received only temporal and conditional acceptance in the type approval process. The black list contains illicit equipment, such as stolen phones or equipment which has not received type approval. Even if a phone were placed on the black list, calls could be switched to the emergency number.

The international version of the EIR is called the Central EIR (CEIR), and is nowadays coordinated by the GSMA under the name *IMEI Database* as indicated in Ref. [61]. The CEIR compiles the information on illicit equipment that is delivered via an international packet-switched network. When information on stolen or illicit equipment is added to the black list of one operator, the information is also passed on to the CEIR and to the EIRs of all operators connected to the CEIR.

2.4   UMTS/HSPA

2.4.1   Principles of 3G Security

When the UMTS was specified, the stakeholders already had experience from the GSM networks that helped in the further development of protection mechanisms. The main principles of the UMTS protection are that it is based on the GSM security solution, that it takes into account the noted GSM weaknesses, and that it adds further security for 3G services.

Concretely, the trust principle of the UMTS was enhanced in such a way that it also includes the mutual authentication between the system and the user. This avoids the potential utilization of fraudulent spoof base stations which could capture user communications. There have also been enhancements in the protocols and algorithms, including the EAP-SIM for the 3G-WLAN interworking [3]. In the 3G, the related term is *User Authentication*, whereas the GSM utilizes the term *Subscriber Authentication*. More information on the GSM BTS vulnerabilities can be found in Refs. 1,2,4,5,11,19–21.

The UMTS also enhanced the radio interface encryption by introducing longer keys, publicly verifiable algorithms and multiple radio interfaces between the terminal and base station. Also the functionality of the SIM that is referred to as the Universal SIM (USIM) in the 3G network was enhanced. Furthermore, the encryption and decryption functionality of the radio

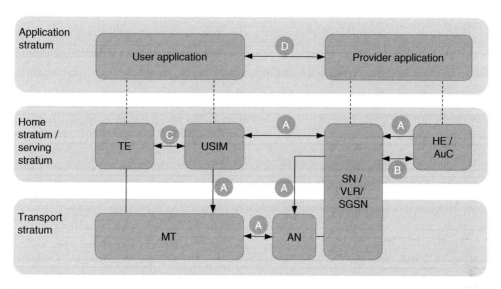

Figure 2.12 The 3GPP security architecture. The symbols of the figure refer to the following: (A) network access security; (B) provider domain security; (C) user domain security; and (D) application security

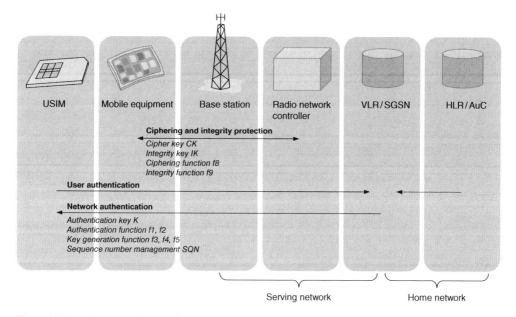

Figure 2.13 The role of the UMTS interfaces in 3GPP security procedures. *Source*: Reprinted from [38] by courtesy of ETSI

interface is moved to a deeper level, now residing at 3G RNC compared to the GSM which locates the decryption functionality at the base station. Figure 2.12 summarizes the 3G security architecture, and Figure 2.13 depicts the role of the UMTS interfaces in the security procedures [17].

The UMTS authentication and key agreement protocols, the UMTS Authentication and Key Agreement (AKA), are applied in the UMTS networks. Furthermore, 3GPP defines EAP-AKA for the authentication of the users that connect via WLAN, and EAP-AKA' to authenticate users that connect via trusted non-3GPP access networks to the LTE core, i.e., Enhanced Packet System (EPS).

The UMTS specifications define three entities related to the authentication: the Home Environment (HE) which means the home network; the Serving Network (SN) which can be the home network or the roaming network; and the USIM (UICC). At the start of the call attempt, the SN assures the user's identity is correct via a challenge–response procedure as is the case in GSM. In the UMTS, the User Equipment (UE) also makes sure that the SN is authorized by the HE to execute this procedure via the mutual authentication.

2.4.2   Key Utilization

The UMTS authentication is based on the user's permanent key K which is stored in the HE database as well as in the USIM. For the encryption and integrity check, additional temporal keys are generated. As soon as the user is identified, the actual authentication takes place via an AKA procedure. For that, an authentication vector is generated by the AuC and stored to the respective Visitor Location Register/Serving GPRS Support Node (VLR/SGSN). At the start of the authentication procedure, the VLR (in the case of a circuit-switched (CS) call) or SGSN (for packet-switched (PS) calls) sends a user authentication request to the UE accompanied by two parameter values for RAND and Authentication Token (AUTN). The USIM of the UICC receives these parameters. The USIM already has the permanent UMTS key K which is used to calculate the authentication vectors prepared previously by the AuC. This happens via multiple algorithms, and the outcome is the assurance if the AuC actually generated the AUTN. Now, if the AuC is noted as correct, a Response (RES) value is sent back to VLR/SGSN which, in turn, compares it with the Expected Response (XRES) generated at the AuC. If the RES and the XRES are the same, the authentication phase is passed and the call can be initiated.

Along with the authentication procedure, the temporal 128-bit Cipher Key (CK) for radio access network encryption and 128-bit Integrity Key (IK) values are calculated by the USIM which transfers them to the mobile equipment for the actual encryption. On the other hand, the CK and IK are transferred from the AuC to the VLR/SGSN. As the encryption and integrity procedure is initiated, these values are transferred to the RNC via the RAN Application Protocol (RANAP) message's Security Mode Command. The RNC thus takes care of the actual encryption of the UMTS. The aim is to assure the user and network that neither the CK nor the IK have been used previously. As a basis for this, the authenticated management field in the HE and USIM have authentication key and algorithm identifiers, and they limit the CK and IK usage before the USIM triggers a new AKA. The pre-requirement for the AKA that the UMTS utilizes is that the AuC and USIM share the user-specific key K as well as the message authentication functions (f1, f1*, f2) and key-generating functions (f3, f4, f5). The AuC is able to generate random integers and sequence numbers while the USIM is able to verify the freshness of these sequence numbers.

Prior to the ciphering, the UE and the RNC agree the version of the encryption algorithm. The algorithm set is enhanced and there is more variety in the UMTS compared to the GSM.

Table 2.1   Variables used by AKA in UMTS

Variable/ function	Description	Length/ bits
AK	Anonymity Key $f5_K$(RAND). The f5 is a key-generating function	48
AMF	Authenticated Management Field. Provides secured channel from HE to USIM in order to define operator-specific options during the authentication procedures	16
AUTN	Network Authentication Token (concealment of SQN and AK optional) SQN $\oplus$ AK ‖ AMF ‖ MAC	128
CK	Cipher Key $f3_K$(RAND). The f3 is a key-generating function.	128
IK	Integrity Key $f4_K$(RAND). Designed for integrity protection of signalling information. The f4 is a key-generating function	128
K	User Key	128
MAC	Message Authentication Code (based on RAND, SQN and AMF) $f1_K$(SQN ‖ RAND ‖ AMF). The f1 is a message authentication function.	64
RAND	Random challenge which the AuC generates	128
RES	Based on RAND, USIM calculates user response $f2_K$(RAND). The f2 is a message authentication function (possibly truncated)	32–128
SQN	Sequence Number	48
XRES	Based on RAND, AuC calculates expected user response $f2_K$(RAND)	32–128

The encryption is performed in the Medium Access Control (MAC) or Radio Link Control (RLC) layer. Each Protocol Data Unit (PDU) increases a relatively small counter. This counter is called the Connection Frame Number (CFN) in the MAC layer, and the RLC Sequence Number (RLC-SN) in the RLC layer. A Hyperframe Number (HFN) counter with a larger value range is also applied. The combination of all these counters is referred to as COUNT-C.

To complement the encryption, some additional inputs are needed: BEARER (radio bearer identity), DIRECTION, i.e., indication of the Uplink (UL) or Downlink (DL) encryption, and LENGTH (size of data to be encrypted). The integrity protection mechanism is also applied in the RRC layer with the IK. The RRC message together with DIRECTION (1 bit), IK (128 bits), COUNT-1 (32 bits) and random number FRESH (32 bits) are fed into the one-way function f9.

The UMTS AKA utilizes the variables and functions listed in Table 2.1. Based on the variables, the result is a set called Quintet which consists of the values of RAND, XRES, CK, IK and AUTN variables.

2.4.3   3G Security Procedures

The UMTS AKA protocol takes into account the compatibility with the previous GSM as widely as possible, yet adds security functionality specific for the 3G system. The original architecture of the GSM for the symmetric challenge–response functionality is thus still valid in the UMTS networks, while the 3G-specific enhancements include (1) authentication of HE to the user; (2) agreement on an IK between user and SN; (3) mutual assurance of freshness of agreed CK and IK between SN and user; and (4) joint specifications between 3GPP and 3GPP2 for providing global 3G roaming.

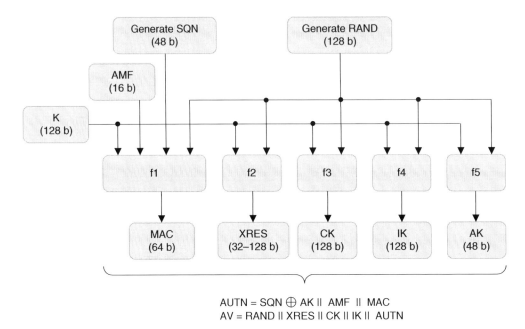

AUTN = SQN ⊕ AK || AMF || MAC
AV = RAND || XRES || CK || IK || AUTN

Figure 2.14   The principle of the 3G authentication vector generation as described in 3GPP TS 33.102

The following description summarizes the 3G security functionality while a more complete UMTS security description can be found in Ref. [64], and a more thorough analysis of the functionality of 3G AKA can be found in Refs. [25,26] which detail the 3GPP methodology for the authentication vectors as listed in Table 2.1. Figure 2.14 depicts the related high-level procedures. As a basis for the 3G AKA, there is a shared user-specific key K in both AuC as well as in the user's USIM. The AKA also uses message authentication functions f1 and f2, and key-generating functions f3, f4, f5.

There are two phases in the 3G AKA, which are the generation of authentication vectors procedure and posteriorly the authentication and key agreement procedure. The first phase relates to the *generation of authentication vectors*. In the initial stage, the HE/AuC receives an authentication data request from an SN which triggers the creation of a set of arrays of Authentication Vectors (AVs) in HE/AuC. The practical number n of these vectors can be around five. Each array of AVs consists of RAND, XRES, CK, IK and AUTN. The HE/AuC then sends this set of n arrays of AVs back to the requesting SN.

The second phase is the *AKA procedure*. The SN, via VLR or SGSN, selects one AV i from the received set of n arrays of vectors. It can be of any from the set of 1 … n. Based on this information, the SN sends RAND(i) and AUTN(i) to the user. Now, as a first step, the USIM makes sure that the AUTN(i) has the correct AUTN. If all is in order, the USIM calculates a respective Response RES(i) back to the SN. As is done in the GSM authentication procedure, the SN now performs a comparison of the RES(i) and XRES(i). In addition, if the results are the same, the USIM also calculates the CK and IK for further ciphering and integrity protection of the radio interface. The procedure for the generation of the authentication vectors is shown in Figure 2.14.

2.5 Long Term Evolution

2.5.1 Protection and Security Principles

The LTE/SAE system is based on IP which means that it may have vulnerabilities that are similar for any other packet network. One of the important aims of the MNO is thus to minimize the illegal opportunities to misuse the network. Since the early days of 3GPP 3G deployments, the security has been identified as an essential part of the service set. The first, Release 99, 3G standards included 19 new specifications by the SA3 Working Group. Ever since, 3GPP has produced advanced specifications for the security, taking into account increasingly the IP domain as the mobile networks are developing towards IMS and all-IP concepts.

The 3GPP SA3 Working Group continues to produce specifications for protecting the LTE/SAE networks. The LTE system provides *confidentiality* and *integrity protection* for the signalling between LTE-UE and MME. The confidentiality protection refers to the ciphering of the signalling messages. The integrity protection, in turn, assures that the signalling message contents may not be altered during the transmission.

All the LTE traffic is secured by the PDCP in the radio interface. In the control plane, the PDCP provides both encryption and integrity protection for the RRC signalling messages that are delivered within the PDCP packet payload. In the user plane, the PDCP performs encryption of the user data without the integrity protection. It should be noted that the protection of the internal LTE/SAE interfaces like *S1* is left as optional.

2.5.2 X.509 Certificates and Public Key Infrastructure (PKI)

The digital certificates are used to authenticate communication peers and to encrypt sensitive data. They are essential for Transport Layer Security (TLS) and for IP Security (IPSec) support. The X.509 certificates contain a public key that is signed by a commonly trusted party. Via this method, the receiving end trusts in the correctness of the public key, as long as a trusted party has confirmed the matching identity by using its digital signature as part of the certificate. This idea of the trusting parties and certificates forms a trust chain.

The essential challenge of the trust chain is to get the keys delivered into the place at the stage when the security does not yet exist. The most secure way would be to install the keys physically at the site the element is produced or installed. This solution is feasible for a low number of sites, or for a new network which is going to be commissioned in any case. Nevertheless, for a large network such as LTE, this is extremely challenging because the certificates have a limited lifetime and they need to be replaced from time to time.

In order to cope with this challenge, the Certificate Management Protocol (CMP) is a feasible option that industry is utilizing in practice. This standardized protocol provides the capability to retrieve, update and revoke certificates automatically from a centralized server. The initial authentication (when operator certificates are not yet in place) is done based on vendor certificates (which are installed in the factory) which are trusted by the operator's Certification Authority (CA). As a result, the eNB PKI can be introduced as a flat hierarchy, having one root CA in the operator's network.

The eNB can thus support secure device identity where a vendor certificate is installed to the eNB in the factory. The vendor certificate is used to identify the eNB in the r CA and to receive an operator certificate for the eNB. This functionality can also be used as a part of

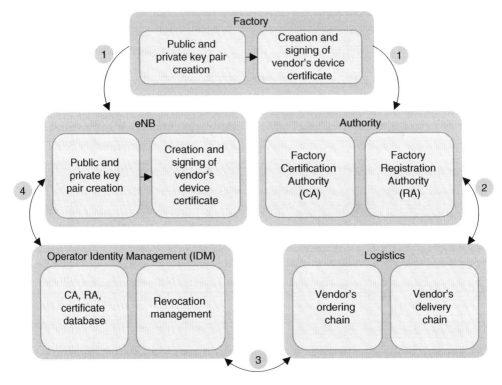

Figure 2.15   The principle of the vendor certificate process

SON LTE BTS Auto Connectivity type of features and in order to support the automated enrolment of operator certificates for base stations.

Figure 2.15 outlines an example of the interaction between the operator and vendor. The process begins in the factory, where the public and private key pair is created, and the vendor's certificate is created and signed. From the factory, the vendor's device certificate is stored to the eNB, and it is also delivered to the factory CA and factory Registration Authority (RA) as shown in the delivery chain (1) of Figure 2.15. Next, the product information consisting of the module serial numbers and vendor's root CA certificate is placed to the vendor's ordering and delivery chain. At this point, the eNB is shipped to the operator. Following the process, the vendor's device serial numbers and vendor's root CA certificate are delivered (3) to the operator's Identity Management (IDM). Next, the IDM creates an operator node certificate to assure the authenticity of the eNB (4). At this point, the operator node certificate substitutes the previous vendor's node certificate. It is now possible to ensure that the equipment is genuine by observing the serial number and vendor's device certification.

2.5.3   IPsec and Internet Key Exchange (IKE) for LTE Transport Security

The eNB follows the rules established by the Network Domain Security/IP Security (NDS/ IPSec) architecture. 3GPP introduces Security Gateways (SEGs) on the borders of security domains to handle the NDS/IPSec traffic. All the NDS/IPsec traffic passes through the SEG

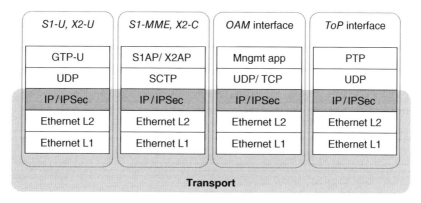

Figure 2.16 The eNB protocol stacks with embedded IPSec layer

before entering or leaving a security domain. The SEGs are responsible for enforcing security policies for the interworking between networks. In this role they also provide the IPsec functions.

It is possible to implement the SEG functionality in dedicated hardware (external SEG) or to integrate it in existing nodes (internal SEG). From the eNB point of view it is not relevant if SEGs are external to the peer entities or integrated, i.e., they are not visible to the eNB. At the eNB side, the IPSec function is integrated into the eNB. Therefore, the eNB represents a security domain of its own and can act as an SEG.

The following logical interfaces can be protected by means of IPSec: *S1_U i/f* (user data transport, or U-plane, between the eNB and S-GW via GTP-U tunnelling), *S1_MME i/f* (signalling transport, or C-plane, between the eNB and MME via the *S1AP* protocol), *X2_U i/f* (user data transport, or U-plane, between eNB nodes during handover via GTP-U tunnelling), *X2_C i/f* (signalling transport, or C-plane, between eNB nodes via the *X2AP* protocol), *O&M i/f* (transport of Operations and Maintenance, i.e., O&M data, or M-plane, between the eNB and O&M system), and *ToP i/f* (transport of Timing over Packet (ToP) synchronization data, or S-plane, between the eNB and ToP Master). Figure 2.16 depicts the eNB protocol stacks with the embedded IPSec layer.

2.5.4 Traffic Filtering

2.5.4.1 Firewall

It is possible to have the eNB element support a firewall function with the capabilities to ingress IP packet filtering, ingress rate limiting, egress rate limiting, and to contain DoS countermeasures.

2.5.4.2 Filtering for Site Support Equipment

The eNB may provide access to site support equipment (e.g. battery backup units) via additional Ethernet interfaces. Typically, this type of IP-based equipment does not provide its own IP packet filter or firewall. Thus, the site support equipment would be directly accessible if there were no packet filter at the eNB side. Therefore, the eNB provides an IP packet filter service that protects the site support equipment from harmful network traffic, and also protects the network from unintended traffic via this interface.

2.5.5   LTE Radio Interface Security

The following sections summarize information about the key hierarchy and the Key Derivation Function (KDF) as a part of the Access Stratum (AS) protection. The key derivation is relevant for normal operation, i.e., call establish and further in the case of a handover (i.e., inter eNB handover). A more detailed description of the LTE security can be found in Ref. [65].

2.5.5.1   Key Hierarchy

Figure 2.17 depicts the LTE key hierarchy concept. It presents the EPS key hierarchy which is valid for steady state when no handover takes place. Nodes are represented by frames, and keys are represented by boxes. An arrow represents a KDF. If a key is derived at one node and transmitted to another, then its corresponding box is located on the border of the frames which corresponds to the involved nodes.

The part of the key hierarchy which is known to the eNB is called the *eNB key hierarchy*. It comprises all AS keys, i.e., an AS base key K_{eNB}, and three AS derived keys which are K_{UPenc} for the UP encryption, K_{RRCint} for the RRC integrity protection and K_{RRCenc} for the RRC encryption.

K is the only permanent key in the LTE network. All other keys are derived on demand via the KDF which is controlled by a key derivation procedure. The existence of a key depends on

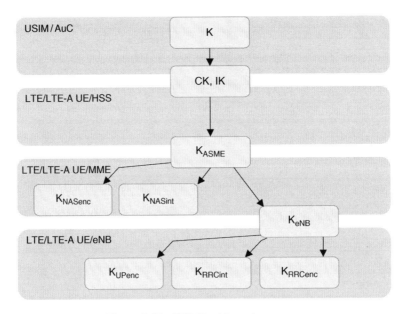

Figure 2.17   LTE Key hierarchy concept

the state in the following way: (1) K exists always; (2) Non-Access Stratum (NAS) keys CK, IK, K_{ASME}, K_{NASenc} and K_{NASint} exist in the EMM-*Registered* stage; (3) AS keys K_{eNB}, K_{UPenc}, K_{RRCint}, and K_{RRCenc} exist in the RRC-*Connected* stage.

2.5.5.2 Key Derivation Functions

The key derivation works with a KDF, which for the EPS is a cryptographic hash function with Ky derived from the KDF as a function of Kx and S. Ky calculates a hash with the key Kx from the string S. The hash value becomes the derived key Ky. In this context, Kx is a superior key which is located at a higher level within the hierarchies (besides the handover case which allows the same level). Furthermore, Ky is the derived inferior key whereas S is a concatenation of several substrings and may be classified as follows: (1) *Bound*, which refers to a string representation of parameters to which Ky will be bound to. Typically, these parameters describe a part of the environment such as the cell identifier, and Ky is only valid while these parameters do not change. (2) *Fresh* refers to a string representation of parameters which will ensure different 'fresh' Ky, also if all other parameters are unchanged. Usually, these parameters are unique for each instant of calculation, such as is the case for random numbers.

A cryptographic hash function provides a result of a fixed size and it is not reversible, i.e., it is not feasible at the current state of the art to derive unknown parameters if the result and the other parameters are known. In particular, it is not feasible to solve Kx even if Ky and S are known.

2.5.5.3 Key Establishment Procedures

There are three basic key establishment procedures. The first procedure refers to the Authentication and Key Agreement (AKA) which establishes CK, IK, and Key for Access Security Management Entity (KASME) in USIM and UE, and the same set is also established in the AuC, Home Subscriber Server (HSS) and MME. The AKA is an NAS procedure and does not have any prerequisites apart from the availability of the permanent key K. Please note that the MME is the Access Security Management Entity (ASME) for the EPS. The second procedure refers to the NAS Security Mode Command (NAS SMC) which establishes the NAS keys K_{NASenc} and K_{NASint} which are required for the NAS message encryption and integrity protection. The NAS SMC is an NAS procedure and needs a valid KASME as a prerequisite. In addition, the NAS SMC activates the NAS security. The third procedure refers to the AS Security Mode Command (AS SMC) which establishes the AS keys, K_{UPenc}, K_{RRCint} and K_{RRCenc}, which are needed for the UP encryption, RRC integrity protection and RRC encryption. The AS SMC is an AS procedure and needs a valid K_{eNB} as a prerequisite. In addition, AS SMC activates the AS security.

The establishment procedure for K_{eNB} depends on case. In the first case, at change to RCC-*Connected*, K_{eNB} will be derived in the MME and transmitted to the eNB by the S1AP *Initial Context Setup Request* message. In the second case, at active intra-LTE mobility, K_{eNB} will be derived by a procedure which is shared by source eNB and target eNB.

In the case of intra-LTE handover, the key hierarchy differs temporarily, because the K_{eNB} for the target eNB will be derived from the K_{eNB} of the source eNB.

2.5.5.4 Key Handling in Handover

Figure 2.18 depicts the key handling in handover procedure. The boxes represent keys and the arrows the KDFs. All keys of the same row are derived in a single chain of KDFs starting from an initial K_{eNB} or Next Hop (NH) parameter. These chains are called forward chains.

At the beginning of the context set up, an initial K_{eNB} key is derived from the KASME at the MME. This triggers the first forward chain (referred to as NH Chaining Counter, or NCC, which is set to 0). The initial K_{eNB} key is transmitted to the eNB and becomes its K_{eNB}.

At the handover instance, a transport key $K_{eNB}*$ and finally a fresh target K_{eNB} will be derived. Because the derivation uses a cryptographic hash function, it is not feasible to derive the source K_{eNB} from the fresh target K_{eNB}. Therefore a target eNB cannot expose the security of a source eNB. This is referred to as perfect backwards security.

However, if the derivation happens on one forward chain, i.e., if the $K_{eNB}*$ is derived from the source K_{eNB}, then the target eNB keys are not secret for the source eNB because their derivation is now known. This is true in a recursive manner. Therefore, all keys of the same forward chain which are located to the right hand side of any K_{eNB} are not secret to this key owner. In other words, there is no forward security in these cases.

In order to obtain forward security, the current forward chain has to be terminated and a new one started. This is done by deriving $K_{eNB}*$ from an NH parameter which was derived from KASME and not only from a source K_{eNB}. In the case of the *S1* handover, the NH parameter is transported by an *S1AP Handover Request* message and it applies for that specific handover instance. Therefore, forward security is reached in that handover, and is called forward security after one hop (which may also be referred to as perfect forward security). In the case of the *X2* handover, the NH parameter is transported by *S1AP Path Switch Acknowledgment* and may apply only the next time, but not for this specific handover because the new keys are already determined at this point of time. Therefore, forward security is reached at the next handover. This is called forward security after two hops.

If the NCC reaches its maximum value of 7, it will wrap around and start repeating at the next increment. Please note that 3GPP dictates that the first forward chain is skipped the first time round of NCC after the initial context set up.

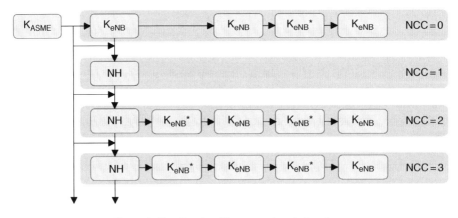

Figure 2.18   Key handling procedure in handover

2.5.5.5   Security Handling of RRC Connection Reestablishment

If the UE decides to initiate RRC *Connection Reestablishment*, the following two steps take place regarding the security: (1) The UE transmits an RRC *Connection Reestablishment Request* message on SRB0 to the cell it selected for requesting the reestablishment. This cell is called the *requested cell*. The message contains a UE-Identity to inform the eNB, which was the cell that triggered the RRC reconnection. This cell is called the *serving cell*. At the RRC *Connection Reestablishment*, the network adapts its serving cell assumption to the UE. (2) The RRC *Connection Reestablishment Request* message cannot be authenticated by the PDCP MAC-I integrity protection, because it transmits across SRB0. Instead, the requested eNB ensures the authentication of the received UE-Identity by comparing a received authentication code (shortMAC-I, which is included in the Initial UE-Identity, or IE) with an authentication code calculated by the network. Each cell enabled to authenticate the RRC reestablishment request applies a dedicated shortMAC-I because this code is bound to a cell. Now, if the serving cell is controlled by the same eNB as the requested cell, the authentication on the network side is an internal matter of the requested eNB and may happen based on need. If this is not the case, then if the serving cell is controlled by a different eNB as the requested cell, the authentication on the network side is handled between two eNBs, and the calculation of the network's authentication code takes place on the serving eNB because it requires the RRC integrity protection key (K_{RRCint}) of the serving cell. The comparison of the authentication codes takes place on the requested eNB because it received the shortMAC-I from the UE, and the calculation of a set of shortMAC-I and the delivery to another eNB happens in the course of a handover preparation.

The requested cell is controlled by the same eNB as the serving cell if the UE-Identity is related to the requested eNB, i.e., if the physical cell identity reported by the UE-Identity belongs to the requested eNB. If the RRC connection reestablishment request is accepted, both UE and eNB refresh their AS key hierarchy in the same manner as happens in the handover procedure from the serving cell to the requested cell, although always with the security algorithms of the serving cell. The possible cases are the same as for the authentication procedure, so as a first option, if the serving cell is controlled by the same eNB as the requested cell, the reestablishment procedure is an eNB internal matter on the network side. In this case, the key refresh happens as for the intra eNB Handover (HO) or, if the requested cell is equal to the serving cell, by intra cell AS security maintenance. As a second option, if the serving cell is controlled by a different eNB from the requested cell, the reestablishment procedure is a matter between two eNBs on the network side. The key refresh happens now as for inter eNB HO according to the HO type which prepared the target cell. In the case of *X2* HO, the source eNB needs to calculate a dedicated K_{eNB*} for each cell which will support a reestablishment and to signal it to target eNB in the course of handover preparation. This is similar to the shortMAC-I provision described above. In the case of *S1* HO, the target eNB derives the fresh K_{eNB} from an NH parameter received from MME. Because the NH is independent from the cell, no additional support for reestablishment is needed.

The UE needs to know the NCC parameter for key refresh. It is thus signalled by the RRC *Connection Reestablishment* message. This message is unprotected, because it transmits across SRB0. If the *X2* interface is not protected by IPSec then the *X2AP* messages including the keys are transferred in plaintext. The SRB1 and all posterior RRC connection reconfiguration procedure bearers will apply the fresh keys immediately.

2.5.6 *Authentication and Authorization*

In the authentication and key agreement procedure, the HSS generates authentication data and provides it to the MME element processing. There is a challenge–response authentication and key agreement procedure applied between the MME element and LTE-UE.

The confidentiality and integrity of the signalling is assured via the RRC signalling between LTE-UE and LTE (E-UTRAN). On the other hand, there is NAS signalling between LTE-UE and MME. It should be noted that in the *S1* interface signalling, the protection is not LTE-UE specific, and that it is left optional to implement the protection in *S1*.

As for the user plane confidentiality, the *S1-U* protection is not LTE-UE specific. There are enhanced network domain security mechanisms applied that are based on IPSec while the protection is optional. The integrity is not protected in *S1-U*.

Figure 2.19 depicts an example of a signalling flow for the authentication procedure. In the initial phase of the authentication, the MME element evokes the procedure by sending the IMSI as well as the Serving Network's Identity (SN ID) to the HSS of the home network of

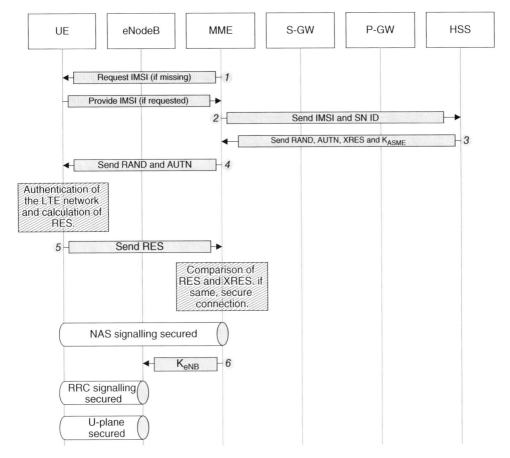

Figure 2.19   The mutual authentication procedure of LTE

the subscriber (2). If the MME does not have information about the IMSI code at this stage, it is requested first from the LTE-UE (1). The IMSI is delivered over the radio interface in a text format, which means that this procedure should be used only if no other options are available.

After the MME user authentication request to the HSS, the HSS responds with an EPS authentication vector containing RAND, AUTN, XRES and the ASME key (3). Upon the MME receiving this information, it sends the RAND and AUTN to the LTE-UE (4). Now, the LTE-UE processes this information and authenticates the network according to the mutual authentication concept. Based on the received information and its own key, the LTE-UE also calculates a RES and sends it back to the MME (5).

Both the LTE-UE and the HSS store the same algorithm which is utilized for the calculation of the response with the same inputs. The MME now compares the RES of the LTE-UE and the XRES that the HSS calculated previously. If they match, the LTE-UE is authenticated correctly and the NAS signalling will be secured. The eNB key K_{eNB} is calculated and delivered to the eNB in order to encrypt the radio interface for all further signalling and data transmission (6).

The normal procedures that are applied in the 2G and 3G environments for the physical protection of the customer subscription data, charging record data and other confidential information are also utilized in the LTE/SAE environment. The fraud prevention and monitoring applied in the previous generations are applicable also in LTE/SAE.

2.5.7 LTE/SAE Service Security – Case Examples

The LTE/SAE is currently changing the concept of the mobile communications clearly into an all IP environment. Along with its benefits, it also may open potential security holes. The motivations of such illegal activities exploiting the vulnerabilities of the systems include financial, destructive and political reasons, or the aim could also be simply to demonstrate the skills of the hackers.

As an example, the base station equipment has traditionally been well protected in 2G and 3G networks as it has been located physically inside the site with only limited access. Along with the home eNB concept, the equipment may be placed in public areas and homes which exposes the HW for potential malicious intentions. At the same time, the attack methods are getting more sophisticated, and advanced hacking tools are widely available via the Internet. The following sections present examples of the protection mechanisms for this new environment.

2.5.7.1 IPSec

The IPSec together with the PKI are utilized as a standardized LTE security solution. The PKI is applied to authenticate network elements and to authorize network access while the IPSec provides integrity and confidentiality on the transport route for the control and user plane. The IPSec concept is based on the certificate server which is the registration authority within the operator's infrastructure. It takes care of the certificates which provide secured IPSec routes between the elements via the migrated Security Gateway (SecGW), as shown in Figure 2.20.

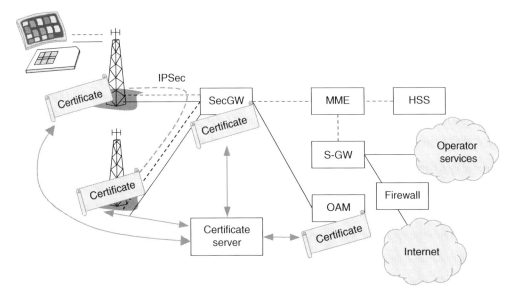

Figure 2.20 The architecture of the combined IPSec and PKI. The light dotted line indicates signalling, and solid line represents user plane data flow. The thick dotted line symbolizes the IPSec tunnel. The communication between SecGW as well as Operations Administration and Maintenance (OAM) can be done via Transport Layer Security (TLS) or Secure HTTP (HTTPS)

Security Gateways with high availability such as Juniper or Cisco are examples of scalable platforms terminating the IPSec traffic from the eNBs and anticipate fast-growing performance figures for the next few years. Insta Certifier is a PKI platform for issuing and managing digital certificates both for users and machines. It provides Certification Authority (CA) and Registration Authority (RA) functionality which are manageability for very large PKI environments by introducing centralized management of authentication keys with support for scalable revocation and key updates.

The LTE/SAE network elements possess identity and are able to authenticate. There are two kinds of authorities in the security solution. The first one is the Factory Registration Authority, which requests vendor certificates, centralized vendor-wide CA issues and keeps the certificates in a database. The second one is the Operator's Certificate Authority which recognizes the vendor's certificates and authorizes requests. This authority issues and manages operator certificates for the network elements. Figure 2.21 summarizes the SW architecture and interfaces for the PKI solution.

2.5.7.2   IPSec Processing and Security Gateway

The LTE standards define IPSec capability at the eNB level, including the authentication procedure based on X.509 certificates. The support of this functionality is mandatory, but for the trusted networks (considered reliable by the operator), the actual utilization of IPSec is left optional. In practice, the eNB may include both the IPSec and the firewall together with the authentication procedure via the X.509 certificates. The combination of the IPSec and the

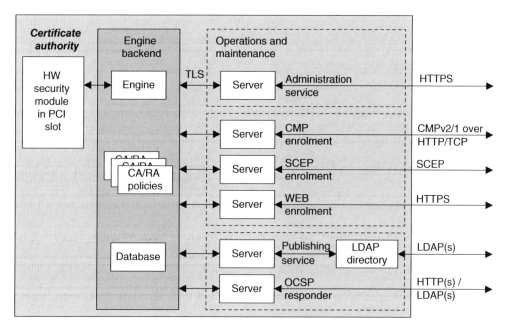

Figure 2.21 The PKI design with the architecture and interfaces

firewall provides the possibility to integrate the respective firewall rules with the rest of the network management which requires only minimum manual configuration.

In the following scenario, the SecGW is a complementary solution placed alongside the Aggregation Router (AR) with both incoming and outgoing interfaces connected to the AR. The advantage of this scenario is that only a few changes on the existing network are required. Furthermore, the scenario allows the aggregation of all interfaces between the AR and SecGW into one logical link. Once aggregated, various types of traffic can be separated on Layer 2 by defining corresponding Virtual LAN (VLAN) interfaces on the SecGW. This setup provides better flexibility and resilience against single link outages. Figure 2.22 depicts an integration example.

The SecGW may provide several options to achieve traffic separation. The first solution is the use of *virtual routers* which allows the separation of routing domains into logical entities and separated routing table. Each virtual routing entity handles its directly connected networks and static or dynamic routes. The second option is based on *VLANs* that are used to separate the traffic within physical links. All security concepts work with physical and logical sub-interfaces. The third option is the definition of dedicated *security zones* as shown in Figure 2.23. These zones are used to logically separate network areas and provide more granular traffic filtering and traffic control by defining access control and filter policies.

The Virtual Private Network (VPN) design of the SegGW can be based either on a single tunnel set up, or a multiple tunnel set up. For the *single tunnel set up*, the traffic of all planes is encrypted with the same encryption set, i.e., either a single IKE SA or a single IP-SEC SA. The set up can be based on the dedicated tunnel interface per eNB, which means

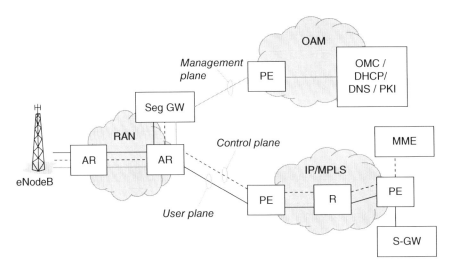

Figure 2.22   An integration example for the gateway attached to the access router

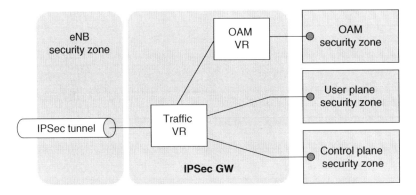

Figure 2.23   The security zone principle

that each eNB has its own tunnel interface on the SecGW. The set up can also be based on the shared tunnel interface, which means that all the eNBs share one tunnel interface on the SecGW.

For the *multiple tunnel set up*, the traffic for each plane is encrypted with different encryption sets (1 IKE-SA/3 IP-SEC SA). The multiple tunnel set up can be based on the dedicated tunnel interfaces per eNB, or on the shared tunnel interfaces.

2.5.7.3   Single Tunnel with Dedicated Tunnel Interfaces

The benefit of this solution is that it provides a persistent tunnel interface per eNB. All routes to one eNB point to the same interface, which makes the design easier. The third benefit is the small amount of Security Associations (SAs). As a drawback, the solution requires a large amount of tunnel interfaces.

2.5.7.4   Single Tunnel with Shared Tunnel Interfaces

The benefit of this design is that only one tunnel interface is required per chassis. The eNB inner routes can be aggregated on the SecGW. As in the previous case, only a small amount of security associations is needed. The drawback of this solution is that the scalability of the design is linked to the IP address concept.

2.5.7.5   Multiple Tunnels with Dedicated Tunnel Interfaces

The benefit of this solution is related to the dedicated tunnel interfaces per plane per eNB. A drawback is that three tunnel interfaces per eNB are required. Another issue is the larger amount of security associations. Due to these drawbacks, this is the least feasible solution in practice.

2.5.7.6   Multiple Tunnels with Shared Tunnel Interfaces

The benefit of this solution is that only one tunnel interface per plane per chassis is needed. In addition, the eNB inner routes can be aggregated on the SecGW. As drawbacks, larger amounts of security associations are required. Also, an additional IP network for the VPN next-hop table is required if the eNB inner routes cannot be aggregated per plane. Furthermore, the scalability is limited by the IP address concept.

2.5.8   Multimedia Broadcast and Multicast Service (MBMS) and enhanced MBMS (eMBMS)

2.5.8.1   General

The MBMS offers broadcast and multicast user services over cellular networks. The MBMS specifications define four types of services: (1) streaming services; (2) file download services; (3) carousel services; and (4) television services.

The MBMS is suitable in many environments for data reception. It applies to consumer use cases like audio/video contents streaming as well as to M2M use cases like a car's navigation system map updates. Furthermore, the MBMS contents delivery can be varied based on location and device type.

The MBMS introduces a Point-to-Multipoint (PTM) service into 3GPP 2G and 3G systems. Originally standardized in 3GPP Rel. 6, the enhanced MBMS (eMBMS) via LTE provides solutions for delivering a variety of broadcast/multicast contents. One of the requirements of the MBMS User Service is to be able to securely transmit data to a given set of users. In order to achieve this, there needs to be a method of authentication, key distribution and data protection for an MBMS User Service as defined in 3GPP TS 33.246 (Rel 12). This means that MBMS security is specified to protect MBMS User Services, and it is independent of whether multicast or broadcast mode is used. The security refers to the protected delivery and utilization of the contents, e.g., by limiting the access to closed channels for authorized user groups (pay channels, special groups). In a wider context, security also may refer to the assurance of the contents delivery. As the MBMS is based on downlink data there is no need for a separate return channel so all the contents are broadcasted/multicasted without acknowledging the success of the reception (which is the benefit of MBMS as it does not reserve additional resources per user). The MBMS includes a Forward Error Correction (FEC) mechanism and file repair functionality. The level of these functionalities is parametrized, and the feasible selection of the values is one of the optimization tasks of the MNO.

The delivery method of MBMS leads to different challenges for securing the connection and contents compared to the security of Point-to-Point (PTP) services, like eavesdropping on non-public contents over a wide service area, or a valid subscriber may circumvent the security solution, e.g., by publishing the decryption keys enabling non-subscribers to consume broadcast content.

The MBMS contains two data transmission modes which are broadcast mode and multicast mode. The main principle of the MBMS broadcast service is that it is a unidirectional PTM service. This service delivers the data stream from the network to all capable 3GPP User Equipment (UE) within the respective broadcast service area. In this way, broadcast services can be received by all the users within the service area if they have activated the reception of the MBMS service in their UE.

The MBMS utilizes the radio and core network resources in an optimal way as the data is transmitted over the common radio channel, and it is flexible for the data rate. The transmission thus adapts to variable radio access network capabilities as well as for varying availability of resources. The adaptation happens by adjusting the MBMS streaming bit rate accordingly. As there is no return channel present for error recovery purposes for the broadcast mode, the correct reception of MBMS cannot be guaranteed. Nevertheless, as FEC functionality is incorporated, the receiver may be able to recognize data loss and try to recover the data according to the FEC principles.

In addition to the broadcast mode, MBMS also contains the multicast mode. In this case, the data is transmitted over a unidirectional PTM connection from a server to multiple subscribers that are present in the multicast service area. Compared to the broadcast mode, multicast services can only be received by users that have subscribed to a certain multicast service and have joined the multicast group associated with the specific service. The basic principle is that the broadcast mode is open by default whereas the multicast mode typically requires a subscription to multicast groups prior to the user joining the multicast group.

The MBMS user services can be divided into three groups which are download delivery, streaming delivery and carousel service. The download delivery service delivers files (binary data) via the MBMS bearer. The UE, i.e., the MBMS client activates the related application and consumes the received data. In this solution, it is important that the data is received in a reliable way which makes the FEC method essential.

The streaming delivery service provides a continuous stream of, e.g., audio and video. Supplementary information via text or still images can also be important in this service. This provides an interactive way of using additional services. As an example, the received text may include web links that lead to additional information related to the contents. In this scenario, the link received via broadcast delivery, the user can access the additional content simply by clicking the link, and by creating a dedicated PTP via any available access method.

The carousel service is able to repeat a broadcast transmission of, e.g., a file or set of files repeatedly in such a way that the customer can receive the same contents every now and then. This can be, e.g., an update to the map data of location-based service. The carousel service combines aspects of the download and streaming services.

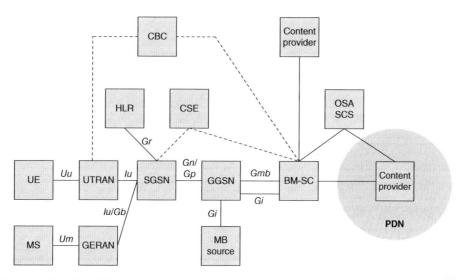

Figure 2.24   The MBMS reference architecture. *Source*: Reprinted from [38] by courtesy of ETSI

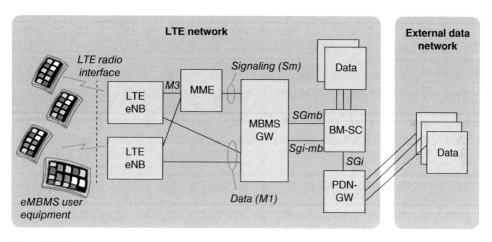

Figure 2.25   The eMBMS reference architecture. *Source*: Reprinted from [38] by courtesy of ETSI

In order to provide MBMS bearer services, cellular network elements like GGSN, SGSN, RNC or Base Station Controller (BSC) take part in various MBMS related functionalities. Figure 2.24 shows the reference architecture of MBMS.

In practice, the basic MBMS service has not taken off commercially. The upgraded eMBMS architecture meant for the LTE networks as defined in 3GPP Rel 9 and beyond is depicted in Figure 2.25. In this solution, the Broadcast Multicast Service Center (BM-SC) is connected to the MBMS gateway, which in turn, connects directly to eNB elements for the user plane contents transfer, and via MME for the initial PTP signalling.

The BM-SC provides functions for the MBMS user (both MBMS and eMBMS) service provisioning and delivery. It may serve as an entry point for content provider MBMS transmissions, used to authorize and initiate MBMS bearer services within the Public Land Mobile

Network (PLMN) and can be used to schedule and deliver MBMS transmissions. The BM-SC shall be able to generate charging records for content provider transmitted data, and supply the GGSN with transport associated parameters such as QoS and MBMS service area in order to initiate and terminate MBMS bearer resources for following transmission of MBMS data. The BM-SC also needs to accept content from external sources and transmit it using error resilient schemes, schedule MBMS session retransmissions, and label each MBMS session with an MBMS Session Identifier to allow the UE to distinguish the MBMS session retransmissions. Furthermore, the BM-SC provides service announcements for multicast and broadcast MBMS user services, and transfers data on separate MBMS bearer services for 2G or 3G coverage, typically having different QoS.

The UE supports functions for the activation and deactivation of the MBMS bearer service and security functions as appropriate for MBMS. The UE should be able to receive MBMS user service announcements, paging information (non-MBMS specific) or support simultaneous services. The MBMS Session Identifier contained in the notification to the UE shall enable the UE to decide whether it ignores the forthcoming transmissions of the MBMS session.

The UTRAN/GERAN network is responsible for efficiently delivering MBMS data to the designated MBMS service area. Efficient delivery of MBMS data in multicast mode may require specific mechanisms in the UTRAN/GERAN. Intra-RNC/BSC, inter-RNC/BSC mobility of MBMS receivers shall be supported. The UTRAN/GERAN shall be able to transmit MBMS user service announcements, paging information and support other services parallel to MBMS.

The SGSN within the MBMS architecture performs user individual MBMS bearer service control functions and provides MBMS transmissions to UTRAN/GERAN. The SGSN provides support for intra-SGSN and inter-SGSN mobility procedures. The SGSN generates charging data per multicast MBMS bearer service per user. The SGSN is also able to establish *Iu* bearers (SGSN-UTRAN signalling) and Gn bearers (SGSN-GGSN signalling) shared by various users on demand when data is transferred to the users. This should be done upon notification from the GGSN.

The GGSN within the basic MBMS architecture serves as an entry point for IP multicast traffic as MBMS data. Upon notification from the BM-SC the GGSN shall be able to request the establishment of a bearer plane for a broadcast or multicast MBMS transmission. Furthermore, upon BM-SC notification the GGSN shall be able to tear down the established bearer plane. Bearer plane establishment for multicast services is carried out towards those SGSNs that have requested to receive transmissions for the specific multicast MBMS bearer service. The GGSN shall be able to receive IP multicast traffic from BM-SC and to route this data to the proper GPRS Tunnelling Protocol (GTP) tunnels set up as part of the MBMS bearer service. The GGSN may also receive IP multicast traffic from other sources than the BM-SC. However, the MBMS bearers are not used to forward this traffic from non-BM-SC sources. The GGSN shall collect the charging data.

2.5.8.2 Solutions for MBMS Security

The LTE network attach/context activation procedure limits the possibilities of non-LTE subscribers to misuse the contents. For example, television type of contents delivery, protecting a potential separate, content-related decryption key publishing might require, e.g., frequent updating of decryption keys in a non-predictable manner while making efficient use of the radio network. The open contents do not basically require specific secure technologies apart from the normal network access procedures. The consumption of MBMS contents can thus be done via a streaming app. Nevertheless, if the contents need to be limited based on, e.g., subscription, time and quality, the standard MBMS procedures might not suffice, but additional solutions could be used, like more thorough app-based management of contents consumption. It should be noted, though, that the pure software might be vulnerable to attacks.

When selecting the proper security mechanism for MBMS contents delivery and utilization, the key task is to balance fluent user experience and adequate protection level. The LTE MBMS provides a highly efficient way to deliver broadcast and multicast contents over the designed service area by maintaining the network load under control as the amount of receiving customers does not increase the data transmission. Straightforward solutions for protecting the contents are based on application level reception of the contents as the customer can be authenticated and authorized for the subscription via the mobile network's procedures. Nevertheless, additional and more precise protection of the contents per customer (e.g., pay-TV during certain hours/occasions like a football event) may benefit from more sophisticated methods like the TEE. The following sections summarize some of the most logical security solutions for MBMS.

2.5.8.3   Security Principles

The Stage 1 requirements for MBMS include security aspects such as the following, according to 3GPP TS 22.146, Release 12. Only those users who are entitled to receive a specific MBMS service may do so. It should be possible to choose whether a given MBMS service is to be delivered with or without ensured group security. If a terminal supports MBMS, then it shall support UICC based key management and all the functions and interfaces required for it. In addition, Mobile Equipment (ME) key management shall be supported. If the UICC is capable of MBMS key management, ME key management shall not be activated; otherwise, ME-based key management is applied.

The Stage 2 requirements include, according to 3GPP TS 22.246 (Release 12), rules dictating that any user-modifiable MBMS service data (e.g. storage of deliveries in the UE, data type and format specific behaviours) shall only be modified by the authenticated user. Stage 3 details further the MBMS security as described in 3GPP TS 33.246 (Release 12), section 4. According to it, the MBMS User Service must securely transmit data to a given set of users by using methods for authentication, key distribution and data protection. This means that the MBMS security is specified to protect the MBMS User Services, and it is independent of whether multicast or broadcast mode is used.

In addition to ME and UICC-based key management that are based on the 3GPP specifications, the eMBMS could also be combined with other solutions that provide additional layers of security, such as TEE or OMA DRM.

2.5.8.4   ME-based Security

The ME procedure as described in 3GPP TS 33.246 relies on the security of the LTE infrastructure, see Figure 2.26.

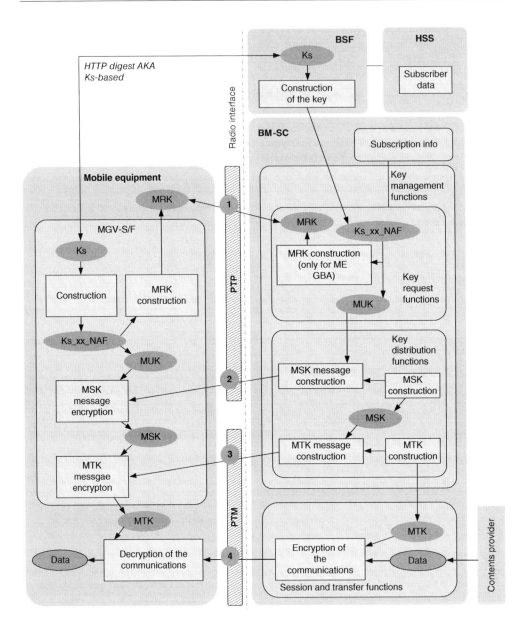

Figure 2.26 The elements and key management procedures for ME-based eMBMS security as described in 3GPP TS 33.246. The events in the radio interface are the following: (1) HTTP Digest authentication with the MRK key; (2) MIKEY MSK key distribution which is protected with the MUK key; (3) MIKEY MTK key distribution which is protected by the MSK key; and (4) user data which is protected via the MTK key. *Source*: Reprinted from [38] by courtesy of ETSI

2.5.8.5   UICC-based Security

The UICC functions for the establishment of the LTE connection which in turn provides the eMBMS utilization as any other LTE service. When more detailed contents security procedures for the eMBMS are done via UICC, it may require additional development efforts such as UICC Operating System support for key derivation functionality, Generic Bootstrap Architecture (GBA) support, support for eMBMS-specific functionalities like the Authenticate command, and support of odd values of instruction bytes (ODD INS bytes). In addition, compliance with the ETSI secure messaging requirements is needed.

2.5.8.6   TEE-based Security

The TEE is based on Trusted Applets within the Secure World, which is isolated from the Normal World. The secure services, or *trustlets* can run within the TEE which protects the contents and is separated from the general operating system, that is, Rich OS, which can be for example Android. Trustlets can be provisioned and managed over the air via TEE-TSM (Trusted Service Manager). The TEE is described in more detailed in section 6.3.3.

2.5.8.7   Digital Rights Management (DRM)

DRM refers to a set of methods for access control technologies for copy protection designed/applied by hardware and software manufacturers, publishers, copyright holders and any other party wanting to control the use of digital content and devices. The early DRM software was intended to control copying while the more mature variants of DRM aim to control executing, viewing, copying, printing and altering of works or devices.

The key benefit of DRM is that it provides an additional level of protection to prevent unauthorized sharing. Its weakness is that with time and sufficient efforts, there may be ways to bypass it. Also, user experience may be limited as DRM can sometimes be non-optimal for purchase portability between devices.

As for feasible DRM solutions, the Open Mobile Alliance (OMA) Mobile Broadcast Services Enabler Suite (BCAST) is an open global specification for mobile television and on-demand video services which can be adapted to any IP-based mobile and Point-to-Point (PTP) content delivery technology, end-to-end service and content protection based on the OMA BCAST smartcard and OMA DRM profile. Designed to support broadcast technologies such as DVB-H, 3GPP MBMS, 3GPP2 mobile unicast streaming systems and ATSC-M/H, the OMA BCAST 1.0 relies on the XML structure to specify a set of features such as electronic service guide, file and stream delivery, service and content protection using the smartcard or DRM profiles, terminal and service provisioning, as well as interactivity and notifications.

2.5.8.8   Comparison of Security Options

As reasoned in the previous section, the eMBMS may be used as standalone solution with its own integrated security, or with an additional security layer. Table 2.2 summarizes some pros and cons of the security options presented previously.

Table 2.2 Comparison of MBMS security solutions

	App-based	TEE	DRM	Terminal-based	UICC-based
Principle	Streaming and contents selection via pre-installed or downloaded software	Trusted applet within TEE's secure world which is based on TSM	DRM via various methods	Embedded functionality within the device	Software embedded into UICC or embedded secure element
Pros	Straightforward to develop and deploy	Good protection for streaming and displaying	Set of established methods	Standard solution for mobile device hardware	Provides highest security level via tamper-resistant element
Cons	Security level and functionality may be limited (e.g., only for initial access)	Limited terminal availability and networks supporting TEE/TSM	Possibly limited end-user experience	Terminal/chipset availability supporting eMBMS	May require OS enhancements and support for key derivation, GBA and eMBMS authenticate command

Figure 2.27   The protocol layers of FLUTE

2.5.8.9   MBMS Error Correction

The File Transport over Unidirectional Transport (FLUTE) is a transport protocol that is designed to deliver files from a set of senders to a set of receivers over unidirectional systems. The FLUTE was developed by the Internet Engineering Task Force (IETF), and it serves for the file transmission over wireless and point-to-multipoint systems, such as eMBMS. The FLUTE is specified on top of the Asynchronous Layered Coding (ALC) protocol instantiation of the Layered Coding Transport (LCT) building block. The FLUTE uses the UDP/IP layer and is transported via respective packets. It is thus independent of the IP version and underlying link layers. The FLUTE uses a File Delivery Table (FDT) to provide a dynamic index of files. Optionally a Forward Error Correction (FEC) can be used for improving the reliability of the downlink data transfer. A Congestion Control (CC) can also be optionally selected in order to enable Internet friendly bandwidth use. See Figure 2.27.

2.5.8.10   MBMS Charging

According to the specifications for the charging, broadcast and multicast modes as stated in 3GPP TS22.146 (Rel. 12), it shall be possible to collect charging information for the transmission of broadcast services to enable billing of broadcast services providers, such as billing third parties for advertising. The charging information therefore needs to be secure to ensure correct billing. Some examples of types of charging information are the broadcast usage duration broadcast volume of contents, multicast session duration and time when joining and leaving a multicast subscription group. Furthermore, the charging information may capture the duration of membership to a multicast subscription group, and multicast session volume of contents. It is also possible to collect subscriber's charging information in home networks and when roaming based on the security procedures for the key management, in order to bill based on the reception of broadcast data on a per-broadcast service basis. Furthermore, the MBMS User Service shall support charging mechanisms as stated in 3GPP TS 22.246 (Rel 12), which describe the charging on a subscription basis as well as the charging for keys that allow the user access to the data.

2.6   Security Aspects of Other Networks

2.6.1   CDMA (IS-95)

Interim Standard 95 (IS-95) is a US-driven digital CDMA cellular system, known also as cdmaOne or TIA-EIA-95. IS-95 belongs to the 2G cellular systems. The IS-95-based networks are utilized widely in North America, and it is also supported in various areas outside the United States. The first specification set for the system was called IS-95A, and the enhanced specification IS-95B was released later. The system is meant for voice and data services. The data transfer rate for IS-95A is up to 14.4 kbps, and for IS-95B up to 115 kbps.

IS-95 is based on Direct Sequence Spread Spectrum (DSSS) technology. It functions efficiently to hide the signal in the radio spectrum below the thermal noise level, which makes the detection and jamming of the signal challenging; in fact, for that reason, the technique was first utilized in the military environment well before its commercial phase.

The basic CDMA system, i.e., IS-95A and IS-95B, later evolved towards the 3G system. The resulting 3G system is called commercially as CDMA2000. This system includes CDMA2000 1x and CDMA2000 1x EV-DO (Evolution Data Only/Data Optimized). There has also been a theoretical variant called CDMA2000 1x EV-DV (Evolution Data and Voice).

CDMA is both a radio interface definition as well as an access method for cellular systems. It is based on the separation of users by different codes, dedicated for each user within the same area. To complement the whole network system, the core part is based on the Time Division Multiple Access (TDMA) principles in a similar manner as in the GSM system. There are thus considerable similarities in the practical solutions of the upper layers to the air interface, including Radio Resource Management (RRM) and Mobility Management (MM).

The idea of the DSSS is to multiply the data stream with another one that is using a considerably higher data rate. This spreading code widens the frequency bandwidth for the actual data transmission. The original data stream can be detected when the very same spreading code is applied in the reconstruction of the data. As the spreading codes and data stream of a certain user is seen as background noise to outsiders, it is thus possible to utilize the same, wide frequency band for several users in such a way that they are not interfered by the others. This phenomenon can be compared with a cocktail party where participants speak in different languages (comparable to scrambling codes). The participants interpret the unknown languages merely as background noise, and even if the noise level rises, the ones that understand a common language can interpret the respective contents from far away. The correlation of the codes for scrambling and unscrambling thus provides the possibility to apply multiple access entries in the system.

Via spread spectrum techniques like the DSSS, CDMA utilizes the same, relatively wide frequency band for sharing the data transmission for all the users located in the area. The essential parts for making the CDMA system work are the transmit power control and error correction codes. There are also other functionalities which increase the performance of the system, like the RAKE receiver for combining multi-path propagated signal components of the same signal, variable data rates for data transmission and voice service, and a soft handoff mechanism. The complete set of different functionalities provides the possibility to lower the requirements of the carrier per interference (C/I) level which in turn enhances the spectral efficiency.

Each IS-95 carrier contains logical channels. As everything is transmitted and received within the same frequency band of the carrier, these channels are separated from each other by a variety of spreading codes. When the signal is processed, it goes through several steps including the coding itself and interleaving, which results in the symbols to be transmitted over the radio interface. In order to do this, the symbols are modulated with Walsh codes.

The security of the IS-95 radio interface is based on the CDMA signal itself (being below the noise level), and Cellular Authentication and Voice Encryption (CAVE) algorithm [56]. The CAVE is based on a 64-bit A-key which is used together with an Electronic Serial Number

(ESN) and a random number in order to generate a 128-bit long Shared Secret Data (SSD). Furthermore, the SSD is divided into two 64-bit blocks from which block A is meant for authentication and block B for encryption. The principle uses a challenge–response technique for the authentication of the subscription while the random number is used to create a session key for the encryption of voice and data traffic. Ref. [57] claims that there is an inherent signal feature on the plaintext in the IS-95 CDMA system, and has been able to prove the functioning of a ciphertext-only attack method to solve the initial phase of the key sequence by eavesdropping on a 20 ms ciphertext frame. The described attack method is based on the exploitation of the linear relations between the key sequence and the state of the long code generator which has resulted in an algorithm for decoding the private mask, showing that the voice encryption of IS-95 system is unsecure against ciphertext-only attack.

2.6.2   CDMA2000

CDMA2000, which also is known as International Mobile Telecommunications-Multi Carrier (IMT-MC), belongs to the 3G cellular systems together, e.g., with UMTS. CDMA2000 standards consist of CDMA2000 1X, CDMA2000 EV-DO Rev. 0, CDMA2000 EV-DO Rev. A, and CDMA2000 EV-DO Rev. B. These variants comply with the ITU IMT-2000 requirements, and furthermore, CDMA2000 is backwards-compatible with the North American 2G system, IS-95. In the United States, CDMA2000 is a registered trademark of the Telecommunications Industry Association (TIA). The security risks for CDMA2000 are comparable with the ones described for UMTS.

The CDMA2000 1x EV-DO, often abbreviated as EV-DO or EV, is a telecommunications standard for the wireless transmission of data through radio signals, typically for broadband Internet access. It uses multiplexing techniques including CDMA as well as TDMA to maximize both individual users' throughput and the overall system throughput. It is standardized by the 3GPP2 as part of the CDMA2000 family of standards and has been adopted by many mobile phone service providers around the world – particularly by the ones in the previous CDMA network. It is also used on the Globalstar satellite phone network.

The radio interface of the CDMA2000 system differs from the one used in UMTS. Nevertheless, the core network has synergies between these and other systems. As an example, the TIA packet core network was specified by TIA TR45.6., and the work was based strongly on IETF solutions [7].

CDMA2000 belongs to the ITU IMT-2000 and complies thus with ITU-defined 3G standards. CDMA2000 is based on the CDMA IS-95 system providing an evolution path. The other entity specifying CDMA2000 is the 3GPP2, which assures that the core solution is compatible with 3GPP systems.

Ref. [58] summarizes the principles of CDMA2000 for the protection of system resources against unauthorized use. The security of the system is based on the terminal authentication which needs to prevent fraudulent use of the network. There is thus a proof of subscription identity used, proof of sender identity and message integrity. In addition, CDMA2000 includes network authentication, comparable to the UMTS mutual authentication principle, which prevents false base station attacks on user information. The successful authentication is a prerequisite for authorization, and the service access rights subscription data are passed from the home system, i.e., Home Location Register (HLR) or Authentication, Authorization and Accounting (AAA) to the serving system.

IS-2000/C.S0001-0005 is based on symmetric keys for the encryption. The system relies on root a authentication key for the security association, and the session keys are derived from the root key during authentication. IS-856/C.S0024, in turn, is based on a public-key agreement to establish airlink session keys. It uses symmetric keys for RADIUS authentication.

Ref. [58] further summarizes the procedures for IS-2000 authentication which is based on the challenge–response principle. For Rev B and earlier variants, the authentication is based on IS-95 for legacy reasons whereas for Rev C and later, the AKA principle is used for the authentication in a similar fashion as it is deployed in UMTS. In addition, the latter variants may use optionally a User Identity Module (UIM) authentication procedure for proving presence of a valid UIM, preventing rogue shell attacks. CDMA2000 also includes checking the message integrity, based on the keyed SHA-1 hash of the message contents. Furthermore, there is a Cryptosync based on time and other data defined in order to prevent replay attacks.

e IS-2000 also contains identity privacy via the TMSI. For the user data privacy, IS-2000-B and later variants rely on the 128-bit Rijndael algorithm (AES). The IS-2000 encryption uses 64-bit keys from legacy authentication, and 128-bit keys from AKA.

Ref. [59] has concluded that the security mechanisms in CDMA2000 represent a major improvement over the 2G mechanisms. The system provides an efficient one-pass challenge–response mechanism including mutual authentication that tackles the threat of false base station attacks. Furthermore, the confidentiality algorithm of CDMA2000 is noted to be stronger than the one used in 2G systems, and Ref. [59] also claims that it is more efficient than the one used in the 3GPP UMTS system.

2.6.3   Broadcast Systems

The broadcast systems are typically used for delivering radio and television programmes to the public. The security of the systems is related to the assurance of the confidentiality in the closed channel delivery. It can be related to, e.g., pay-TV contents that need to be purchased prior to their consumption.

The global distribution of digital television deployments is based on mainly the following main standards: Digital Video Broadcasting (DVB) (Europe, Africa, Asia and Australia), Advanced Television Systems Committee (ATSC) (North America), Terrestrial Integrated Services Digital Broadcasting (ISDB-T) (South America, Japan) and Digital Terrestrial Multimedia Broadcast (DTMB) (China). The Digital Rights Management (DRM) and Conditional Access (CA) function as a basis for the controlled programme delivery and for providing security for terrestrial television and radiobroadcast networks.

In addition to the PTP links that can be utilized for audio and video streaming reception in the mobile communication environment, the mobile networks may contain built-in functionalities for providing broadcast type of services. Examples of such services are Cell Broadcast (CB) and MBMS. Furthermore, the CB and MBMS (including the evolved version eMBMS) can be utilized in the normal information delivery as well as a base for warning systems.

2.6.4   Satellite Systems

As Ref. [16] states about modern Satellite Communications (SATCOM), proper protection of satellite systems for privacy and security is increasingly important especially for special environments such as military applications. At present, though, the securing is typically done merely via upper layer protocols as can be interpreted from [22,23]. Ref. [16] furthermore concludes that there may be alternatives for multi-beam secure satellite communications such as the ones through physical layer security techniques which refers to joint power control and beamforming schemes and related individual secrecy rate constraints.

2.6.5 Terrestrial Trunked Radio (TETRA)

TETRA is a radio system meant for professional, global use which is aimed at police, fire brigades, defence forces, rescue departments and other entities requiring safe and closed communication for emergency-type of operations [7]. The TETRA is an open standard developed by ETSI. The main purpose of the TETRA standard is to define a series of open interfaces, as well as services and facilities, in sufficient detail to enable independent manufacturers to develop infrastructure and terminal products that would fully interoperate with each other as well as meet the needs of traditional Private Mobile Radio (PMR) user organizations. The TETRA has already been deployed in many regions and nations within and outside Europe [7].

The PMR networks are highly needed because public networks cannot adequately guarantee required RF coverage or Grade of Service (GoS) during emergency situations. In addition to the basic communications, public networks cannot typically provide specialized voice services such as fast call set up, Direct Mode Operation (DMO) and high levels of secure encryption for voice and data.

The original TETRA standard, TETRA I, was known as the TETRA Voice plus Data (V + D) standard. In addition to the set of TETRA-specific functionalities as well as functionalities familiar from the public mobile communications, there are many security enhancements that provide both clear and encrypted options for advanced and fast group call services, individual calls, Short Data Services (SDS), and Packet Data Services (PDS). The evolved version is referred to as TETRA II, and it contains additional functionalities such as Trunked Mode Operation (TMO), range extension, enhanced voice codecs and TETRA Enhanced Data Service (TEDS).

The TETRA security is extensive as it needs to provide varying levels ranging from the basic level comparable to commercial networks up to the level that complies with the requirements for the national public safety network. The security mechanisms are covered through authentication, Air Interface Encryption (AIE) and end-to-end encryption that provide the shielding mechanisms for the attacks against confidentiality, authenticity, integrity, availability and accountability. The standard-based services are further expanded by the Security and Fraud Prevention Group (SFPG) of the association. The TETRA utilizes the TETRA Subscriber Identity Module (TSIM) as defined in Ref. [19].

Similarly, as in the UMTS network, the mutual authentication procedure of the TETRA ensures that the network can control the access to it while the radio terminal can trust that the network is authentic, as a basis for the voice and data connections. In the TETRA, as in most other secure systems, the authentication is the basis for the overall network security and can also be used to ensure the correct billing in public access systems. It also provides the foundation for a secure distribution channel for sensitive information delivery such as the encryption keys.

The standard defines TETRA encryption algorithms TEA1, TEA2, TEA3 and TEA4. There are differences in the intended use and the exportability of the equipment containing these algorithms. As an example, the TEA2 is meant for public safety users within the Schengen and related European countries only, and the others have wider applications ranging from general commercial use to public safety use in the regions where the TEA2 is not used. The encryption is closely bound to the TETRA signalling protocols. The algorithms may be implemented as a software within the radio terminal and base station equipment instead of hardware encryption modules for providing cost-optimization.

The TETRA standard also supports end-to-end encryption based on a variety of encryption algorithms as deemed necessary by national security organizations. The TETRA Association Security and Fraud Prevention Group has worked on a general framework for the end-to-end encryption. The evolution also includes work on the International Data Encryption Algorithm (IDEA) and the newer AES algorithm, which benefits from a larger cryptographic algorithm block size. Custom and indigenous algorithms are also possible with the end-to-end encryption, although these are not recommended for the radio interface encryption due to their need for integration in signalling protocols and availability of standard compliant terminals.

Besides these core security capabilities, the TETRA can also support a wide range of security management capabilities such as those used to control, manage and operate the individual security mechanisms in a network. The most important of these is encryption key management which is fully integrated into the TETRA standard functions. Even though the security functions are integrated in a network, this does not automatically imply that a network is fully secure. However, the security risks are concentrated to specific elements in the network, which can be adequately controlled.

It also should be noted that the characteristics of a trunking system provide enhanced security. As an example, the dynamic and random allocation of channels makes it more difficult for a casual eavesdropper to monitor conversations. Also, the possibilities for abuse are minimized as the identity of all radio users and the time and duration of messages are known and can therefore be traced.

2.6.6   Wireless Local Area Network (WLAN)

As protection technologies evolve, malicious attacks become more sophisticated, and the original methods are not sufficient to protect Wi-Fi access (and thus the contents and systems behind the access point). Currently, Wi-Fi networks using the Wi-Fi Protected Access (WPA2) protocol provide the latest functionality both for the security providing the control of access intentions and privacy protecting the transmissions from others [10]. For the most updated security, is thus recommended to allow in the network only devices that comply with the latest protection technology. The current Wi-Fi certified devices implement WPA2.

In addition to the actual Wi-Fi standards, there also exist various solutions at the equipment and application level. Table 2.3 summarizes the currently available protection mechanisms of Wi-Fi which are detailed further in the following sections.

It should be noted that the major part of Wi-Fi equipment has security disabled by default, the reason being to ease the setting up of the Wi-Fi network. Furthermore, the access points, routers and gateways have typically default Service Set Identifiers (SSIDs) as well as administrative credentials, i.e., username and password for fluent user experience while configurating the connectivity. It is thus of high importance to change the default settings at the earliest convenience on setting up the network [10].

Table 2.3 Current security solutions for Wi-Fi/WLAN connectivity

Method	Term	Description	Standard
WEP	Wired Equivalent Privacy	Currently, WEP is considered as a weak security standard. The password it is based on can often be resolved in only a few minutes by utilizing a standard laptop and software tools available from the Internet. Is based on manual key handling	IEEE 802.11,1999
WPA	Wi-Fi Protected Access	WPA has been an interim solution to improve the security of WEP. It is based on dynamically generated keys, and provides robust security in small network environments	IEEE 802.11, 2003
WPA2	Wi-Fi Protected Access, evolved	The current most updated standard is the evolved variant of WPA, WPA2. Some hardware may require firmware upgrade or replacement for supporting WPA2. It is based on the longer, 256-bit encryption key which improves security over WEP. It is based on manual handling of a pre-shared key and provides robust security in small network environments	IEEE 802.11i, 2004
802.1X	Port-based network access control (PNAC)	Provides authentication mechanism for devices attaching to LAN and WLAN. Involves local machine (supplicant), authenticator and authentication server. Based on RADIUS or Diameter server and dynamically generated keys	Part of IEEE 802.1 networking protocols
WISPr	Wireless Internet Service Provider roaming	Aimed at automatizing the authentication procedure. Allows users to roam between wireless Internet service providers in a similar way to cellphone users roaming between MNOs. RADIUS server is used to authenticate the subscriber's credentials	Wi-Fi Alliance (WFA)
EAP	Extensible Authentication Protocol	Authentication framework which includes EAP-SIM, EAP-AKA, EAP-AKA', EAP-TLS, EAP-TTLS and EAP-LEAP	IETF RFC 4017, 4372, 3579, 3580, 5216, 5281

In all cases, regardless of the access point offered via the public Wi-Fi hotspot, mobile networks or fixed internet access, the security level of the network and applications can be increased by deploying a VPN, as well as additional tools like firewalls and HTTPS for providing PTP protection. In this way, it is not possible to monitor communications via parallel equipment connected to the same access point.

2.6.6.1   Wi-Fi Authentication and Accounting

Before a user may be authenticated and access the Internet, the device must be associated to the Wi-Fi network by scanning the available Wi-Fi networks and by selecting the wanted access point. An alternative manner is based on the storing the desired SSID to the device which is then selected automatically for the connection if found. There are also devices that connect to the available Wi-Fi networks without prior adjustments.

The Wi-Fi networks can be configured for different security levels. The network may be open for all devices, or they may be configured to request an authentication procedure prior to the connection. Especially in the home environment, users may sometimes leave the network open with respective SSID broadcasting publicly. Nevertheless, it is not recommended to leave the network without authentication to avoid potential misuse. Some security breaches include the possibility for outsiders to monitor data and hijack the connection. Thus, in order to avoid security issues, it is recommended that users set up the Wi-Fi access point based on pre-shared keys for accessing and encrypting radio interface data transmission for both access point and device. The keys may be typed manually to the access device, or via automatized methods such as Wi-Fi Protected Setup (WPS). However, the WPS is not recommended as it contain major security flaws.

The very first Wi-Fi encryption method was based on Wired Equivalent Privacy (WEP). It soon became obvious that its protection level was weak, which triggered the adaption of an evolved alternative, (Wi-Fi Protected Access (WPA). It is based on the Temporal Key Integrity Protocol (TKIP) and RC4 ciphering. TKIP generates dynamically a new 128-bit key for each packet which prevents the attack types that compromised the WEP. WPA also optionally supports Advanced Encryption Standard (AES) ciphering.

Nowadays, the WPA2 security protocol has replaced the previously mentioned protocols. It is based on IEEE 802.11i, and includes the Counter-mode Cipher block Protocol (CCMP) (chaining Message authentication code Protocol, or Counter-mode CBC-MAC Protocol) which is an AES based, strong encryption method. At present, there are no successful attacks reported against AES.

The Wi-Fi router can also be configured to hide the broadcasted SSID. Even if this prevents the access point name from appearing to the consumers in normal cases and might thus increase the feeling of security, monitoring of the data traffic can reveal the hidden SSID relatively effortlessly. As another alternative, applying MAC filtering increases the security level, although it requires advanced Wi-Fi skills in the home environment.

The RADIUS protocol is also capable of transmitting accounting data, including, e.g., connection time, data usage and location. Wi-Fi service providers may use this information for billing purposes. The service providers may have a variety of different data models based on flat rate, data usage, time, roaming etc.

In the beginning of the Wi-Fi hotspot era, billing was based typically on the data utilization per megabyte as this was similar to the model used by the mobile network operators. As another example, the Wi-Fi hotspots of hotels provided time-limited payable connections. However, along with the reduced data utilization cost of cellular data subscriptions, Wi-Fi flat-rate subscriptions have increased in popularity. Currently, the advanced MNOs tend to include Wi-Fi usage to the offered mobile data package which makes the radio access technology transparent to the end-users [7].

2.6.6.2 Username and Password in Web Authentication

The public Wi-Fi service operators typically base their service on a layer 3 browser and username/password login. In this mode, the user's device is attached to the Wi-Fi access point to receive an IP address. As soon as the web browser is opened (and any web page is typed), the user is redirected to the web login page which is used to request the credentials, i.e., the username and password. In the roaming use case, the home operator needs to be typically selected from a drop-down menu. It is common to base the login procedure on an HTTPS connection although the data is not protected after the secured login and is thus open to be monitored by any outsider due to the lack of the encryption on the radio interface. Another issue is the fixed IP address allocation for the device once it is attached to the network which wastes IP address resources in providing the addresses to non-customer devices of the operator and might never authenticate to the network.

The WFA has specified a WISPr attribute to ease and automatize the authentication procedure although it is not straightforward to implement. Various mobile network operators have developed operator-branded connection managers to simplify the Wi-Fi authentication process by including cellular connection functionalities into the same software. The web authentication is widely used at present as it can be used in laptops and tablets. Nevertheless, its use in smartphones is challenging as it requires the user to open the browser and to insert a username and password. Thus, a more advanced method is needed in the smartphone environment which the following sections summarize.

2.6.6.3 802.1X

An enhanced Wi-Fi authentication solution is based on the standardized IEEE 802.1X, a port-based access control protocol, that can be used in either wireless or wired environments. It executes between the Mobile Terminal (MT) and the Wi-Fi Access Point (AP) in order to authenticate the user. The AP also has a RADIUS client function that initiates the RADIUS protocol between the home Wi-Fi networks' RADIUS server. IEEE 802.1X brings some of the cellular network features, like encryption, to the Wi-Fi networks.

IEEE 802.1X provides an authorization framework that controls the traffic through a port and access network resources. It enhances the authentication in such a way that prior to a successful authentication there is only a port open for the EAP packets to authenticate the user while all remaining traffic is blocked until the authentication is completed successfully, which triggers the allocation of an IP address for the user together with access rights to the Wi-Fi network. Then, the successful authentication triggers the AP to send a RADIUS Accounting Start message to the RADIUS server. The benefit of this mechanism is the reduced reservation of the IP addresses.

The IEEE 802.1X framework contains three key elements, which are Supplicant, Authenticator and Authentication Server (see Figure 2.28). The *Supplicant* is a host software requesting authentication and access to the network. In practice, it is typically a client installed in the terminal. The *Authenticator* is a device controlling the traffic to pass the port. There are two port types, an uncontrolled port and a controlled port. The uncontrolled port allows EAP authentication traffic to pass through while the controlled port blocks all other traffic until the supplicant has been authenticated. Typically this is a WLAN access point or an access controller, depending on the network architecture. The *Authentication Server* (AS) validates the credentials of the supplicant and notifies the Authenticator about the authorization of the supplicant. The AS may contain a database or it may proxy the authentication request to the appropriate database, e.g., to the HLR in the case of EAP-SIM. The AS is typically a RADIUS server but Diameter servers are also available on the market.

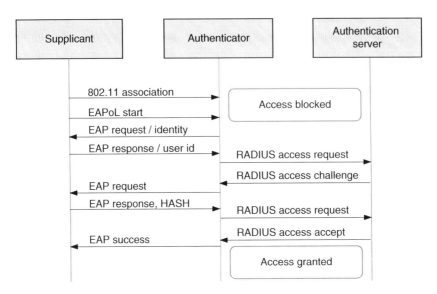

Figure 2.28 The flowchart of successful EAP authentication

The EAPoL protocol (EAP over LAN) is used on the radio interface between supplicant and authenticator. It includes the EAP protocol that is transferred over the RADIUS protocol. The authenticator divides the EAP part of the EAPoL, and sets it to communicate with the AS over the RADIUS. The Diameter protocol can also be used in this interface.

The RADIUS and Diameter protocols both support the EAP framework. It should be noted that the EAP is an authentication framework instead of an authentication mechanism. It provides common functions and negotiation of authentication procedures called EAP methods. The commonly applied modern methods that are capable of operating in wireless networks include EAP-SIM, EAP-AKA, EAP-TLS, EAP-TTLS and Extensible Authentication Protocol-Lightweight Extensible Authentication Protocol (EAP-LEAP). More specific requirements for the EAP methods applied in the WLAN authentication are described in Ref. [44].

The IEEE 802.1X introduces also the key exchange for the data encryption. As stated previously, the WEP encryption has been noted to be very weak. Based on the IEEE 802.1X feature which is designed to deliver dynamically the encryption keys, it is possible to use the enhanced security of WPA and WPA2.

The modern Wi-Fi networks in the MNO environment typically support the IEEE 802.1X. It provides enhanced security, provides the use of enhanced authentication methods as well as new discovery features in Wi-Fi networks [7].

2.6.6.4 EAP-SIM, EAP-AKA and EAP-AKA'

The increased traffic flows along with the success of smartphones have generated an increased need for the data offloading. It may well be that the most essential success factor of the public Wi-Fi services – observed from the user's point of view – is the transparency of the authentication and the method for joining the Wi-Fi network. In the MNO environment, which is mostly based on smartphone utilization, the EAP-SIM [46], EAP-AKA [47] and EAP-AKA' [45] authentication methods have been noted to be highly practical and useful. These methods are based on the SIM or USIM for authentication of users. On the network side it uses the Mobile Network subscriber information that resides in the HLR or HSS. It relies on the same data that is used for the 2G/3G/4G network authentication but in an enhanced way.

The key benefit of the EAP-SIM in the smartphone is that it includes the SIM card along with the utilization of the already existing user database (HLR/HSS). As a result, the MNO can re-use the existing infrastructure to provide a transparent authentication to the Wi-Fi network, and the end-user does not need to set any separate username and password on the client side. Additionally, the Wi-Fi charging can be based on the cellular data so that the user may be charged via a single bill for the data utilization.

The EAP-SIM functions with the older SIM cards and is based on GSM authentication. The newer EAP-AKA and EAP-AKA' require 3G USIM cards to function, and they rely on the 3G network AKA authentication in an enhanced way to authenticate the Wi-Fi users. The EAP-SIM authentication uses IMSI as a mobile identity, and the authentication in the newer ones is based on Temporary IMSI (TMSI) for increased security [7].

2.6.6.5   EAP-TLS

The EAP-Transport Layer Security (EAP-TLS) is defined in Ref. [48]. It provides strong security and uses a certificate and password for the user authentication. It is supported natively in several operation systems, including Windows, Windows Mobile, Mac OS and iOS. Most EAP-TLS implementations require a unique client certificate which has reduced the adoption of the EAP-TLS in the wider markets. It is thus used typically in the enterprise environment for laptops. In the normal case, the certificate is delivered on a smartcard by inserting it into the laptop's smartcard reader, or by installing the certificate onto the laptop prior to its delivery.

A typical use case is an enterprise Wi-Fi network based on IEEE 802.1X and EAP-TLS. Once the laptop is equipped with an individual certificate (either smartcard or pre-inserted certificate), and it is connected to the office Wi-Fi network, it is possible to authenticate it automatically towards the company's authentication server. The password can be given separately, or alternatively, a single-sign-on feature of the operation system may be used. As a result, the user is authenticated, data of the air interface is encrypted and the user may access the Intranet via the single process which is beneficial for the employees and employers [7].

2.6.6.6   EAP-TTLS

The EAP-Tunneled Transport Layer Security (EAP-TTLS) is an EAP protocol that extends the TLS as defined in Ref. [49]. The difference with the EAP-TLS is that the EAP-TTLS does not require individual client certificate in order to authenticate to the server. In the authentication process, the server is first authenticated to the client securely (and optionally, the client is also authenticated to the server). Then, the server can establish a secure tunnelled connection to authenticate a client. The credentials (username and password) are transferred to the authentication database on that secure tunnel. The tunnel provides protection against eavesdropping and Man in the Middle (MITM) attacks.

There are two versions of EAP-TTLS which are the original EAP-TTLS (EAP-TTLS v0) as described in Ref. [49], and EAP-TTLS v1. The EAP-SIM, EAP-AKA and EAP-TTLS are all feasible and practical authentication methods for the Wi-Fi service providers since they do not require unique certificate to the device whereas the EAP-TLS provides a user-friendly authentication method to the Wi-Fi networks in the managed corporate environment [7].

2.7 Interoperability

This section presents key aspects of the interoperability of 3GPP and non-3GPP networks, Wi-Fi Offload, Simultaneous Voice and LTE (SVLTE), signalling during location updates and in typical roaming scenarios. This section also presents examples and identifies potential security issues during the interoperability procedures. In the beginning of LTE/LTE-A deployments, other RATs were already offered by the operators. As the LTE/LTE-A coverage area is typically evolving and by the first network launch the service area is probably less than the offered 2G and/or 3G services, thus the inter-working with the LTE/LTE-A and the operator's own RATs is essential to assure the fluent continuum of both voice and data calls [7].

2.7.1 Simultaneous Support for LTE/SAE and 2G/3G

One important item to be considered by the operators is related to LTE/SAE interaction with the existing 2G and 3G networks. It depends on the network support as well as device capability. The support for 2G/3G is by default built into LTE/SAE devices so that they are able to switch to 2G/3G whenever LTE/SAE is not available. However, it is not very straightforward to build a device which needs to be connected to both LTE/SAE and 2G/3G networks at the same time. This would require two separate radios functioning simultaneously, both connected to the core network. Likely it also requires two SIM cards.

Having this kind of device would solve a number of issues related to problems of handover between LTE/SAE and 2G/3G. For example, voice-related functionality such as CSFB or Single Radio Voice Call Continuity (SRVCC) would not be needed since the LTE/SAE device would always be connected to the 2G/3G network, meaning the native CS voice service of 2G/3G would be available without any need for the handover. Nevertheless, due to the respective complexity of the phone design, e.g., size, cost, battery consumption and inter-system interference caused by two simultaneously used radios, it is unlikely to that such devices will be deployed at the least in the initial phase of the networks.

The 3GPP specifications define the states of the LTE-UE and state transitions, including inter-RAT procedures. The states have been divided into an RRC_CONNECTED state (when an RRC connection has been established) and an RRC_IDLE state (when no RRC connection is established). The RRC states of the LTE-UE are characterized according to the following.

2.7.1.1 RRC Idle State

The RRC Idle state means that the LTE-UE controls the mobility and takes care of the monitoring of the paging channel in order to act when there is an incoming call waiting. In this state, the LTE-UE takes care of the monitoring of the system information exchanging.

Furthermore, for the LTE terminal models that support the Earthquake and Tsunami Warning System (ETWS), the terminal monitors the respective notifications that can be delivered via the paging channel. In addition to the paging channel monitoring, the LTE-UE performs neighbouring cell measurements, cell selection and cell re-selection procedures, and in general, is able to acquire system information from the LTE/SAE network.

2.7.1.2 RRC Connected State

The LTE-UE is capable of delivering unicast data in downlink and uplink in the RRC Connected state. The mobility is controlled by the LTE/SAE network, meaning that the network takes care of the handover and cell change procedures order, possibly with additional support of network (NACC, Network Assisted Call Control) to 2G radio access network (GERAN). In this state, the LTE-UE still monitors the paging channel and/or System Information Block Type 1 contents in order to detect system information changes. As in the Idle state, the LTE-UE also monitors the ETWS notifications if the terminal is capable of supporting the system. Furthermore, the LTE-UE monitors control channels that are associated with the shared data channel to determine if data is scheduled for it.

2.7.1.3 Mobility Support

The mobility of the LTE-UE between LTE and 2G is illustrated in Figure 2.29, and the same idea is presented for 3G and CDMA2000 in Figure 2.30 and Figure 2.31, respectively. In the latter case, HRPD refers to High Rate Packet Data.

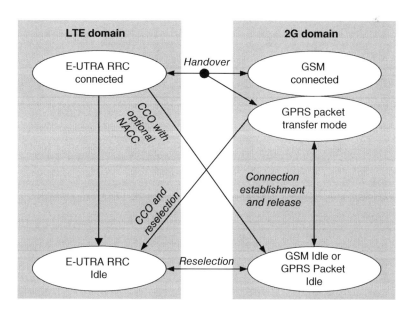

Figure 2.29 The LTE-UE states and the inter-RAT mobility procedures with the GSM network as interpreted from Ref. [38]. *Source*: Reprinted from [38] by courtesy of ETSI

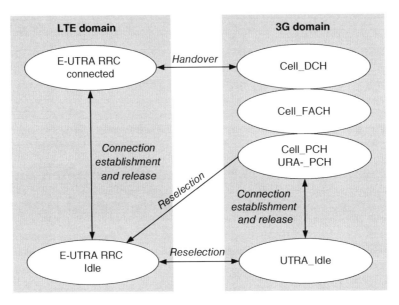

Figure 2.30 The LTE-UE states and the inter-RAT mobility procedures with the UMTS network as interpreted from Ref. [38]. *Source*: Reprinted from [38] by courtesy of ETSI

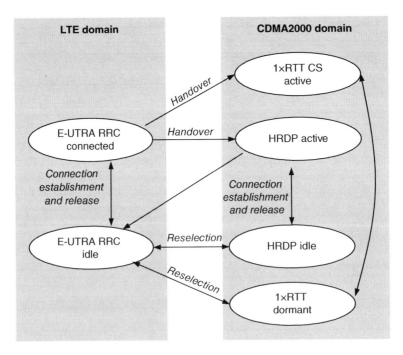

Figure 2.31 Mobility procedures between E-UTRA and CDMA2000 as interpreted from Ref. [38]. *Source*: Reprinted from [38] by courtesy of ETSI

2.7.2   VoLTE

The network development typically follows the device evolution. The first LTE/LTE-A networks were used merely as a big bit-pipe providing customers with better bandwidth via data dongles. In the more advanced phase, the LTE/LTE-A operators need to deploy additional core network elements and upgrade the CS core to offer functions such as CSFB, VoLTE and SRVCC. As indicated by GSMA and Next Generation Mobile Networks (NGMNs), the CSFB is considered to be the common interim solution for the voice service in LTE/SAE networks.

Since the VoLTE is a long-term target, it will not be widely available in a commercial environment. This is because the IMS-based IP voice deployment requires major work to be implemented due to the complexities of the IMS core system, related application servers, SRVCC support, QoS support in LTE/SAE RAN and Policy and Charging Control (PCC) architecture that are all likely to be required for a realistic CS voice replacement service. A related issue is the support for the CSFB and/or VoLTE for the inbound roamers.

2.7.3   CS Fallback

Even though the LTE/SAE is an all-IP network, it also contains interfaces to the legacy networks. One example of this is the CSFB for voice service as defined in Ref. [62]. The CSFB means that the 2G/3G CS network is used to deliver the voice call for an LTE/SAE user. In the LTE/SAE, the delivery of the voice traffic would require some IP-based solution like VoLTE or Over the Top (OTT). If these other mechanisms are not (yet) deployed, the feasible way to deliver voice traffic in LTE/SAE is via the existing 2G/3G network and the CSFB. 3GPP has also defined 'Short Message Service (SMS) fallback', which refers to SMS over the *SGs* interface, and allows the LTE/SAE device to send and receive short messages via the MME. Figure 2.32 depicts the procedure.

Also the operator community of GSM Association and the NGMN are interpreting the CS Fallback as the common intermediate step towards the target solution of VoLTE, so therefore it likely is a part of many vendors' roadmap [63].

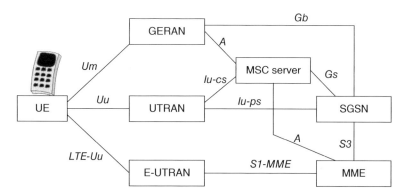

Figure 2.32   Enhanced Packet System (EPS) architecture for CSFB and SMS over *SGs* interface

2.7.4   Inter-operator Security Aspects

In the LTE/SAE environment, it is important to consider the inter-operator aspects, namely roaming and interconnection. The GSMA document IR.77 describes the general guidelines for the GPRS Roaming Exchange/IP Exchange (GRX/IPX) network, which are also valid when deploying, for example, LTE/SAE roaming regardless of the used service/application.

The main issues relate to the fact that GRX/IPX is a completely separated network, and in fact, it is not visible or accessible from the Internet. This requires core network nodes that are able to access both GRX/IPX and the Internet. They need to support multi-homing or be capable of having two completely separate interfaces: one for the Internet and the other for the GRX/IPX. The IP addressing used for these interfaces needs to be separate as one cannot reuse the same IP address in GRX/IPX that is already utilized in the Internet.

Also additional guidance is required especially if another network such as the Internet is used for accessing other operators. This particularly relates to the native level of security offered by the network; the GRX/IPX can be considered to be secure as it is only accessible to operators that are part of the trusted parties whereas the Internet is open to everybody. So the Internet connectivity requires additional security-related functions such as Session Border Controller (SBC) in order to guard incoming access and the IPSec tunnel for securing the traffic. Being essential parts in the inter-operable environment, the security solutions also need to be simple and manageable to provide efficient fault, mis-usage and fraud discovery.

2.7.5   Wi-Fi Networks and Offload

Most mobile devices that are cellular data-capable also include integrated Wi-Fi capability. The importance of Wi-Fi hotspots is increasing especially in dense Internet usage locations like airports, hotels and city centres. The operators are thus facing a challenge in the converged Wi-Fi and cellular networks for ensuring secure and seamless handovers between the systems in such a way that the user experience is as fluent as possible.

Wi-Fi Offloading provides a solution to this challenge. As identified in Ref. [24], the standardization has focused on both tight and loose coupling between the mobile communications networks and the Wi-Fi access, depending on the forum. One of the 3GPP approaches is the Enhanced Generic Access Network (EGAN) architecture, which is based on tight coupling via rerouting the cellular network signalling through Wi-Fi access networks. This results in Wi-Fi being one of the 3GPP radio access networks. Figure 2.33 depicts the Wi-Fi Offloading architecture.

There is also the loose coupling approach in 3GPP for Wi-Fi via Interworking Wireless LAN (IWLAN) architecture. In this option, IP data can be delivered between mobile devices and the operator's core network through a Wi-Fi access. Here the mobile communications network and Wi-Fi are handled in a separated way, and the client application decides the network selection. IWLAN architecture is based on the VPN or IPSec tunnel between the user device and dedicated IWLAN server that resides in the operator's core network. Users can thus access the operator's internal services or gateway that provides connection to Internet.

Only a few user equipments support native IPSec connectivity. Such user equipment requires an additional client. As Ref. [24] has noted, the impact of installing the extra

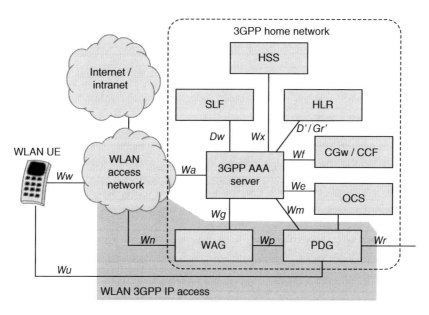

Figure 2.33 Wi-Fi Offload architecture

client and its behaviour is a concern for new implementations. The simplest offload method for directing data to the Wi-Fi network is via a public Internet connection relying on the non-coupling option. In this case, there is no need for interworking standardization.

The Access Network Discovery and Selection Function (ANDSF) of 3GPP provides an advanced solution for the control of offloading between 3GPP and non-3GPP access networks like Wi-Fi hotspots. ANDSF has been designed to offer assistance for the discovery of access networks and to provide policies for prioritizing and managing connections.

The focus of ANDSF is to provide assistance to User Equipment (UE) in order to discover non-3GPP access networks, e.g., Wi-Fi and WiMAX that are suitable for data communications at the location in question in addition to 3GPP access networks. Furthermore, ANDSF is designed to provide the UE with rules policing the connection to these networks. Operators can list the preferred networks and provide automatically respective policies via ANDSF. ANDSF thus offers the possibility for carriers to enable Wi-Fi hotspots with secure connectivity and seamless experience in locations where roaming between cellular and Wi-Fi networks is controlled by operators. The combination of ANDSF and Hotspot 2.0 is an efficient enabler for enhanced and fluent user experience across Wi-Fi and cellular networks. In order to maintain the adequate QoS level, Hotspot 2.0 provides a first-step solution for roaming.

As identified in Ref. [24], there are three basic schemes for the initiation of the mobile offload to Wi-Fi: WLAN scanning initiation (user device periodically performs WLAN scanning); user initiation (user selects the network technology); and remotely managed initiation (network server initiates the offloading).

2.7.6   Femtocell Architecture

The femtocell is a small mobile communications base station which has been designed primarily for the home or small business environment. The coverage area of a femtocell is limited to some tens of metres. It can be connected to the service provider's network via broadband connectivity of, e.g., xDSL or cable. Figure 2.34 shows the femtocell architecture.

At the moment, there is support for accessing typically 2–4 active mobile devices in the home environment and 8–16 active mobile devices in the business environment. The benefit of the femtocell is the extended radio coverage indoors assuring that possible outage areas can be covered. At a smaller scale, the enhanced coverage also has a positive impact on battery duration along with lower output power levels of user devices. Femtocells also provide capacity enhancement and enhanced QoS, e.g., for voice calls. The femtocell concept has been mainly designed for Wideband Code Division Multiplexing Access (WCDMA) but it is valid for other mobile communications standards like GSM, CDMA2000, TD-SCDMA, WiMAX [37] and LTE. Furthermore, the concept allows the operator to design additional pricing strategies, e.g., customers may benefit from the utilization of the femtocell coverage areas.

In the femtocell deployment, it is worth noting that the concept works with existing handsets on the market as the equipment operates in the licensed spectrum. The Home NodeB (HNB) refers to the WCDMA femtocell of 3GPP systems, while HeNB refers to the femtocell deployed in LTE/LTE-Advanced networks.

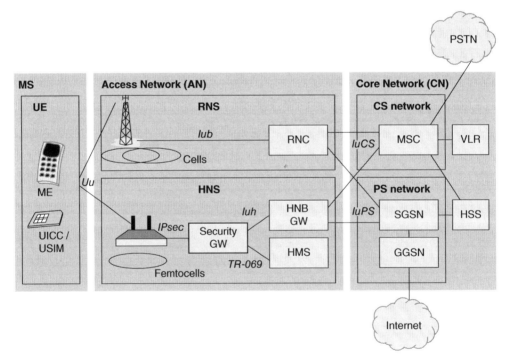

Figure 2.34   Femtocell architecture

References

[1] I. Androulidakis, D. Pylarinos and G. Kandus. Ciphering indicator approaches and user awareness. *Maejo International Journal of Science and Technology*, **6**(3):514–527, 2012.

[2] Aftenposten. The spoof GSM base stations revealed in Oslo, 16 December 2014. http://www.aftenposten.no/nyheter/iriks/Secret-surveillance-of-Norways-leaders-detected-7825278.html (accessed 4 July 2015).

[3] 3GPP TSG SA WG3 Security – SA3#25 S3-020557, 8–11 October 2002. http://www.3gpp.org/ftp/tsg_sa/wg3_security/tsgs3_25_munich/docs/pdf/S3-020557.pdf (accessed 4 July 2015).

[4] *Wired*. GSM spoof BTS demo, 31 July 2010. http://www.wired.com/2010/07/intercepting-cell-phone-calls (accessed 4 July 2015).

[5] *Forbes*. GPRS relay, 19 January 2011. http://www.forbes.com/sites/andygreenberg/2011/01/19/smartphone-data-vulnerable-to-base-station-spoof-trick/ (accessed 4 July 2015).

[6] *Forbes*. Information security of automotives, 8 April 2014. http://www.forbes.com/sites/andygreenberg/2014/04/08/darpa-funded-researchers-help-you-learn-to-hack-a-car-for-a-tenth-the-price (accessed 4 July 2015).

[7] J. Penttinen. *The Telecommunications Handbook*. John Wiley & Sons, Inc., Hoboken, NJ, 2015.

[8] *Forbes*. Security breach of vehicles. 8 April 2014. http://www.forbes.com/sites/andygreenberg/2014/04/08/darpa-funded-researchers-help-you-learn-to-hack-a-car-for-a-tenth-the-price/ (accessed 19 April 2015).

[9] Verizon Security Breach Report, 2015. http://www.verizonenterprise.com/DBIR/2015/ (accessed 19 April 2015).

[10] Wi-Fi security description by Wi-Fi Alliance, 2015. http://www.wi-fi.org/discover-wi-fi/security (accessed 13 June 2015).

[11] BBC. Mass snooping fake mobile towers 'uncovered in UK', 10 June 2015. http://www.bbc.com/news/business-33076527 (accessed 14 June 2015).

[12] 2016 Data Breach Investigation Report. Verizon, 2016.

[13] Intel Curie, 2015. https://iq.intel.com/tiny-brain-wearables-cute-button/ (accessed 15 June 2015).

[14] M. Green. A few thoughts on cryptographic engineering, 14 May 2013. http://blog.cryptographyengineering.com/2013/05/a-few-thoughts-on-cellular-encryption.html (accessed 4 July 2015).

[15] 3GPP TS 55.216 V6.2.0 (2003-09). Technical Specification, 3rd Generation Partnership Project; Technical Specification Group Services and System Aspects; 3G Security; Specification of the A5/3 Encryption Algorithms for GSM and ECSD, and the GEA3 Encryption Algorithm for GPRS; Document 1: A5/3 and GEA3 Specifications. (Release 6).

[16] J. Lei, Z. Han, M.A. Vazquez-Castro and A. Hjørungnes. Secure satellite communication systems design with individual secrecy rate constraints, 2011.

[17] M. Walker. On the security of 3GPP networks. Eurocrypt 2000.

[18] Elad Barkan, Eli Biham and Nathan Keller. Instant ciphertext-only cryptanalysis of GSM encrypted communication. Advances in Cryptology – CRYPTO 2003. Lecture Notes in Computer Science Volume 2729: 600–616, 2003.

[19] ETSI TS 100 812-2, V2.4.1. Terrestrial Trunked Radio (TETRA); Subscriber Identity Module to Mobile Equipment (TSIM-ME) interface; Part 2: Universal Integrated Circuit Card (UICC); Characteristics of the TSIM application, 3 August 2005.

[20] Jen Valentino-Devries. Stingray phone tracker fuels constitutional clash. *Wall Street Journal*, 22 September 2011.

[21] WPG Harris. Harris Wireless Products Group catalog, page 4. 25 August 2008. https://www.documentcloud.org/documents/1282631-08-08-25-2008-harris-wireless-products-group.html (accessed 4 August 2015).

[22] L. Liang, S. Iyengar, H. Cruickshank and Z. Sun. Security for the flute over satellite networks. Proceedings, International Conference on Communications and Mobile Computing, Kunming, China, January 2009. Pp 485–491.

[23] M. Mahmoud, N. Larrieu and A. Pirovano. An aeronautical data link security overview. Proceedings, IEEE/IAII Digital Avionics Systems Conference, Orlando, USA, October 2009. Pp 4.A.4-1–4.A.4-14.

[24] 4G mobile broadband evolution. Release 10, Release 11 and beyond, HSPA+, SAE/LTE and LTE-Advanced. 4G Americas. October 2012.

[25] 3GPP TR 33.902. V3.1.0. January 2000. 3rd Generation Partnership Project; Technical Specification Group Services and System Aspects; 3G Security; Formal Analysis of the 3G Authentication Protocol (3G TR 33.902 version 3.1.0 Release 1999).

[26] 3GPP TS 33.102. 3G security; Security architecture. V. 12.2.0, December 22, 2014.

[27] C. Gabriel. Managing the new mobile data network. The challenge of deploying mobile broadband systems for profit. 2012. Rethink Technology Research.

[28] Broadband technology overview. Corning, white paper, 2005.

[29] Realistic LTE performance. From peak rate to subscriber experience. Motorola, white paper, 2009p.

[30] J. Markendahl and Ö. Mäkitalo. Analysis of business opportunities of secondary use of spectrum. The case of TV white space for mobile broadband access. 22nd European Regional ITS Conference. Budapest, 18–21 September 2011.

[31] P. Croy. LTE backhaul requirements. A reality check. Aviat Networks, white paper, 2011.

[32] Realistic LTE performance. From peak rate to subscriber experience. Motorola, white paper, 2009.

[33] New wireless broadband devices. Understanding the impact on networks. 4G Americas, May 2012p. http://www.4gamericas.org/UserFiles/file/White%20Papers/4G%20Americas%20White%20Paper%20New_Wireless_Broadband_Applications_and_Devices%20May%202012.pdf (accessed 28 February 2014).

[34] Signaling considerations of apps. http://www.seven.com/mobile-signaling-storm.php (accessed 9 March 2014).

[35] Traffic and market data report. Interim update. Ericsson, February 2012.

[36] Launch of the 2014 Manual for Measuring ICT Access and Use by Households and Individuals. 11th World Telecommunication/ICT Indicators Symposium (WTIS-13). ITU Document C/23-E. Mexico City, México, 4–6 December 2013.

[37] IEEE 802.16m technology introduction. White paper, Rohde&Schwarz, 2010.

[38] 3GPP TS 36.331.

[39] Limits of human exposure to radiofrequency electromagnetic fields in the frequency range from 3 kHz to 399 GHz. Safety Code 6. Environmental Health Directorate, Health Protection Branch. Publication 99-EHD-237. Minister of Public Works and Government Services, Canada, 1999.

[40] J. Penttinen. *The Telecommunications Handbook*. John Wiley & Sons, Inc., Hoboken, NJ, 2015.

[41] www.icnirp.de (accessed 14 September 2013).

[42] J. Penttinen. The DVB-H radio network planning and optimisation. Doctoral thesis, Aalto University, School of Electrical Engineering, 2011.

[43] K. Zhao. Interaction between the radiation of LTE MIMO antennas in a mobile handset and the user's body. Masters' Degree Project, Kungliska Teknska Hogskolan, Stockholm, Sweden, June 2012.

[44] IETF RFC 4017. Extensible Authentication Protocol (EAP) method requirements for wireless LANs. March 2005.

[45] IETF RFC 4372. Chargeable user identity. January 2006.

[46] IETF RFC 3579. RADIUS (Remote Authentication Dial In User Service) Support for Extensible Authentication Protocol (EAP). September 2003.

[47] IETF RFC 3580. IEEE 802.1X Remote Authentication Dial In User Service (RADIUS) Usage Guidelines. September 2003.

[48] IETF RFC 5216. The EAP-TLS Authentication Protocol. March 2008.

[49] IETF RFC 5281. Extensible Authentication Protocol Tunneled Transport Layer Security Authenticated Protocol Version 0 (EAP-TTLSv0). August 2008.

[50] Will Wi-Fi relieve congestion on cellular networks? GSMA, 5 May 2014.

[51] The Android SDK. http://developer.android.com/sdk/index.html (accessed 2 March 2014).

[52] Android OS statistics. http://blogs.strategyanalytics.com/WSS/post/2014/07/30/Android-Captured-Record-85-Percent-Share-of-Global-Smartphone-Shipments-in-Q2-2014.aspx (accessed 16 November 2014).

[53] The iOS SDK. https://developer.apple.com/devcenter/ios/index.action (accessed 2 March 2014).

[54] The BlackBerry app development. http://developer.blackberry.com (accessed 8 June 2014).

[55] The Windows Phone SDK. http://www.microsoft.com/en-us/download/details.aspx?id=35471 (accessed 2 March 2014).

[56] D. Tipper. IS-95. Graduate Telecommunications and Networking Program, University of Pittsburgh. http://www.sis.pitt.edu/~dtipper/2720.html (accessed 31 December 2015).

[57] D. Chen, M. Liu and X. Liu. Attacks and enhancement on security architecture of IS-95. 14th International Conference on Communication Technology (ICCT). IEEE, Chengdu, China, 2012. Pp. 1143–1148.

[58] F. Quick. Security in CDMA2000. ITU-T workshop on Security, Seoul, Korea, 13–14 May, 2002.

[59] L. Ertaul, S. Natte and G. Saldamli. Security evaluation of CDMA2000.

[60] ITU statistics. Mobile phone penetration, 3 January 2016. http://www.itu.int/en/ITU-D/Statistics/Pages/stat/default.aspx (accessed 3 January 2016).

[61] IMEI Database of the GSMA, 3 January 2016. https://imeidb.gsma.com/imei/login.jsp (accessed 3 January 2016).

[62] 3GPP TS 23.272.

[63] CSFallback.www.ngmn.org/news/ngmnnews/newssingle2/article/ngmn-alliance-delivers-operatorss-agreement-to-ensure-roaming-for-voice-over-lte-357.html (accessed 3 January 2016).

[64] V. Niemi and K. Nyberg. *UMTS Security*. John Wiley & Sons, Ltd, Chichester, 2003.

[65] D. Forsberg, G. Horn, W.-D. Moeller and V. Niemi. *LTE Security*, 2nd Edition. John Wiley & Sons, Ltd, Chichester, 2012.

3

Internet of Things

3.1 Overview

The Internet of Things (IoT) environment is as versatile as the word 'things' just sounds. It basically refers to a vast set of all kinds of equipment connected to the Internet, including surveillance cameras, refrigerators, printers, self-driving cars and telematics boxes which are able to communicate and perform actions. Remarkably, IoT also refers to yet-to-be-seen devices that may be so innovative that we might not have a clue about their future form. The key idea of IoT is the facilitation of useful functions via the 'always-on' connectivity everywhere and at any time, providing fluent and seamless user experiences, and making our lives easier in the transition from the information society towards the truly connected society.

This chapter discusses the principles of IoT by interpreting its typical definitions and analysing the suitability of it in different environments. The overall development of IoT is explained by investigating the evolution of M2M solutions and mobile connectivity, the Connected Living concept, and other relevant industry forum, alliance and standardization initiations. This chapter also presents technical descriptions of current key case examples of IoT in environments such as telemetry, automotive and e-health. The aim is to identify trends and development of IoT for novelty and future solutions such as wearable devices, household appliances, industry solutions, robotics, self-driving cars to understand their role and how they relate to IoT.

To complement the overall picture, this chapter also discusses the contributing role and technologies of utilities, i.e., the ways they are relying on wireless technologies and their importance in the electric domain, including generation, transmission, distribution and area networks, mobility and smart grid applications.

Wireless Communications Security: Solutions for the Internet of Things, First Edition. Jyrki T. J. Penttinen.
© 2017 John Wiley & Sons, Ltd. Published 2017 by John Wiley & Sons, Ltd.

3.2 Foundation

3.2.1 Definitions

IoT is a term that has been publicly visible for many years. Even before the appearance of the term 'IoT', it existed as an idea for a very long time, as can be interpreted from early activities involved with connected devices in the 1G and 2G mobile communications environment at the beginning of the 1990s – which represented an era well before the adaptation of the graphical World Wide Web (WWW). With only a limited and text-based Internet, the (highly unusually utilized external) data connectivity of analogue systems like Nordic Mobile Telephony (NMT) and SMS of GSM networks were utilized to test ideas that were in fact quite similar to current IoT thoughts. The early mobile devices were used as remote management tools for a variety of inventions, e.g., providing automatic slow-scanning surveillance video contents that could be triggered via sensors [15], opening garage doors, and even warming up saunas. As important facilitators, the public Internet, the WWW with Graphical User Interface (GUI) and more advanced mobile communications services, finally provided a suitable user-friendly base for all kinds of communications wirelessly and via wired connections.

In general, the term 'IoT' itself indicates the blurriness of the environment, generalizing the presence and utilization of services and connected devices under the highly interpretable word 'things'. A concrete example of this environment is the M2M communications which does not require the intervention of persons, and can be interpreted as belonging to part of, or in many cases, be even a synonym for IoT. In fact, IoT and M2M are often understood to refer to the same environment although M2M is a subset of IoT. This is due to the nature of IoT which, in addition to the pure machines communicating with each other, can be understood to also include communications between machines and human-operated devices [27].

The GlobalPlatform (GP) defines IoT as *uniquely identifiable objects and their virtual representations in an Internet-like structure* [1,2]. IoT can thus refer to the overall trend of increasing amounts of devices that are connected to the Internet, by all means, and especially in a wireless manner. To mention some examples, there is a clear tendency to include IoT functionality to automotive, medical, home and utility devices by enhancing the User Experience (UX) and by automatizing functionalities.

Investigating further the definitions of the GlobalPlatform, IoT devices need to perform *measurements* of physical properties and gather information via *sensors*, and to influence or modify their environment by, e.g., performing measurements via *actuators*. In addition, IoT may include devices capable of *processing* the data obtained by sensors, perform tasks like correlation of measurement data and analysis of the information, possibly to be delivered further to some other entity for post-processing. In practice, a single device may be able to perform one or more of the above-mentioned tasks. Furthermore, IoT devices need to be able to *communicate* with the external world, either via local wireless connectivity such as proximity or vicinity technologies (NFC, RFID, Bluetooth LE, etc.), Wi-Fi, wider range systems such as cellular networks or fixed networks such as Asymmetric Digital Subscriber Line (ADSL). The connectivity may be done via any known standardized or proprietary technology, and via licensed or unlicensed RF bands, as long as the IoT devices are able to deliver their messages to the counterparty, which may be system or other device performing functions or relaying the messages. Figure 3.1 summaries the principle.

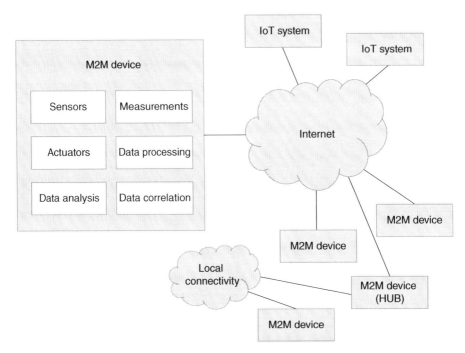

Figure 3.1 IoT consists of devices that are able to perform functions such as measurements and data processing, as stated in Refs. [1,2]. The connectivity can be based on all known data transfer techniques, including mobile communications networks, local wireless and wired networks, and even direct connectivity. IoT may have communications with other consumer devices, and furthermore, part of the devices can act as hubs to connect the local equipment to the Internet

From a multitude of IoT definitions, Ref. [6] summarizes it in the following sentence: *The Internet of Things (IoT) is a computing concept that describes a future where every-day physical objects will be connected to the Internet and be able to identify themselves to other devices. The term is closely identified with RFID as the method of communication, although it also may include other sensor technologies, wireless technologies or QR codes.*

Furthermore, Ref. [6] emphasizes the importance of IoT as an *object* that can represent itself digitally and become something greater than the object itself. In other words, the object will be connected to surrounding objects and database data to be part of a whole environment. This is fascinating because as soon as a multitude of objects are acting in unison, they can be considered to have *ambient intelligence*.

The RFID has been mentioned in various early sources related to IoT. Nevertheless, as the communications is being adapted into the modern IoT environment, the connectivity may actually be any of the known methods, including fixed and wireless technologies. In the most typical cases, wireless radio access technology is used from which the cellular connectivity provides the widest service areas while Wi-Fi is typical for the local solutions. Many other short-range technologies can be used such as low-energy Bluetooth and light-based communications (Li-Fi) that function within limited coverage ranges.

3.2.2   Security Considerations of IoT

IoT provides a vast amount of new possibilities for managing, coordinating, automatizing and benefiting in general from M2M communications without human intervention, as well as by taking human interactions into account when needed. At the same time, new types of equipment will be introduced into the market as a base for innovative services. In addition to such a huge amount of new opportunities, this also opens known and totally new security threats that may compromise the identity of users and confidentiality of information, which in turn may jeopardize the safety of, e.g., economic funds, and in the worst case, threaten even the personal well-being of persons as in compromised medical or traffic control applications. One of the current, highly concrete demonstrations of such life-threatening risks is related to the possibility of taking over remotely the control of connected, self-driven vehicles [7,8].

To ensure protection against the known and future security breaches, IoT needs to take into account the underlying technical mechanisms. As presented by the GlobalPlatform, some of the items that contribute to security are the following [2]:

- The *Secure Element* (SE) has the form of the SIM/UICC, embedded SIM or external card and is protected against physical and logical security attacks. It thus provides a suitable means for hosting multiple stakeholders' applications.
- The *Security Domain* (SD) serves for storing cryptographic content for each stakeholder on the SE. It also provides a means for secure communications between the content and external entity, and for managing the content.
- The *Trusted Service Manager* (TSM) is a service delivery broker which is able to establish business agreements and technical relationships between stakeholders.
- The *Controlling Authority* (CA) manages the confidential post-issuance of new stakeholders onto the SE.
- The *Trusted Execution Environment* (TEE) is a secure and trusted area on the mobile equipment HW, which is used for safe and protected storing and processing of sensitive data by isolating it from the 'normal' world.

These technologies together with other relevant bases for IoT are described and analysed throughout this book.

3.2.3   The Role of IoT

The role of both fixed and wireless telecommunications has increased and provides a platform for essential functionalities in daily life for personal and business environments, as well as government services. The amount of current information flow via data networks is breathtaking. The utilization of the Internet is growing steadily as can be seen in Figure 3.2 [16].

In fact, along with the enhancements of IoT, the information society – with ever evolving technological solutions for more efficient communications – is converting the world to a vast integrated information system [3]. Looking at this groundbreaking evolution, the old saying about the telecommunications network being the biggest machine in the world may indeed need to be reviewed because the Internet now serves as an umbrella for such a great amount of services, telecommunications being in fact only one of its parts in a set of all kinds of subsystems and devices. Thus, it is justified to state that the Internet is nowadays the world's biggest machine.

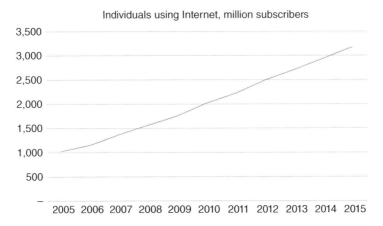

Figure 3.2   Individuals using the Internet [16]

Human-based communications still represents a high share of the total information flow, but the fully automatized M2M communications is taking over with giant leaps. IoT with sensors and actuators embedded in physical objects are preparing us for a totally new era of a socio-techno-economical ecosystem where old business rules and earning models are refreshed on a daily basis. There is plenty of room for enhancing current services via M2M devices as well as by taking totally new innovations into use, via wired and wireless, current and future networks that rely typically on IP connecting the uncountable machines to the Internet.

Along with the increased amount of M2M equipment, the level of data transferred increases, which needs special attention so that MNOs can deploy new networks and enhance the capacity of the current ones to maintain or enhance the quality level required by solutions from background data exchange up to extremely delay-sensitive real-time solutions. The estimation for the growth of M2M and the resulting increase for data utilization is an extremely hard task to perform, but the following summarizes some of the publicly available forecasts. Ref. [4] has noted that according to industry experts, IoT was expected to include more than 15 billion connected devices by 2015 while the estimate for 2020 is 50 billion. This reference mentions that the major part of the devices would not be conventional computers or smart-phones, but small, economical, interconnected and partially autonomously communicating devices like home appliances, security systems, smart thermostats, smart meters, portable medical devices, health and fitness trackers and smart watches. Ref. 4 foresees 26 billion installed IoT units by 2020 without counting personal computers, tablets or smartphones, while ABI Research estimates that the total number of connected devices will reach at least 40 billion IoT units by 2020 [19].

In such a dynamically developing environment, it is most challenging to forecast precise values for the number of IoT devices – and after all, it might not be even necessary or possible to form such statistics as the devices are based on such a variety of systems and communi-cations technologies. Nevertheless, the overall messages from all these efforts indicate that the IoT market will be of extremely high relevance and surpass even today's smartphone sales. It can thus be expected that IoT has big direct and indirect impacts on the telecommunications development as it creates totally new business opportunities and impacts notably on our lifestyle. It could be argued that after the information society, the next big step in human life would, in fact, be the new connected society era.

There will be more and more enhanced and novelty equipment types in the market, like integrated miniature telemetry devices and cameras with short-range wireless communications that are embedded into medical pills. Other examples could be an environment where M2M devices observe processes like animal populations in isolated remote areas, and communicating with food and medical supply chains to provide automatic refills when required. In an urban environment, there are unlimited possibilities for optimizing announcement types via real-time knowledge of the preferences of people present in the area, e.g., by displaying the most relevant messages based on user profiles. The question in this highly developing environment might be whether it will happen too fast. Imagine an example of such personalized announcements in a public environment, the systems knowing exactly the preferences of the users and displaying related information on screens on the streets as the individuals walk by.

The IoT era is thus most fascinating as it counts for more possibilities for data mining than ever before, and, at the same time, there will be need to maintain at least some level of personal privacy, or provide users with the means to opt out from the personalized messaging. The basic problems are who has the right to collect and store information of individuals, who has right to protect the data and, in general, who may protect who and with what cost of privacy. The task is more than challenging in the modern environment with increasing levels of cyber-attacks, which needs to be addressed by increasing cyber surveillance, the entities being not only individuals but also governments and a whole ecosystem of banks, production chains, armed forces, educational institutes, etc. IoT can thus be also interpreted as an elemental component in the cyber world.

3.2.4   IoT Environment

Being used so widely all around the globe, one may wonder what the origin of the term 'Internet of Things' is. Various sources, like Ref. [5], summarize the key steps towards IoT and discuss who should be credited for 'inventing' IoT. In fact, as for the terminology, IoT definitely sounds sufficiently general to express the big impact it is causing on society by blurring the technicalities in such a compact yet intuitive way. Nevertheless, the meaning of IoT – regardless of the popularizing of the term as such – is after all a result of a long chain of innovations related to the theories and technologies such as electricity, communication methods, transistors, microprocessors, modulation and coding schemes, communications devices, protocol stacks, signalling methods, computers, programming, audio and video technologies, wired and wireless systems, measurement and antenna technologies, and of course, the Internet as it has emerged from early ARPANET and TCP/IP. These areas represent only a tiny snapshot of the building blocks that have paved the way to automatized functionality of connected devices in the form of IoT, M2M communications, Internet of Everything, Connected Society, Intelligent Home or any other term we want to express the areas of IoT.

It can be argued that the environment called 'IoT' would not exist without all the efforts made by an impossible-to-estimate number of inventors, technicians, commercial people and other contributors to the standardization, development, deployment and marketing of communications technologies. As for the term 'IoT' itself, according to several public sources like [5,9], it was utilized in this form by sensor researcher Kevin Ashton in 1999 as part of his presentation about early predictions of the development of the Internet. The term has been ever since widely adopted and it seems to be established as a part of the countless abbreviations of the telecommunications technologies. More background information for the development of IoT terminology can be found, e.g., in Ref. [5]

Based on the publicly available information, the very early enabler for the IoT type of environment has been the RFID, which provides the possibility to read basic information from a short distance. Some early examples of the RFID were related to inventory management by reading wirelessly the pre-embedded tags from items or to a supermarket where an RFID reader scans the customer purchases wirelessly in an instant fashion. This automatized identification and tracking of objects containing an RFID tag represents one-way wireless communications between passive objects and could in fact be considered as one of the very initial technology steps leading into the fuller scale IoT.

Nevertheless, the relatively simple data transfer between object and RFID reader represents a rather limited methodology and does not exactly comply completely with the modern IoT environment. Instead of merely passive objects, the current IoT can be understood as a set of both passive and active physical objects that are part of the information network in a fluent and seamless way, by participating in the functioning of whole systems and contributing to the actions of various areas of everyday life such as learning, healthcare, telematics and businesses. In fact, IoT is paving the way to a completely novel society which integrates the 'traditional' physical word, the digital world, and the virtual cyber world, as depicted in Figure 3.3 [5].

In many sources, IoT development is presented as waves, each containing more advanced functionalities of IoT solutions as a function of time. The RFID typically has been seen as an early trigger for IoT, especially as an automatized basis for the supply chain to tackle more efficiently inventory management and optimize deliveries [4]. This phase could be generalized as belonging to the first IoT wave. The second wave includes more versatile means for the communications and has been a base for, e.g., applications of many vertical

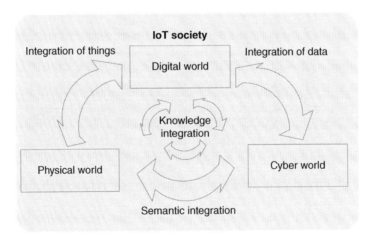

Figure 3.3   The main components of IoT

markets like healthcare, transport and security. The next, forthcoming third wave includes even more advanced means for developing services for locating objects and users. Additional waves could include, e.g., means for monitoring and controlling remote objects via presence and management over the WWW. The respective technology areas, along with all of the IoT waves, develop, thus facilitating further enhancements of the M2M and consumer solutions. In fact, this process can be understood as an iterative one, so that the new technology enablers trigger new types of IoT solutions and respective technologies, which again contribute to the further development of technology areas as depicted in Figure 3.4.

It can be claimed that IoT is an integral part of the overall technological development together with many other areas such as advanced telecommunications networks, sensors, mobile applications and related security solutions such as biometric authentication, cloud-based solutions, gesture recognition and NFC payment. IoT is thus a fairly suitable term to emphasize the inclusion and involvement of such a variety of entities communicating and participating actively in the contribution to the overall 'intelligence' via this always-aware and always-connected environment. Some IoT devices include weather-monitoring tools that can pre-process data for easing the actual predictions, smart meters for monitoring energy consumption, utility networks control and monitoring systems for gas and water consumption, industrial appliances that measure physical, technical and chemical values, home and office automation systems that can control the living environment based on input from the connected sensors, traffic, transport tracking and medical solutions with sensors that facilitate remote diagnostics, among others. The complete list is in fact close to infinity as new innovative devices are introduced into commercial markets, making our life easier and more fluent. One of the highly relevant domains increasing the reliance on the IoT environment is the automotive, including advanced means for self-driving vehicles.

To sum up the high-level picture of the IoT – it can be divided into individual, community and society levels as indicated in Ref. [35]. Some examples of individual IoT devices are smartphones and wearables meant for personal health and money transactions, whereas some community IoT devices include connected cards, health devices and smart homes. Examples of society level IoT devices are smart cities and smart grids.

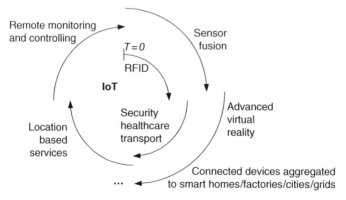

Figure 3.4 The IoT environment is developing along with the technological enablers, each phase or wave influencing the further planning of the enablers in an iterative way

3.2.5 IoT Market

The advanced consumer device markets are indicating that, together with the evolution of computing, the importance of IoT will be much higher relatively soon [4]. As a reference, personal computers resulted in commercial markets of around 100 million units per year, and there are around a billion PCs and related devices in the consumer markets. Furthermore, the mobile communications markets have introduced – in addition to the advanced feature phones – highly powerful hand-held devices like smartphones and tablets that represent the business scale for around a billion units per year. At the moment, forecasts indicate that IoT could reach a computing market of around 10 billion units per year [4]. It is a logical and 'mathematically beautiful' result of the IT evolution, since it fits nicely into the milestones of over 1 million mainframe computers in the 1960s and over 10 million mini computers in the 1970s, followed by hundreds of millions of PCs and desktop Internet devices between the late 1980s and 2000s, which all led to the current status of billions of mobile Internet devices. The next big step is the explosion of the IoT device markets within the forthcoming decade in a range of tens of billions of units. That is in fact one of the most concrete justifications for the deployment of the vast address base of IPv6.

IoT involves a huge number of stakeholders and businesses as well as many connectivity technologies that enable the always-connected devices. Although the device might not be switched on and permanently reserve the communication channels, the term 'always connected' refers to the ability to 'wake up' and exchange information with the network in an autonomic way whether it happens frequently (say, once per second), or only occasionally as needed (like remote low-power telematics devices emptying the measurement data buffers on a monthly basis).

In fact, the ultra-low power devices that operate in physically difficult-to-reach locations have very special requirements for power consumption and energy sources so that they are able to work automatically without human intervention or physical maintenance efforts during extensive time periods. The deep knowledge and new innovations of this energy efficiency are some of the key components, not only for remote devices but also in general for IoT application and device development. The connected devices may work perhaps for several years by relying on modern energy-saving techniques such as solar power panels and radio frequency energy which is harvested from the inducted RF energy. This domain has been noted as important to develop further, as it would benefit all the IoT devices equipped with such novelty energy-saving functionalities.

At the same time, HW and SW developers need to ensure the quality of their products so that the need for maintenance can be minimized, which further optimizes the costly site visits and reparation efforts. This is all a matter of balancing the processing chips, memory types, individual equipment and whole system expenses, quality and performance, interoperability, security and user experiences, among other aspects forming the complete IoT environment.

At the component level, one of the key development areas is the processor power efficiency to ensure an even more feasible base for the connected devices. At the other extreme, on top of the protocol layers, the app developers need to consider the best practices for creating energy-efficient functionalities. One example is the always-on communications based on periodic 'heartbeat signalling' that forces the user equipment to stay in an active stage without a network. It may greatly impact on the device energy and network capacity utilization. This type of heartbeat signalling may make the device respond faster when needed as there is no initial connection setup signalling for the packet data protocol context activation, but reserving the signalling capacity may prevent other users entering the network. Thus, the app developers need to ideally cooperate with the network infrastructure operators to balance accordingly the benefits and drawbacks of such solutions.

In addition to the power consumption of the device, the evolved battery technologies can further optimize the environment. Ref. [13] identifies the potential for the near future solutions related to IoT Systems on Chips (SoCs) by integrating low-energy processors with multi-protocol wireless transceivers and sensor interfaces into a single-chip integrated circuit. The benefit of such integrated solutions are the reduced cost, complexity and power of IoT applications.

3.2.6   Connectivity

The IoT devices rely on a variety of connectivity technologies because a single technology cannot fit optimally into all the imaginable environments. The challenge of this topic is achieving the optimal balance between costs and supported technologies of the IoT devices, so as an example, not all economical IoT devices can be equipped with LTE connectivity. Instead, some other short-range solution would serve better, combined by a gateway device that manages the local IoT network and connects it to the external word via, e.g., Wi-Fi or cellular network access.

Wi-Fi is a suitable method for connecting devices to the Internet as the respective IP protocols match by default, however, a tiny battery-operated IoT device may not be able to provide sufficient power for the longer term Wi-Fi connectivity. The selection of the proper connectivity technology is thus one of the many optimization tasks of the device manufacturers and other IoT stakeholders. Some guidelines may suggest that for applications requiring high data rates and wide coverage areas, LTE/LTE-A or previous cellular systems may serve best while local high data rate applications can be served via Wi-Fi. In fact, the best proofs of the increased consideration of the LTE as a base for IoT are seen in the 3GPP standardization items: the LTE-U for unlicensed operation and LTE-M for the M2M environment, standardized and added as part of the other LTE variants.

For the lower data rate solutions in limited areas, low-energy Bluetooth (BLE) can serve small amounts of IoT devices while ZigBee can manage more devices via mesh networking, both being more energy efficient compared to Wi-Fi or cellular connectivity. High energy efficiency is typically a requirement for battery-powered connected IoT devices that may remotely monitor and control, e.g., security systems. These devices may be based on programmable microcontrollers, embedded sensors and actuators which monitor and respond to environmental phenomena such as an opening door, breaking window, change of temperature, humidity and lightning. Especially in critical environments such as surveillance data monitoring and transmission, controlling and alarming environments with potential life-threatening conditions involved, the assurance of energy as well as connectivity is essential. Also, even less critical environments such as home sensors for controlling temperature or lights benefit from reliable energy and connectivity solutions. One solution for the management of such environments is based on the distributed intelligence of IoT with interconnected devices to optimize the monitoring and delivery of sensor data in real time, combined with cloud computing resources for post-processing and analysing the data. Users may also be able to control the respective IoT devices via smart devices or other remote methods.

The development of smart sensor solutions is thus one of the key areas in IoT, to provide data for the controlling and management of smart homes (varying humidity, temperature and lighting depending on the physical presence of users and user-based profiles etc.), and to provide information for the safety and security, including the health check and alarms related to home appliances that are failing (to act before the incidence happens). There are endless amounts of opportunities in the IoT environment for current and new stakeholders, providing interesting business opportunities that may benefit from partnerships and cooperation.

IoT handles vast amounts of real-time data from the field via always-connected devices such as intelligent sensors, forming part of the 'big data' environment. The IoT devices themselves and respective systems analyse the data and provide processed results as a base for decision making. IoT is thus suitable for managing, controlling and connecting devices such as home appliances, cars, businesses and complete cities in a highly automatized way without the need for human interactions. With so important a share of exposure in data collection and processing, the respective security is becoming one of the most essential elements in this environment with a high level of integration. There is thus room for combining current security solutions in an innovative way to ensure safety of the users, as well as for completely new security solutions.

The IoT environment is transparent for the physical radio (or fixed) connectivity while the messages from the IoT devices such as sensors are delivered or relayed to the destination, and vice versa (for the two-way systems). The wireless IoT connectivity can therefore be based on the licensed or unlicensed bands which are used by, e.g., mobile communications systems, wireless local area networks, low-range systems such as Bluetooth and ZigBee, etc. The typical division of the networks categorized by the operational range: Personal Area Network (PAN), Local Area Network (LAN), Neighborhood Area Network (NAN) and Wide Area Network (WAN), the latter being, e.g., the Internet itself covering the whole globe [14].

As the term suggests, IoT refers to connecting all kinds of devices to the Internet in one or another way. The devices that connect directly to the Internet are based on the IP protocol suite which enables by default the interoperable functions and data transfer between all the other Internet-connected devices. One of the challenges of the other, often varying connectivity technologies is the interoperability – or better said, the lack of it. In fact, in many cases, the internal IoT system might have been designed by utilizing non-IP or even proprietary connectivity technologies in order to simplify the TCP/IP structure – which in many cases is indeed unnecessarily heavy. So, the local network devices may use protocols deviating from the standard IP in the internal communications while connectivity to the Internet of these devices may be done via a gateway which supports the IP connectivity for the communication with the external world, and the non-IP connectivity within the internal network [14]. The physical connectivity of these gateways, or hubs, may be any of the networks able to communicate via IP such as mobile networks.

3.2.7 Regulation

IoT does not yet have a dedicated regulation. Nevertheless, there are such vast impacts foreseen from the deployment of the billions of IoT devices within the forthcoming years that the need to review current regulation is obviously needed. The discussions include especially the principles of storing the data and protecting the privacy of the users.

As Ref. [35] indicates, IoT can be expected to impact on licensing, spectrum management, standards, competition, security and privacy. There are already familiar aspects in current regulations that match with the forthcoming IoT, and that have been managed in previous systems. This is the case especially in the telecommunications and ICT regulation that has been involved with aspects related to competition, privacy and data protection. Also, the regulatory consequences are sometimes quite logical as is the case in the need for very large address spaces in order to identify the billions of connected objects. One of the obvious solutions for this is the IPv6, which requires proper management of the addresses. Ref. [35] also emphasizes the implications that are not necessarily obvious. One of such phenomenon under active investigation by the Federal Communications Commission (FCC) is the additional load that IoT brings to current services such as Wi-Fi and mobile networks, yet it is expected that new spectrum will not be explicitly allocated to the increasing IoT communications. As an example, the European Commission sponsored investigations indicate that a license-exempt model may support IoT development by avoiding contractual negotiations before devices are manufactured and used, which promotes large-scale production of more economical devices. Nevertheless, there is a need to follow up the advances to monitor how the market dominance will develop along with IoT devices and services – which belongs to the role of the competition regulators to ensure the balance between competitive market and innovation.

There is also regulation involved at least indirectly with IoT connectivity, especially related to the high-level frequency strategies in global and regional levels. As an example of the frequency regulation, the additional definition of the LTE/LTE-A bands is a joint effort of 3GPP, 3GPP2, industry and regulatory bodies, and there is obviously a great need to take into account the M2M and IoT traffic which can be seen, e.g., via the inclusion of the LTE-M as a part of the specifications. The identification of the frequencies suitable for the M2M communications has been discussed, and the liberation of previously utilized bands should ease the task. The general principle of the frequency regulation has been to actively identify new possibilities for the band utilization in such a way that the broadband spectrum needs to be flexible. The main body, ITU and World Radio Conference (WRC) events in 2012 and 2015 have identified the clarification of the Digital Dividend (DD) as one of the priority tasks aligned with cross-border coordination.

The DD spectrum, as one of the many options suitable for the M2M traffic, has been allocated to 790–862 MHz. The band was used for commercial LTE deployment in the USA and Germany as of 2010. In the USA, the DD spectrum was auctioned in 2008, and the concrete LTE deployment is advancing via Verizon Wireless that opened commercially the 700 MHz LTE network at the end of 2010. In addition to the already identified DD spectrum, there is also growing interest to extend the band in Europe below the already existing band. The strongest argument for the extension is the previous analogue TV bands that had already been ramped down by the end of 2011 in most European countries.

3.2.8 Security Risks

3.2.8.1 Overall Issues

The basic question in the IoT environment is how to reliably identify IoT devices and ensure secure communications between the devices and systems. One of the early-day principles of the Internet-connected devices is to provide a MAC address that identifies explicitly the devices. This is the case for basically all the IP-connected devices like printers, Wi-Fi cards, etc. The MAC address thus sounds like the most logical way to distinguish between the IoT devices. The issue of the MAC address is that even if it is hard-coded into each device to ensure the uniqueness, it is straightforward to alter in the IP communications via suitable software. Thus, the IoT system cannot for sure conclude if the connected devices really form part of the legitimate equipment or if there are possibly some cloned devices connected into the system which may aim to capture confidential traffic.

The more secure way to identify IoT devices is based on, e.g., a public key structure, or by including some suitable HW-based storage into the device that is hard to tamper with such as a SIM card, embedded SE or TEE. The issue of these solutions is the cost as the business model of the most economical devices would not tolerate such an additional expense. Another, possibly more cost-effective approach for the deployment and user experience may be related to the tokenization although the basic question remains on how to identify the correct IoT devices and how to securely transport and store the tokens into the devices.

Along with the advances of the SIM card and remote subscription management in an inter-operable way – as it is currently under standardization by various organizations – there are new methods presented for these items such as PKI-based developed key distribution and secure channels for accessing the contents of the SE, which refers to the development of the variants of the SCP.

3.2.8.2   Network Element Considerations

Ref. [11] has identified a highly relevant security issue in the growing IoT environment which is related to the embedded systems. These embedded vulnerabilities are a result of years of deployment which has resulted in many kinds of equipment still forming part of the uncountable number of elements in the Internet infrastructure, such as routers and bridges. The issue with these network devices can potentially be such that the security vulnerabilities are not patched automatically as is the case typically with consumer devices like laptops and smart devices. Instead, the security vulnerability may stay unmonitored and open doors for malicious intentions to overload the element, access further the net-work, and scan and modify the traffic within the respective core network. The correct way is obviously not the hiding the problems by the equipment manufacturers and operating entities as the knowledge about these security holes would, once revealed, spread very fast among the interested parties.

The very challenge of this area of potential middleware vulnerability is the difficulty to provide and manage updates for the security holes of the SW. The impact may be considerable as Ref. [11] indicates by referring to the Def Con event where a researcher investigated a set of commercial home routers and managed to break into half of them, including some of the most common brands. One of the basic root causes for these issues is that the respective systems are equipped with computer chips that are typically very economical. Due to the highly competi-tive environment, the chip manufacturers have limited ways to differentiate from each other for features and bandwidth. These chips often contain Linux-based OS and open-source SW components. This environment does not encourage the companies to enhance further the secu-rity aspects of the chip as it may require reservation of the very tight resources from the further manufacturing of the following chip generations.

The routers, bridges and other IP core elements are typically made by OEMs that might not get the brand name visible in the final product and thus do not have too much incentive to exceed the requirements for the chips, while the company representing the brand might have equally low interest in any security enhancements for the final product. As soon as the equipment is commercialized and deployed, it might be equally challenging to get security updates for the chip-related issues because of the low priority for maintaining old variants of the chips and products.

Furthermore, as Ref. [11] emphasizes, the SW might be relatively old even for new devices. Regardless of the relatively low possibility of new security patches developed for the chips, some components might lack patching which leads to security vulnerabilities as the systems age; one reason for this situation being the lack of the original complete source code. If the patch is available, the user normally needs to proactively download and install it. Nevertheless, the issue is that there is not often alerts for these patches, and if the ISP does not have a process for remotely installing the upgrades, the users might not be experienced enough to perform such upgrades for routers, modems, etc. Ref. [11] concludes that there might be a considerable amount of unpatched and insecure devices in the Internet. Some potential security breaches arise thus, e.g., via malicious DNS changer type of attacks against home routers and computers, and mentions an example of a Linux worm that can target routers, cameras and other embedded devices.

Along with the exponential growth of IoT devices, they could also potentially suffer from the same type of issues as a result of possibly low levels of SW patch maintenance, unless it is specifically addressed by the device manufacturers in an early phase. It has already been evident that some of the current and forthcoming variants of routers and modems are under special risks because they are located in the interface between users and the Internet. That makes it almost impossible to simply switch these vulnerable equipment off upon security breach events occurring. They also are increasingly powerful regarding the processing performance and the amount of embedded functionalities which makes them comparable with computers. Furthermore, as they are typically always-on-connected, they are increasingly attractive target for malicious attacks.

3.2.8.3 IoT Security as Defined by the GlobalPlatform

The GlobalPlatform has identified several security concerns for IoT as described in Ref. [2]. One highly relevant observation is related to the fact that the current and expected IoT devices are often used in environments that include critical infrastructure and systems such as transportation and medical devices. The potential security issues in this domain are related to the overall security of the devices as well as the users' privacy as a result of the IoT technology that interacts by definition with the physical world, which in turn may expose security holes compromising the private data. The GlobalPlatform emphasizes the special importance of unattended devices like electricity meters that are able to broadcast sensitive data without our awareness, and that require protection from potential attackers. Not only the M2M traffic but all the involved stakeholders need to ensure IoT security, including consumers, network element and mobile device manufacturers, network operators, service providers and application developers. The old saying about the chain being as strong as its weakest link applies in this environment excellently.

In order for the IoT market to evolve favourably, and to ensure that the development takes into account the security aspects, the GlobalPlatform has identified some key principles. These include the requirement for IoT devices to support a multi-actor environment that facilitates the varying of security and access settings per involved party. Furthermore, service providers should independently have means to remotely manage their own security parameters. It is also important to be able to add services and service providers to a device after it is deployed in the field. The service subscriber must be able to change service providers during the expected lifetime of the device which may be considerable depending on the use case and environment, the automotive being one representative for long-term device utilization. All this basically refers to the evolved subscription management concept which is actively under standardization at the GlobalPlatform as well as by various other organizations. More details of the enhanced, interoperable Subscription Management (SubMan) concept is described in later chapters of this book.

Being involved in many mobile telecommunications development items, the GlobalPlatform also acts in the role of IoT standardization and provides open technical specifications that focus on the interoperability and security of connected devices. The specifications of the GlobalPlatform include features that aim to enhance the privacy and security of IoT, such as the SE which works as a separate chip hardened against physical and logical attacks, and enables secure hosting of applications for various stakeholders. The GlobalPlatform also works on SD which stores cryptographic content for a stakeholder on the SE and provides mechanisms to manage such content and establish secure communications with external entities. Other related items of GlobalPlatform are the Trusted Service Manager (TSM) which is a third-party broker establishing business agreements and technical relationships between different stakeholders in a service delivery, the CA which allows for confidential post-issuance introduction of new stakeholders onto an SE, and the TEE which is a secure area residing on a mobile device that ensures sensitive data is safely stored, processed and protected in a trusted environment on that device [2].

The GlobalPlatform has identified various use cases for emphasizing the need of M2M security including healthcare, automotive, wearable devices and energy. Ref. [2] details how the GlobalPlatform specifications can address the privacy and security concerns for the deployment of IoT and M2M devices by relying on embedded technologies for new forms of secure communication and data transmission. Ref. [2] also presents descriptions of the use cases, introduces the function of IoT devices and details how vulnerabilities in security and privacy can be resolved.

3.2.8.4   Threats and Protection

Ref. [18] identifies some concrete key threats in the IoT environment and potential risks related to the security and privacy of the connected devices. In comparison to the currently typical hacking efforts against companies, with IoT, the hacks will be remarkably personal. In fact, as our homes may contain various IoT devices, they could open security holes for someone hacking into living rooms, baby monitors, smart TVs and other connected consumer devices without our awareness.

According to Ref. [18], the IoT devices are typical representatives of Minimum Viable Products (MVPs). The MVP is a product that needs to be released fast while the concrete enhancement feedback comes from the customers which, in turn, helps to make the product better. In the case of IoT devices, this means that there is no time or resources for ensuring the security and privacy of the equipment especially for the most economic ones. Also, for the simple devices, additional security may result in lower fluency for the user experiences. Thus, a strong password may not be attractive for setting up a new, highly economical IoT device during the out-of-the-box phase.

Ref. [20] summarizes further the reasoning behind such a huge amount of security breaches of IoT so far. The base for these phenomena is the very fast pace of digitally connected devices becoming an essential part of our daily life, including our homes, offices, cars and even the very close proximity in the body area. Along with the growth of the IPv6 and Wi-Fi network deployments, the IoT environment is shaping fast – and can have large amount of potential security holes. The positive side of IoT is definitely the existence of a completely new set of platforms that enable us to perform something totally new, but at the same time the drawback is that IoT is becoming an increasingly attractive target for cybercriminals along with the increased number of connected devices that can expose more attack vectors.

Some of the compromised IoT devices that have been reported in public include IoT baby monitors that may provide a door for criminals to monitor the feeds, to change camera settings and to authorize external users to remotely view and control the monitor. Another widely reported case relates to Internet-connected cars with indications of potential risks for criminals being able to take control of the entertainment system, unlock the doors or even shut down the car in motion. Ref. [20] further emphasizes the increasingly important role of wearables that may become a source of security threats as hackers could potentially target the motion sensors embedded in, e.g., smartwatches to interpret the contents, or to steal personal sensor data from health devices.

In addition to the compromised personal data, the more severe security breaches are related to those cases that may have impact on personal physical security. An example could be medical device monitoring and maintaining vitals such as pacemakers which, in case of any disturbances such as altering the results of the heart-rate functioning, could cause unrepairable damage or even death.

As an example of the industry forums, the IoT Security Foundation is a non-profit industry body responsible for vetting Internet-connected devices for vulnerabilities and offers security assistance to tech providers, system adopters and end-users. The aim of the Foundation is to raise awareness through cross-company collaboration and encourage manufacturers to consider security of connected devices at the HW level [31]. There are also efforts by many companies to set up platforms enabling large networks of IoT devices to identify and authenticate each other in order to provide higher security and prevent data breaches. Other examples can be found from various research works to enhance IoT security through device and smartphone linking.

Some simple guidelines are often sufficient in the initial phase of IoT device deployments. From the consumer's point of view, upon installing a new IoT device such as an Internet-connected toaster, baby monitor or any other 'always-connected' device, the rule of thumb is to change the default password immediately before the equipment is used. For the manufacturers, it would be ideal to totally isolate personal health-related functions from the open Internet, even if it were well protected via passwords. Furthermore, the protected data (cloud, devices, bitlocker) is justified for both industries and consumers as it is highly risky to leave any plain data (texts, images) if the contents could be accessed by non-authorized persons.

For the device manufacturers, the assurance of the data storage either via the utilization of HW-based secure elements, as much as reasonable cost-wise, guarantees the highest protection level. Solely SW-based security is typically more vulnerable than, e.g., micro-SD, embedded SE, removable UICC or TEE based methods.

It is always recommended to check the weakest links in the end-to-end security. Even if the IoT device looks relatively innocent, and the home environment is otherwise well protected, a device such as a non-protected sensor-containing freezer that is connected to the same home network could open a surprisingly easy security breach door for criminals. It is also a good idea to keep up to date with security news such as the latest advances in the DNS protection, firewalls and virus protection SW for all the relevant devices.

3.2.9   Cloud

Due to its usefulness for many consumer and M2M environments, the cloud can be considered as an important – if not integral – part of IoT. Cloud computing is based on the sharing of resources in order to obtain capacity gain, to distribute processing resources and to achieve economies of scale. The drawback of dedicated data delivery networks is that their typical load is only a fraction of the dimensioned maximum capacity due to the fact that there are occasional load peaks that demand considerably more capacity. Instead of all the capacity operators managing this type of highly over-dimensioned networks for the sake of the serving of these capacity peaks, the whole delivery infrastructure may, instead, be concentrated into a single, large IP cloud. As the capacity peaks of different operators or areas within a single operator typically won't occur at exactly the same time according to statistical probability characteristics, the total offered capacity of the cloud can be considerably lower than the sum of the individual networks.

This concentration of capacity into the cloud is one of the benefits of cloud computing in telecommunications [10]. Furthermore, the uploading and downloading of cloud data in thee mobile environment is especially fluent via LTE/LTE-A as they provide fast data rate and low latency. Taking into account the charging of data transmission over air via mobile networks, one of the typical solutions is the utilization of Wi-Fi Offloading together with the LTE/LTE-A radio access. Cloud computing is a solution that enables on-demand network access to a set of shared computing resources that are configurable. The initial evolution of cloud networks and services has been driven by the needs of large companies, but it can be assumed that cloud computing is also becoming an increasingly important daily tool for smaller companies and end-users.

As the cloud concept is still relatively new, there might be concerns about the connectivity control, security and privacy as the functionality and contents are distributed outside the previously highly local, much more isolated and thus more manageable environment. The important task for the network service providers is thus to address these concerns and make the respective measures to minimize potential security risks. The proper encryption and protection of user accounts are some of the priority tasks of service providers.

An important need for cloud computing is to outsource IT tasks to cloud services available over the Internet. The benefit of this evolution is that an increasing amount of resources can be managed via a single point and utilized by applications and services based on need. Naturally, it may take years for the large-scale support of cloud services. The server virtualization is one of the ways to create internal and external cloud networks and services. There are also other areas that need to be initiated and developed like the meta-operating system that eases the management of distributed resources as a single computing pool. The meta-operating system is a virtualization layer between applications and distributed computing resources. It takes advantage of the distributed computing resources in order to manage and control applications and related tasks such as error control. In addition, there is also a need for a service governor for the decision making about the final allocation and prioritizing of the computing resources for applications.

Cloud computing offers dynamic network structuring. An example of this functionality is SW-based networking as part of cloud computing access via low-cost and flexible adjustments by the cloud customers. Cloud networks also provide safety features for essential network functions like IP routing, address management, Network Address Translation (NAT), authentication and QoS.

Via SW-based networking of cloud computing, the network architecture of the customers can be replicated regardless of the location. The replication includes changes to topologies and policies, making cloud networking a logical platform for disaster recovery in a dynamic and cost-effective way.

Specifically for the IoT environment, it is essential to facilitate the secure communications with the cloud as well as the storage without exposure to external users. Some cloud-related use cases include secure payment via HCE and cloud storage for IoT telematics and traffic analysing service. The cloud concept is highly useful as it makes an abstraction to IoT communications and can give added value in many type of environments, e.g., a tokenization entity for banking and safe storage for telematics.

3.2.10 Cellular Connectivity

3.2.10.1 RF Band Deployment Scenarios

The cellular RF bands form one of the most useful bases for IoT devices that require connectivity within a wide range. With only a small set of quadruple bands, GSM can provide global functioning. Even by adding the latest bands such as 450 MHz, GSM connectivity is relatively straightforward to deploy by device manufacturers as well as by the network operators as it is part of such a robust technology. The same principle applies to 3G technologies such as UMTS, which provides global functioning with only five RF bands. The benefit of 2G technologies, especially for GSM, is the widely available service area. The networks are typically deployed on 900 and 1800 MHz or 850 and 1900 MHz bands, which are able to provide very large coverage for IoT devices far into future assuming the systems are maintained in a parallel fashion along with the 3G and evolving 4G systems – and of course, during the ITU-R compliant 5G systems that will be deployed as of 2020.

The current environment is much more diverse with the pre-4G LTE and fully ITU-compliant 4G LTE-A as there are also considerably more RF bands available. At the same time, it ensures a sufficient set of options for the stakeholders (terminals and networks optimized per region), on the other hand, a single device can hardly support all the possible frequency options if it is supposed to be techno-economically feasible. The selection of the most optimal frequencies by the device manufacturers and operators is thus one of the most important tasks in the device selection, deployment and operation of the networks, and it clearly depends on each market. The following sections summarize the global status for the LTE/LTE-A bands in order to present the challenges of RF band fragmentation.

3.2.10.2   LTE/LTE-A Spectrum in Practice

The high number of possible LTE and LTE-Ad bands provides a large set of options per ITU region, continent and country. There is such a variety of e possibilities that regulators, operators and device manufacturers need to form their strategies with great care. One example of the diversity can be seen in Figure 3.5 that indicates the LTE Frequency Division Multiplex (FDD) and Time Division Multiplex (TDD) frequency deployments or plans in the Latin America region. Figure 3.6 shows the typical, assumed LTE/LTE-A band deployment and carrier aggregation scenarios for the rest of the world. The information is based on various publicly available indications about regional frequency plans.

As can be noted from Figures 3.5 and 3.6, there is a very wide set of potential LTE/LTE-A RF band variants in the world, and many differences even within smaller regions. The planning

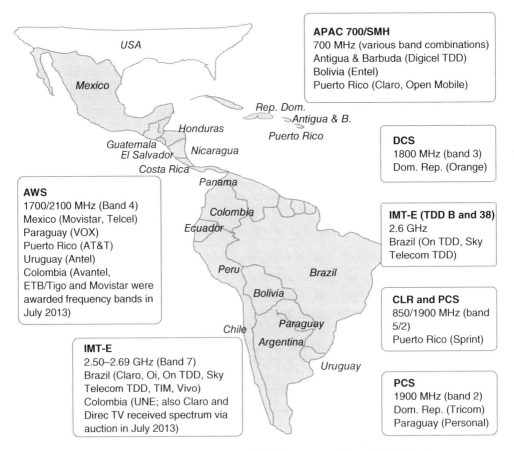

Figure 3.5   An example of the potential LTE spectrum plans of Latin America

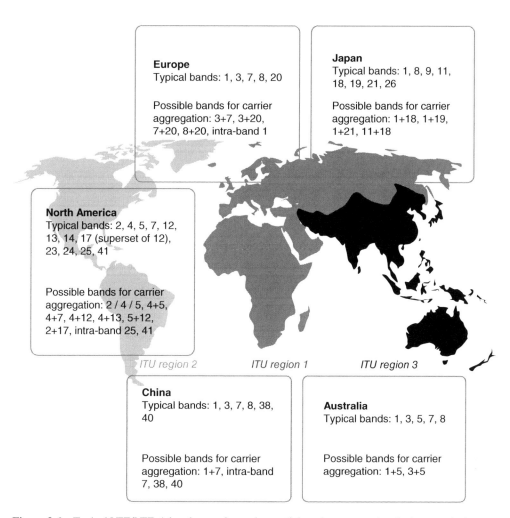

Figure 3.6 Typical LTE/LTE-A band scenarios and potential carrier aggregation deployment in the rest of the world

of the LTE/LTE-A devices – both consumer devices as well as IoT devices – is thus much more complicated than with previous mobile communications systems. The optimization of the device's band support depends on many aspects like target markets, device category and size. Chapter 6 discusses in more detail the methodology the device vendor may consider for optimizing the device RF band set.

The first commercial LTE networks were deployed at the end of 2009, and the initial significant launches were seen in the USA and Japan by the end of 2010. In addition to the FDD frequency band, the TDD spectrum also provides increasing opportunity for LTE/LTE-A developments, despite the fact that traditionally the TDD spectrum has been less attractive to operators than the FDD spectrum. Based on recent developments, it can be expected that the role of TDD will increase significantly during the forthcoming years.

The LTE networks have been implemented on many frequency bands. It can already be seen that the fragmentation of the RF bands is a reality, which is noted by the industry, vendors, operators and chipset developers. The availability of options offers possibilities but also interoperability challenges which need to be optimized.

In addition to the various bands, LTE/LTE-A also supports various frequency bandwidths depending on the frequency and region. The range may vary between 1.4 MHz, 3 MHz, 5 MHz, 10 MHz, 15 MHz and 20 MHz. The flexibility ensures that the operators are able to select the most adequate strategies for LTE/LTE-A deployment in align with the already deployed GSM and UMTS RF bands especially in 850, 900, 1700/2,100, 1800, 1900 and 2100 MHz bands. This flexibility provides highly dynamic re-farming in such a way that while the capacity of the previous systems lowers, the customers may be directed gradually to utilize more efficiently the LTE/LTE-A system. In the initial phase, the utilization of the narrowest bandwidths 1.4 MHz and 3 MHz may often make sense in order to introduce LTE/LTE-A into the commercial market although the achieved data rates would logically be relatively low. As the situation develops, LTE/LTE-A may reserve more bandwidth up to 20 MHz, and with carrier aggregation up to 100 MHz when applicable which provides the highest data rates.

In the frequency bands without previous legacy systems, such as 700 MHz and 2600 MHz, LTE/LTE-A may be deployed with the maximum possible bandwidth from scratch. Especially the 700 MHz band, as well as the latest addition of 450 MHz, may be highly suitable for wide rural areas. It should be noted though that the antenna size of the lowest range, e.g., on the 450 MHz band, needs to be relatively large in the UE in order to fully exploit the benefit of the lower frequency propagation characteristics. For small-size equipment like hand-held smart devices, the small antenna may compensate the benefit of the wide propagation due to the relatively high antenna loss, so the 450 MHz band probably suits best for larger-size equipment like integrated LTE/LTE-A tablets.

A feasible solution for both device and network would be the support of combined low bands (selected frequency around 700–900 MHz) and high bands (selected frequency from 1800 MHz up to 2.6 GHz or even 3.5 GHz). The low bands provide large coverage while the high bands provide more capacity and highest data throughput.

Even if the 3GPP list of frequencies looks overwhelming, it allows the selection of the most suitable network deployment scenarios at the regional level – as long as the network vendors and OEMs ensure the availability of sufficient sets of equipment per region. As an example, the new band 2600 MHz is increasingly important in many regions like Latin America, Europe, Asia (Pacific), the Middle East and Africa, as can be interpreted from spectrum acquisitions and publicly available near-future plans of these regions. Also the reorganization of the television frequencies due to the ramping down of the analogue systems brings new opportunities for e LTE/LTE-A deployments in the DD spectrum, and the re-farming scenarios in 850/900 and 1800 MHz bands are also logical options for the LTE/LTE-A networks. The 700 MHz frequency spectrum especially has triggered much discussion and concrete steps for deployments. The Asia-Pacific (APAC) band 28 has been identified as one of the most potential candidates in the Latin America region. Japan is also active in the initial deployment of 2100 and 1500 MHz bands. The benefit of 2300, 2500 and 2600 MHz bands is the wide availability of the spectrum.

The new LTE/LTE-A options include s new opportunities for the deployment in the DD spectrum and in the 2600 MHz band, as well as on many other new frequency bands like the L-band and 1800 MHz band. The first concrete steps in the deployment of DD spectrum bands were taken in the USA and Germany, and continued by more auctions. The deployment of the 2600 MHz band is especially popular in Europe, and there seems to be increasing interest in the 1800 MHz band in Europe and Asia Pacific. There are also some concrete ideas for further enhancing LTE/LTE-A spectrum utilization via cognitive radio and the white space concept. The development of the white space concept is driven by the USA and the UK, and might offer considerable benefits as part of the band strategies.

3.2.10.3   Advanced RF Utilization

Cognitive radio is another idea for more efficient LTE/LTE-A spectrum utilization. Nevertheless, the concept is still at an early stage and commercial deployments are not as yet taking concrete steps. The first deployments are US-driven along with the utilization of white space spectrums. As the idea is still relatively new, it must also need regulative bodies to make efforts.

As for the more concrete LTE-A items, the utilization of Carrier Aggregation (CA) is growing fast in order to increase the capacity. CA is one of the major items for paving the way to the ITU-compatible 4G era and respective higher data rates. The additional benefit of CA is that it is able to take advantage of the otherwise isolated frequencies. Furthermore, CA may be used for unidirectional frequencies. An example of this option is the reassignment of the former US MediaFLO band 29, which is only defined for Downlink (DL), to LTE/LTE-A.

3.2.11   WLAN

WLAN with various variants is currently one of the most popular Internet access methods. Along with the general development of wired Internet access methods and packet core networks, WLAN solutions have gone through major enhancements. As a result, the bit rate has increased exponentially since the 1990s, and the functional area of the networks grows constantly. The first-phase WLAN was formed by the early IEEE 802.11 standards, which are being gradually complemented. Table 3.1 summarizes the current most relevant WLAN variants [17].

As is the case for the mobile networks, security is an important aspect in private and business WLAN environments. In practice, if the radio interface of the WLAN is not protected by a password, it is basically available for public use. In the worst case, the Access Point (AP) can be used for illegal purposes by external attackers. Setting up the server as a delivery element for image, music or video contents without the permission of the content owners is only one of the countless examples on how the open access can be misused.

The most elemental and easy option for increasing the security level of the WLAN networks is to activate the access code request upon establishing the connections with the AP. Insecure WEP is used for the initial WLAN access restriction, and it has been enhanced via several versions. Chapter 2 details the modern methods for the WLAN protection.

3.2.12   Low-Range Systems

While the cellular systems provide the widest coverage area for IoT (apart from the satellite systems which are probably too costly for general use), the local connectivity methods can be

Table 3.1   The key WLAN IEEE 802 standards

Version	Name	Frequency band	Bit rate (maximum theoretical)
IEEE 802.11 (legacy)	WLAN	2.4 GHz	1 Mb/s–2 Mb/s
IEEE 802.11a	WLAN (Wi-Fi)	5 GHz	54 Mb/s
IEEE 802.11b	WLAN (Wi-Fi)	2.4 GHz	11 Mb/s
IEEE 802.11 g	WLAN (Wi-Fi)	2.4 GHz	54 Mb/s
IEEE 802.11n	WLAN (Wi-Fi)	2.4 and 5 GHz	300 Mb/s
IEEE 802.11 ac	WLAN (Wi-Fi)	5 GHz	1 Gb/s (total area) and 500 Mb/s (single station)
IEEE 802.11ad	WiGig	60 GHz (and backwards 2.4/5 GHz)	7 Gb/s
IEEE 802.15.1	Bluetooth	2.4 GHz	1 Mb/s
IEEE 802.15.3/3a	UWB	Various bands	10–500 Mb/s
IEEE 802.15.4	ZigBee	2.4 GHz, 915 MHz (America), 868 MHz (Europe)	250 kb/s
IEEE 802.16	WiMAX	10–66 GHz	120 Mb/s
IEEE 802.16a/e	WiMAX	2–11 GHz	70 Mb/s
IEEE 802.16 m	WiMAX 2	Licensed IMT-Advanced bands	100 Mb/s, 1 Gb/s
IEEE 802.20	WMAN/WAN	3.5 GHz	1 Mb/s
IEEE 802.22	Wireless Regional Area Network	VHF/UHF TV bands	19 Mb/s

used for information sharing between the devices within the area, or via hubs which could be further connected to, e.g., the cloud. There are various methods for the local connectivity available in the commercial market, and each solution has pros and cons depending on the use case, mobility, required data rate and maximum needed coverage. Some of the key technologies for the connectivity are:

- **Bluetooth**. The popularity of this method has been increasing along with the technical advances since the first versions. The low-energy variant of Bluetooth (BLE) is especially feasible in the IoT/M2M environment as it requires only low power consumption yet offers sufficient coverage for local coverage.
- **Wi-Fi**. The importance of Wi-Fi has been increasing gradually as more hotspots become available in public places. Wi-Fi is typically a method to connect to Internet services as an affordable alternative to cellular connectivity. Typically, the SW updates of smart devices are increasingly done via Wi-Fi as an alternative to the SW downloading via a laptop/PC that is connected to the mobile device. Operators are also getting interested in offering wider Wi-Fi Offloading solutions to the consumers to balance the cellular network load.
- **Near Field Communications (NFC)**. This is the closest range technology that has been gradually appearing into new markets since 2012. It can be used in many solutions via a tap gesture of the consumer, including information sharing (similar to RFID), establishing connections for audio/video and performing secure payment.
- **Radio Frequency ID (RFID)**. This is based on readable and optionally writeable tags. It represents a very basic connectivity of the IoT mobile device, although it can be integrated into the overall functionality of the device/SIM.

- **Wireless USB**. A short-range, high-bandwidth wireless radio communication protocol designed by the Wireless USB Promoter Group. It is maintained by the WiMedia Alliance. Wireless USB was based on the Ultra-Wide Band (UWB) platform of the WiMedia Alliance, providing 480 Mb/s within a 3 m distance, or 110 Mb/s within a 10 m coverage range. The bandwidth was 3.1–10.6 GHz. The Current implementation is W-USB, which is able to form USB systems including host, devices and interconnection support. It is based on the USB hub-spoke model that allows a maximum of 127 wireless devices connected wirelessly via PTP links, or spokes, with the host, or the hub. There is a single host controller in the system. The topology is comparable to a star network.
- **Ultra-Wide Band (UWB)**. A wireless technology based on very low energy and short-range distance to be utilized in high-bandwidth communications over a large portion of the radio spectrum. Typical use cases for UWB are short-distance radar imaging, sensor data collection, precision locating and tracking. UWB was evaluated as a base for PAN networks, and it was present in the IEEE 802.15.3a draft PAN documentation. Nevertheless, the IEEE 802.15.3a task group was dissolved and the development was taken over by the WiMedia Alliance and the USB Implementer Forum. Although technically functional, the un-ideal progress and techno-economic suitability have limited the use of UWB in consumer products.
- **ZigBee**. An IEEE 802.15 standard designed to carry small-scale data within short distances, and it uses only a low amount of power. Unlike Wi-Fi, ZigBee represents a mesh networking standard. The nodes of the ZigBee network are thus connected to each other. It provides a fixed data rate of 250 kb/s.
- **6LoWPAN**. Combines the IPv6 and low-power PAN, providing IP-based wireless transfer for very small and low-processing powered devices.
- **Symphony Link**. A wide-area, low-power wireless system. End-users may employ a long-range radio activity module for the communications with the gateway. It is based on star topology networking, i.e., the end nodes communicate with a single gateway. The gateway manages the stars and interacts with the modules in a flexible way.
- Optical reader technologies for barcodes (variants: 1D, 2D, 3D).

There also are (practically) obsolete legacy systems like Infrared (IR) which could be used to create a link between two devices (e.g., mobile to mobile, or mobile to laptop) for information sharing, including photos, contacts and other contents. The importance of this method has dramatically decreased in recent years, and the modern devices no longer include IR. Nevertheless, the method is still useful for the older devices that support IR. In those cases, the security aspects need to be considered especially in open communications as the traffic may be eavesdropped within the visible line of sight.

Figure 3.7 summarizes the currently most relevant local connectivity technologies that are especially suitable for the IoT environment, and the following sections detail further the principles of the NFC, barcode, RFID and Bluetooth.

3.2.12.1 Bluetooth

Bluetooth replaces wires between equipment like computers, their peripherals and mobile devices. Bluetooth has low power consumption and a relatively short functional range which

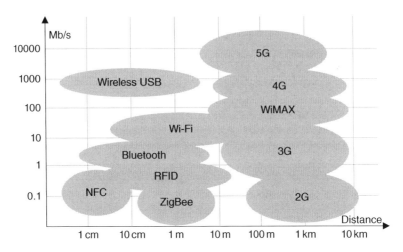

Figure 3.7   High-level examples of wireless connectivity solutions with respective coverage and data rate

Table 3.2   The theoretical distances of Bluetooth devices per class

Class	Max power (mW)	Max power (dBm)	Distance (m)
1	100	20	100
2	2.5	4	10
3	1	0	5

depends on the standardized power classes. Bluetooth provides low-cost transceiver technology for many types of devices like wireless headsets and microphones. As the interface is based on the RF, one of the benefits over older technologies like IR is that no line of sight is required as long as the received power level is high enough. Table 3.2 presents the power classes of Bluetooth [17].

Bluetooth profiles can be defined as general behaviours through which Bluetooth-enabled devices communicate with other devices. This means that in order to be able to connect the devices together via Bluetooth technology, both must support and understand the common Bluetooth profile in use. The Bluetooth profile describes the possible applications that can be used in the connection, and how Bluetooth is used. As an example of the profile, a File Transfer profile defines how devices should use Bluetooth in order to transfer files between devices, which can be physically, e.g., mobile device and a Personal Digital Assistant (PDA).

In order for the Bluetooth devices to connect with another, both devices must share at least one of the same Bluetooth profiles. An example would be a Bluetooth headset that is utilized via a Bluetooth-enabled cell phone. Both the headset and the mobile device should have and use the Headset (HS) profile, which basically defines the way to initiate, maintain and release the connection between, e.g., headsets and mobile devices.

Various Bluetooth profiles have been developed. The manufacturer of the Bluetooth device assigns a set of Bluetooth profiles for the device to a certain set of applications that work with other Bluetooth devices. According to the Bluetooth standards, all the Bluetooth profiles should include as a minimum the following set of information: (1) dependencies on other profiles; (2) recommended user interface formats; and (3) particular parts of the Bluetooth protocol stack used by the profile. To perform the planned functions, each profile uses particular options and parameters at each layer of the stack. This may include an outline of the required service record, if applicable. Most Bluetooth devices are given just a few profiles. For example, a Bluetooth headset will use the HS profile, but not the LAN Access profile which defines how devices use Bluetooth technology to connect to local area networks.

Bluetooth pairing can be done via strong but optional pre-shared key authentication and encryption algorithms. The Bluetooth security depends largely on the length and randomness of the password for pairing Bluetooth devices. This procedure performs mutual authentication when done for the first time, and sets up a link key for later authentication and encryption. anther parameter related to the security is the visibility setting of the Bluetooth device, so it is recommended to make devices visible only when needed. Furthermore, the optional user authorization gives additional protection in the case of incoming linking requests.

The potential security risks of Bluetooth are related to the lack of centralized administration and security enforcement infrastructure. As Ref. [32] has identified, the Bluetooth specification is highly complex and supports more than two dozen voice and data services. The challenge arises from the fact that devices and services supporting Bluetooth are based on a variety of chipsets, devices and operating systems, as well as different user interfaces, security programming interfaces and default settings that are not constantly the same for all. The respective Bluetooth attacks include identity detection, location tracking, DoS, control and access of data and voice channels without permission, to mention just a few. One of the simplest ways for security breaches is based on a weak default password that may be, e.g., '0000'. Also, the support for the HS profile that provides telephony signalling commands may open means for misuse of the device via Bluetooth. More information about the Bluetooth security aspects and protection can be found in Ref. [33], and Ref. [34] which detail examples of real-world Bluetooth attacks such as unauthorized downloading of phone books and call lists, the sending and reading of SMS messages from the attacked phone, and breaches from relatively long distances (more than 100 m).

3.2.12.2   RFID

RFID technology belongs to the Automatic Identification and Data Capture (AIDC) methodology. The AIDC, or Auto-ID, is in turn a set of methods for identifying objects, collecting object data and entering respective data into computer systems in an automatized way. In addition to the RFID, the AIDC technologies include barcodes, biometrics technologies, magnetic stripes, Optical Character Recognition (OCR), smartcards and voice recognition [21].

RFID is based on wireless data transfer in order to automatically identify and track tags that are attached to objects. The tags identify the respective objects based on electrically stored information. Unlike the case of optically readable barcodes, the RFID reader does not need to be located in Line-Of-Sight (LOS) with the tag as the radio waves also propagate through material.

There are a variety of RFID tags available. They can function in standalone environments without an external power source based on the electromagnetic induction of magnetic fields upon reading the tag, while other RFID types harvest electromagnetic energy from the interrogating radio waves in a passive transponder mode. RFID may also be based on its own power source, e.g., a battery, which also provides larger coverage area of the tag compared to the induction-based passive mode.

RFID tags work on a set of unlicensed frequencies such as 3–8 MHz, 13 MHz, 27 MHz, 433 MHz, 902–928 MHz, 2.4 GHZ and 5.8 GHz. There are also various use cases for the tags. They are useful in tracking objects during the production, storing, logistics and after-sales market. In addition to goods and products, they are equally useful in identifying living objects such as pets via an embedded RFID chip under the skin, or runners in a sports event with the tag attached to clothes.

Figure 3.8 depicts the RFID system architecture which consists of the RFID tag attached to the object, antenna, reader and host computer which in turn is connected to the application such as Enterprise Resource Planning (ERP).

According to Ref. [22], the potential security breaches related to RFID are currently under control, or at an acceptable level. The source indicates that the current data protection provided by the EPCglobal generation 2 protocol represents an advance over previous protocols and is acceptable for still rather limited RFID deployments within the supply chain. While the IP communication between RFID readers and the network is secure due to standard IP network security, the threat in this domain is related to the RF communication between tags and readers which needs to be taken into account in the further development of protocols.

The potential future RFID security threats may include clone tags and unauthorized readers, as well as malicious intentions to intercept reader data via external devices. There is also – although still only a theoretical – possibility to deliver viruses via RFID tags. Ref. [22] thus suggests that the future deployments need updated security and 3G protocols along with the increasing importance of RFID deployments in the consumer domain.

Ref. [23] also discusses RFID security claiming that security breaches can happen at the RFID tag, network or data level. One of the potential issues in adopting sufficiently protective solutions and standards is the very low cost and light functionality of the tags. This means that

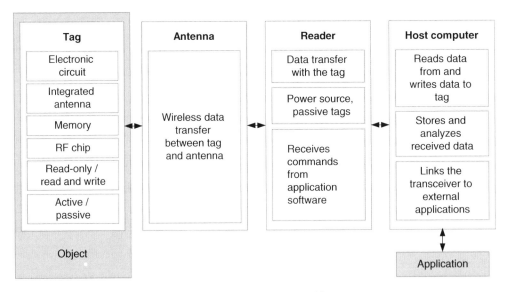

Figure 3.8   The RFID system architecture

the functional and improved security tools do not make production of the tags feasible cost wise, and additional encryption within a tag would impact negatively on the tag's processing power. Nevertheless, the industry is considering these issues. The EPCglobal UHF generation 2 protocol is expected to work with ISO 18000-6C RFID wireless interface specifications. Furthermore, EPCglobal has enlisted security vendor VeriSign Inc. as its infrastructure provider.

Ref. [24] further mentions the DoS attacks. In the case of RFID, a relatively simple radio jammer or signal oscillator set on the RFID tag frequency may interfere with the communications located nearby. The result may be significant in environments such as mobile payment or security applications. The requirements of very close RFID tag proximity to the reader reduces these risks but it would not prevent completely the effects of a powerful jammer located near the reader. Ref. [24] also mentions the possibility to eavesdrop on the communications between reader and tag, and the cloning of RFID devices. Furthermore, strong encryption methods require more memory and more complex chip designs than the typical RFID tags support, which results in weak encryption and thus vulnerabilities for malicious attacks.

3.2.12.3   Barcodes

The barcode is an optical machine-readable representation of data relating to the object to which it is attached. The original 1D form of barcodes represents data via different widths and spacing of parallel lines. The further development of barcodes has brought 2D formats that contain a variety of optical representations such as rectangles, dots and hexagons. The barcodes can be interpreted optically by using barcode readers as well as with equipment capable of digital imaging such as smart devices with respective barcode scanner apps.

The typical use cases for any form of barcodes include the retail product tagging for the price information, as well as goods tagging in warehouse and logistics supply chains. As is the case with RFID, barcodes also belong to the AIDC. The benefit of the barcode is the simplicity, universal use and low cost. The drawback is the need for LOS.

Barcodes are not considered as subject to important security breaches although it may be possible to embed instructions to the image that some smart devices may execute, such as the displaying of the IMEI. It is not typical that user devices are capable of executing some other, more harmful code based on the barcode, such as initiation of a phone call upon scanning the code. Nevertheless, barcodes can present web links that can be entered automatically by the smart device; it is thus a matter of ensuring protection mechanisms in the device side to prevent the browser establishing connections to malicious web pages.

3.2.12.4   NFC

NFC is a short-range wireless communication technology which enables the exchange of data between devices like hand-held mobile phones and readers such as Point of Sales (POS) equipment of retail stores. The communications of the NFC is based on the high-frequency radio interface that provides functional connections within a maximum distance of about 10 cm between the NFC-enabled devices. For payment solutions, there may be separate certification processes that dictate different requirements for the useful distance. NFC is a subset of the RFID domain, and is based on a proximity range frequency of 13.56 MHz.

This frequency range is dominated by the ISO-14443A, ISO-14443B, FeliCa and ISO-15693 tag standards.

It should be noted that ISO-14443A, ISO-14443B and ISO-15693 do not define security architecture. The ECMA-340 standard is meant for information exchange between devices that have more capabilities than merely simple memory storage. It is based on the stack utilized in the ISO-14443A standard, but it allows more functionality in addition to the reading and writing memory. Nevertheless, it does not contain security architecture so the protection needs to be ensured, e.g., in the application level.

3.3   Development of IoT

The following sections summarize some of the key concepts in the IoT domain such as the GSMA Connected Living concept and other industry forum, alliance and standardization initiations.

3.3.1   GSMA Connected Living

The GSMA Connected Living concept is planned to enable IoT, creating an environment in which consumers and businesses can enjoy rich new services, connected by an intelligent and secure mobile network [12]. The focus of the Connected Living programme is to ease the task of mobile network operators in adding value to the services as well as to accelerate the delivery of connected devices and respective services in the M2M market. The ways of work of this programme are based on industry collaboration, regulation, network optimization and development of key enablers to support the growth of M2M in the near future and longer term. The programme takes into account the safe communications of the IoT devices and applications via the mobile network.

As the number of IoT devices is predicted to grow considerably, one of the most essential tasks of the MNOs is to plan and optimize the networks accordingly in order to support thousands of simultaneously communicating devices. For that reason, GSMA is developing guidelines for efficient, trusted and reliable IoT services – not only the operators and service providers but also the app developers are in a key position to create such solutions that do not waste the valuable capacity of the networks. One example of such dangers is the heartbeat signalling to keep the PDP context alive even if there are no actual communications; this does not cause too much harm in cells with low amounts of devices, but as the numbers of equipment increase there could be potentially hundreds or thousands of such devices fighting for the signalling resources – and those highly loaded signalling areas would not be able to serve all the devices without app-level optimization. With the help of the Connected Living guidelines, the IoT device and application developers are able to ensure a common approach and fair utilization of the precious resources in such a way that they can scale as the IoT market grows.

In practice, GSMA cooperates with IoT ecosystem partners in the creation of the guidelines that describe how the IoT equipment can communicate within mobile networks efficiently. This approach is elemental as there are hardly better ways to guarantee the fluent functioning of the connectivity of huge amount of IoT devices in a scalable network, so the commonly created rules by the stakeholders ensures a fair share of the efficient connectivity.

3.3.2   The GlobalPlatform

The GlobalPlatform discusses the current development and potential next steps of IoT in Ref. [2], arguing that the ongoing IoT environment is still relatively young which means that existing proprietary solutions will suffice for current use cases. Nevertheless, along with the growth of the number of IoT devices in the market, the level of security and privacy concerns increase also increase. The GlobalPlatform has noted that this is an important threat to general public and critical infrastructures.

Thus, the GlobalPlatform has reasoned that open standards are essential for ensuring interoperability between connected devices in such a way that along with the development of IoT devices the respective security level they offer is sufficiently high. One of the important tasks of the GlobalPlatform is thus to enhance the respective specifications and engage industry participants to guarantee that the needs of the IoT market are met [1].

As a concrete step, the GlobalPlatform has an IoT Task Force which is open to its members and which facilitates the discussion of forthcoming business requirements for network-capable objects. It also aims to identify ways for progressing GlobalPlatform technology. In addition, it collects feedback from the industry for contributing efficiently to the IoT market.

3.3.3   Other Industry Forums

There are many IoT-related organizations and industry forums that are actively seeking solutions for the overall functionality, performance and security of always-connected devices. One of these setups is the IoT Forum, which is considering the future of IoT networks [25]. The IoT Forum recognizes that the GSMA is working to establish common capabilities among mobile operators to enable a network that supports value creation for all stakeholders, including security, billing and charging and device management. The IoT Forum recognizes that all these items can enhance IoT by enabling the development of new services, and that through the provision of the respective value added services, mobile network operators can – in addition to providing cellular connectivity – act as a trusted partner for the end-users. The IoT Forum notes that the operator capabilities need to be tailored for the emergent M2M business model, building a trusted infrastructure that all stakeholders can rely on and profit from.

Some of the concrete means for making this happen is the remote M2M provisioning, and the related GSMA Embedded SIM Specification, which is meant to accelerate growth and operational efficiency for the M2M environment. The GSMA's embedded SIM delivers a technical specification to enable the remote provisioning and management of embedded SIMs to allow the Over-the-Air (OTA) provisioning of an initial operator subscription and the subsequent change of subscription from one operator to another.

The IoT Forum also considers IoT business enablers for M2M services. The Forum reckons that the growing IoT provides socio-economic benefits, but there is a big need to develop devices, applications and services in such a way that consumers can trust their data is secure. The GSMA supports a constructive policy and regulatory framework that can unlock the benefits of M2M services for consumers and businesses, building confidence and network capability. The GSMA thus advocates a sustainable M2M environment that will enable operators to unlock the consumer and business benefits of new and profitable services.

3.4   Technical Description of IoT

3.4.1   General

IoT is involved with secure execution of code as well as secure transport of confidential data between devices and systems, including keys. There are several ways to handle these procedures, based on current and future solutions. The IoT device may need to store the data confidentially which can be done, e.g., with HW-based SE or TEE. As the GlobalPlatform defines, SE can be a removable or non-removable semiconductor device with form factors such as a smartcard, SIM/UICC and an embedded, permanently installed secure device. The role of the SE is to host sensitive data and applications securely for service providers, application providers and other relevant stakeholders.

The idea of the SE is to provide a data storage and means for sharing among only the allowed parties and it functions as a logical base for remote and post-issuance secure management such as subscription data. The management of the SE can happen directly by the respective parties or via third-party solutions such as a TSM in order to provide remote updates within the SE. Figure 3.9 depicts the principle of the TSM.

The GlobalPlatform defines an SD which is basically a dedicated area within the SE reserved for a certain stakeholder in such a way that others have no visibility to the contents outside of their own SD. Figure 3.10 depicts the principle of the SD.

The same physical card can contain several SDs, and they can be used, e.g., for issuers, CAs and secure application providers. The SD is in practice an application on the SE that can be used to store credentials and as a base for managing securely the SE contents via secure

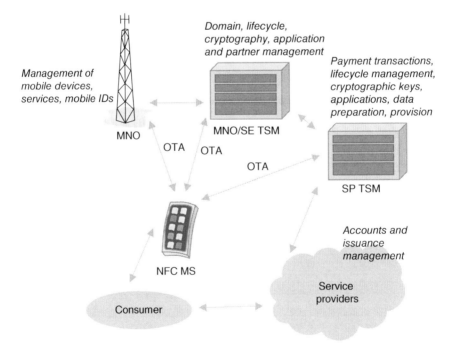

Figure 3.9   The principle of the TSM

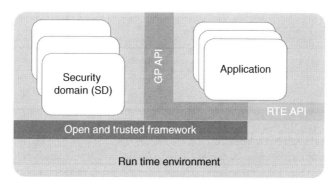

Figure 3.10   The principle of the SD

channels and Application Protocol Data Units (APDUs). As there may be multiple stakeholders, each reserving their own SD (also called 'tenants'), the SDs can be organized in a form of a hierarchical structure by varying the privileges independently from each other. The actual management may happen via all the supported bearers like SMS although the most cost-efficient communications are done over IP connectivity, e.g., via HTTP. The applications within each SD have own Application ID (AID), and they can be individually installed, provisioned and deleted via the TSM.

The TSM is one way of managing the SDs. It is a trusted broker establishing business agreements and technical relationships for the stakeholders such as MNOs, OEMs and such as payment institutions for mobile wallet utilization in the consumer domain, or utility meter reading in the M2M environment. The TSM can be used to manage the card for uploading, installing, updating and deleting data and other remote procedures in both consumer and IoT environments.

One of the important components in this environment is the CA. Its role is to facilitate SE post-issuance procedure for new stakeholders by maintaining the confidentiality. More concretely, the CA can create keys for the new SDs on behalf of other parties in such a way that the SE owner does not have visibility to the contents of the service.

In addition to the SE, the confidentiality of the data within the device, such as a smart device or IoT device, can be handled by the TEE which refers to the secured part of the main processor itself. It can thus be utilized without a separate SE based on, e.g., SIM/UICC although nothing prevents the joint utilization of it with the TEE. The TEE provides area for trusted app execution by ensuring protection, confidentiality, integrity and access rights. One example of the benefits of TEE is the protection of the display and PIN code entries so that they cannot be recorded via malicious SW.

3.4.2   Secure Communication Channels and Interfaces

The secure channels are essential to transport data between entities such as SIM/UICC and respective servers. As an example, GSMA defines the Secure Channel Protocol (SCP) suite that contains, e.g., SCP03, SCP80 and SCP81. These are discussed in more detail in section 4.11.1 along with descriptions of current (e)UICC Subscription Management solutions.

3.4.3 Provisioning and Key Derivation

The base for the consumer and IoT device and application initialization is to establish crypto-graphic keys for the device and the service the device is using. The initial settings can be included in the manufacturing process, but the challenge arises from the dynamic nature of the modern subscription data; the data might change, new data could appear and old data might need to be removed during the lifetime of the device, so the remote subscription management definitely eases the updating of such devices especially if the respective SE is embedded into the device. Also, it might be challenging to even know about the initial services and operators involved with the device such as automotive IoT products that may end up in different countries along with the cars they have been embedded into. As an example, the GlobalPlatform enables trusted SE issuers to perform preliminary provisioning of the credentials into the IoT devices yet maintain the confidentiality between the stakeholders.

The key derivation is needed to establish secure communication channels between the consumer and IoT devices, and respective services. The procedure includes authentication, authorization and encrypted communications. The keys can be divided between a long-lasting 'master key' and short-term 'session key'. The master key needs to be securely stored, e.g., into a tamper-resistant SE to provide highest protection, and the varying short-term key can be derived from it for each session. The maximum security can be furthermore ensured by dedicating a separate master key independently for each stakeholder, e.g., in the case of multiple 'tenants renting' SDs.

3.4.4 Use Cases

This section describes some typical or expected examples of the IoT environment.

3.4.4.1 Telemetry

Telemetry is one of the most logical environments for IoT devices such as intelligent sensors which are able to collect data autonomously from the field, pre-process the results and forward the data to the IoT systems. As the telemetry is utilized increasingly by industry as well as in the everyday M2M environment (such as controlling thermostats and basing billing to the telemetry data received from electricity consumption), so the respective security is becoming increasingly important. The security relates to the privacy of the consumer and system data as well as to the protection of fraudulent intentions to alter billing data.

3.4.4.2 Connected Security Systems

According to Ref. [26], connected security systems typically rely on sensors to monitor events such as opening doors or windows or movement within the monitored location. Also, automatization systems may include elements at home that can be controlled remotely, e.g., via smartphones for locking and unlocking doors, switching on and off the lights and audio systems etc. These systems may rely largely on the cellular network connectivity between the home security system and controlling smartphone with respective control app. These type of security systems are able to inform the user about the event, such as a text message telling parents that children have returned home. More complete systems may include video surveillance cameras installed at the user's property.

3.4.4.3   Automotive

According to Ref. [28], the GlobalPlatform has identified the automotive industry as being one of the most important areas moving towards SW-based control, giving increasing authority to computer-based systems. There can be a multitude of supported systems such as maintenance, location and entertainment services with respective means for SW updates. At the same time, the connectivity of the automotive systems is taking form between each other and external entities so that there could be various connectivity solutions activated at the same moment. This trend may also open unknown security holes for potential malicious attacks. There are solutions tailored for the vehicular environment such as Vehicle-to-Vehicle (V2V) and Vehicle-to-Infrastructure (V2I), together denoted as V2X which employs public key cryptography to authenticate OTA messages. V2X is defined in the USA via IEEE 1609 standards set and ETSI ITS G5 defines it in Europe. The standardization bodies have selected Elliptic Curve Cryptography (ECC) as a basis for the solution, the benefit being the small size of signatures and keys. The signatures are based on the Elliptic Curve Digital Signature Algorithm (ECDSA) by applying 224-bit or 256-bit key length. Furthermore, the vehicles are equipped with private and public key pairs that are changed frequently for added protection level. The public keys are distributed by certificates that are based on Certificate Authority (CA) [36].

The GlobalPlatform presents a practical use case in Ref. [28] involving a car manufacturer, Secure Element Issuer (SEI), several Application Providers (APs) and IoT Device Manufacturer (DM). In this scenario, the DM integrates the SE and/or TEE into the device. The DM may make SE patches originated from the Secure Element Supplier (SES) and integrates the SE into the Onboard Unit (OBU) of the car. The OBU may have applications and provide connectivity for several services like location-based services and maintenance service.

Now, the owner of the vehicle (IoT Service Subscriber, SS) may be subscribed into several application services such as remote diagnostics and location-based service. It also might be that the user does not want these application providers AP-1, AP-2, …, AP-n tpo know about the data transferred between the user and any of the other APs, or the user might want only a limited set of the APs to have visibility to all the data.

Ref. [28] details that in this case, each AP may ask the TSM previously selected by the car manufacturer to install its application in the OBU. The SE issuer and car owner in turn authorize the action, with the SD of the CA, and optionally the CA, assisting in setting up the confidential keys.

3.4.4.4   E-health

Ref. [28] presents a use case for e-health which contains a sensor gateway with an integrated SE. The healthcare system provider procures medical sensors and gateways to be integrated into Remote Patient Monitoring (RPM) equipment. The medical sensors can communicate with the RPM gateway by establishing the connectivity and security parameters. The RPM equipment is sold or rented to healthcare providers (AP). Finally, the medical personnel of the healthcare provider gives the RPM equipment to a patient to be utilized at home. To set up the equipment, the AP contacts the TSM selected by the SE issuer for the installation of the new SD on the gateway on behalf of the AP. The SD of the CA, and optionally the CA, takes part in establishing the AP's secret keys in the SD. The AP can now provision the needed parameters like IP address and public key by using the secure channel. The connectivity from the patient's side can be based on a cellular network which makes the MNO a logical option to serve as the Network Provider (NP). As the AP takes care of the connectivity expenses, it is a logical option to be a Network Subscriber (NS). In this case, either party or even a third party can act as a Service Provider (SP) while the patient is an SS. Please note that because the patient does not own the RPM equipment, the patient by default is unable to change the SP. Nevertheless, the patient may own other types of e-health IoT equipment such as a blood pressure measurement device via other SPs.

3.4.4.5 Utilities

The utilities have an increasingly important contributing role in the IoT environment, being one of the first types of M2M machines in realistic field deployments [29]. The utilities are also developing along with modernized technologies, such as smart grids. The utilities may rely on various types of wireless technologies although the most logical connectivity for these equipment is based on cellular radio technologies. As the amount of transmitted and control data is typically very low and the requirements for the real-time transfer are not too strict, even the most basic cellular technologies such as GSM SMS or packet data are useful for this environment. In fact, as the 2G networks are well established with relatively large radio coverage areas, and there is such a large amount of utilities such as power meter readers connected in the 2G network, it may be one of the most relevant reasons for not ramping down the 2G networks straightaway, even if the 3G and 4G networks are more spectral efficient. The role in the electric domain, including generation, transmission and distribution is also increasingly relying on the IoT solutions for better understanding the local near real-time power consumption.

The Smart Grid (SG) is one of the big raising items in the energy domain. The SG merges the EPS with energy and information technology, end-user applications and loads as defined in IEEE standard 2030-2011. Figure 3.11 depicts the respective graphical representation of IEEE while the NIST presents the SG via a higher level conceptual model. There are also other intentions to describe the SG as for the smart infrastructure, smart management and smart protection system [30].

As can be seen in Figure 3.11, the area types for the IEEE SG include the Home Area Network (HAN), Business/Building Area Network (BAN) and Industrial Area Network (IAN). The communications methods of SG include cellular networks (GPRS, 3G, LTE and 4G), Wi-Fi (IEEE 802.11), fixed Ethernet, WiMAX (IEEE 802.16), fibre optics, xDSL, PLC, WSN/WPAN (IEEE 802.15.4), ZigBee and DASH7. In Figure 3.11, the terminology is the following: NAM refers to the Neighborhood Area Network, AMI to Advanced Metering Infrastructure, EAN to Extended Area Network, FAN to Field Area Network, HAN to Home Area Network.

As the importance of the SG is increasing, and the power systems are highly strategic, the related security vulnerabilities need to be minimized. It is not hard to imagine the magnitude of damage a security breach may cause if the connectivity exposes security holes for, say, switching off the power system of a complete city even for a short time period.

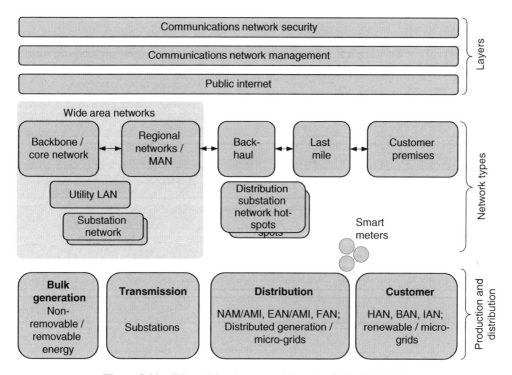

Figure 3.11   SG model as interpreted from the IEEE 2030-2011

The GlobalPlatform presents a use case for the utilities in Ref. [28]. The GlobalPlatform scenario is about a smart meter manufacturer (Device Manufacturer, DM) which produces smart meters with an integrated SE. The SE is provided by an SE supplier. An energy provider (which in this case acts as an SEI and SP) obtains and installs the smart meter in a house while an energy distribution network provider or grid operator company (acting as an AP) requests its TSM to install on the device an application providing measures and management of the line quality. In order to protect this information the cryptographic keys are already installed during the deployment. Another AP (local energy provider) typically bills the consumer for energy usage and will load its own application on the meter to monitor energy usage.

In addition, the house owner or tenant (in a role of SS) may be able to select an energy supplier which differs from the grid company so the user may have the opportunity to optimize energy costs depending on living styles that best map with the selected SS. Now, these various suppliers involved may need different information for calculating the rate, and involved energy distribution network operators may want to monitor different parameters to optimize their grid management.

The use case of Ref. [28] is related to the event of the homeowner changing energy supplier. The new service provider needs to contact the TSM associated with the utility grid company to be able to replace the old supplier's rate information in the smart meter, as well as the secret keys used for protecting the transmitted records. Understandably, the keys should be maintained confidential from the competition. The SD of the CA, and optionally the CA itself, is thus involved. In one scenario, the grid company can become the NP by using power line communications for backhaul while the new energy supplier may be the network subscriber paying the grid company for its connectivity costs. In alternative scenario, where the smart meter supports cellular connectivity, an MNO may act as NP which means that the grid operator would be a logical NS and charge the energy supplier for its connectivity costs. Finally, in the third scenario, the homeowner (being in the service subscriber role) could also have their own application activated in the device for measuring energy consumption which means that the homeowner would be another AP.

References

[1] GlobalPlatform, IoT. http://www.globalplatform.org/medguideiot.asp (accessed 24 January 2015).

[2] GlobalPlatform,Leveraging GlobalPlatform to improve security and privacy in the Internet of Things. Whitepaper, May 2014.

[3] B. Meynert. The Internet of Things. 24 December 2012. http://www.sagevita.com/business/the-internet-of-things/ (accessed 26 July 2015).

[4] T. Tuttle, Silicon Labs. The Internet of Things: The Next Wave of Our Connected World. Embedded Systems Conference '15. http://www.embedded.com/design/connectivity/4430102/The-Internet-of-Things–the-next-wave-of-our-connected-world (accessed 26 July 2015).

[5] A brief history of Internet of Things. http://postscapes.com/internet-of-things-history (accessed 26 July 2015).

[6] Technopedia. Definition of Internet of Things. 2015. http://www.techopedia.com/definition/28247/internet-of-things-iot (accessed 27 July 2015).

[7] A. Greenberg. Hackers remotely kill a jeep on the highway – with me in it. 21 July 2015. http://www.wired.com/2015/07/hackers-remotely-kill-jeep-highway/ (accessed 27 July 2015).

[8] C. Thompson. 14-year-old hacked a car with $15 worth of parts. 19 February 2015 http://www.cnbc.com/2015/02/19/14-year-old-hacked-a-car-with-15-worth-of-parts.html (accessed 27 July 2015).

[9] *Newsweek*. Meet Kevin Ashton, Father of the Internet of Things, 23 February 2015. http://www.newsweek.com/2015/03/06/meet-kevin-ashton-father-internet-things-308763.html (accessed 7 November 2015).

[10] Cloud networks. http://www.networkworld.com/news/2008/111208-private-cloud-networks.html (accessed 9 September 2012).

[11] *Wired*. The Internet of Things is wildly insecure – and often unpatchable, 1 June 2014. http://www.wired.com/2014/01/theres-no-good-way-to-patch-the-internet-of-things-and-thats-a-huge-problem/ (accessed 15 November 2015).

[12] GSMA Connected Living. http://www.gsma.com/connectedliving/ (accessed 15 November 2015).

[13] T. Tuttle. Internet of Things: The next wave in computing. eMedia, 2014.

[14] G. Reiter. Wireless connectivity for the Internet of Things. Texas Instruments, 2014.

[15] D. Bjorklund, J. Rautio and J. Penttinen. NMTImage. DMR (Digital Mobile Radio) Conference, Stockholm, Sweden, June 1994.

[16] ITU statistics, end-2015 estimates for key ICT indicators, 16 November 2015. http://www.itu.int/en/ITU-D/Statistics/Pages/facts/default.aspx (accessed 16 November 2015).

[17] J. Penttinen. The Telecommunications Handbook. John Wiley & Sons, Inc., Hoboken, NJ, 2015.

[18] M. Marjapuro. 7 reasons why IoT device hacks keep happening, 2 November 2015. https://iot.f-secure.com/2015/11/02/7-reasons-why-iot-device-hacks-keep-happening/ (accessed 22 November 2015).

[19] The Internet of Things will drive wireless connected devices to 40.9 billion in 2020. 20 August 2014. https://www.abiresearch.com/press/the-internet-of-things-will-drive-wireless-connect/ (accessed 22 November 2015).

[20] B. Dickson. Why IoT security is so critical, 24 October 2015. http://techcrunch.com/2015/10/24/why-iot-security-is-so-critical/#.i8ddwze:sh4q (accessed 22 November 2015).

[21] The GS1. Fundamental concepts of AIDC and RFID. http://www.gs1us.org/DesktopModules/Bring2mind/DMX/Download.aspx?command=core_download&entryid=51&language=en-US&PortalId=0&TabId=785 (accessed 27 November 2015).

[22] ThingMagic. Security breaches of RFID. http://www.thingmagic.com/index.php/rfid-security-issues (accessed 27 November 2015).

[23] *InformationWeek*. RFID's security challenge, 11 November 2014. http://www.informationweek.com/rfids-security-challenge/d/d-id/1028389? (accessed 27 November 2015).

[24] Enterprise Risk Management. RFID; Great benefits also come with a security risk. http://www.emrisk.com/sites/default/files/newsletters/ERMNewsletter_July_2010.pdf (accessed 27 November 2015).

[25] IoT Forum, 27 November 2015. http://iotforum.org/ (accessed 27 November 2015).

[26] GSMA. The impact of the Internet of Things; The Connected Home. http://www.gsma.com/newsroom/wp-content/uploads/15625-Connected-Living-Report.pdf (accessed 27 November 2015).

[27] GSMA. Understanding the Internet of Things (IoT). July 2014.

[28] GlobalPlatform. Leveraging GlobalPlatform to improve security and privacy in the Internet-of-Things. White paper, May 2014.

[29] C. Lima. Enabling a smarter grid, September 2010, Silicon Valley. Smart Grid Series, Smart Grid Communications.

[30] D. Bakken (editor). Smart Grids: Clouds, Communications, Open Source, and Automation.

[31] The IoT Security Foundation. https://iotsecurityfoundation.org/ (accessed 27 November 2015).

[32] NSA. Bluetooth security. https://www.nsa.gov/ia/_files/factsheets/i732-016r-07.pdf (accessed 27 November 2015).

[33] Bluetooth security and protection. http://blog.bluetooth.com/bluetooth-security-101/ (accessed 27 November 2015).

[34] Bluetooth attacks. http://www.trifinite.org (accessed 27 November 2015).

[35] ITU. Regulation and the Internet of Things, 6 November 2015. https://itunews.itu.int/en/6024-Regulation-and-the-Internet-of-Things.note.aspx (accessed 4 January 2016).

[36] Auto-talks. V2X Security Portfolio, v. 1.3. White paper, 2016. http://www.uato-talks.com (accessed 24 May 2016).

4

Smartcards and Secure Elements

4.1 Overview

This chapter presents key technologies related to smartcards and SEs, especially as they relate to the wireless systems. First, an introduction discusses why smartcards are still a useful and perhaps even indispensable anchor for security in the era of IoT. Nevertheless, the new environment may require updating of the systems and ways of utilization of the SEs. In fact, both the IoT – with its vast amount of M2M connected devices – and the consumer markets are starting to require physically much smaller SEs than we have typically seen to date in the form of traditional SIM cards. As a result of consumers' small-sized wearable devices and huge amounts of low-cost M2M equipment all over the field, the principles for the subscription management may see a radical evolution from the traditional SIM OTA methods, which typically have been vendor-specific.

The industry requires wider interoperability in order to ensure fluent and global mobility from which one example is the automotive environment; once the car has been built, it may end up in any country. So, if the respective car-embedded communications system for cellular networks contains an initial subscription, it should be possible to change it to any operator's system – not only for the initial activation but also for several changes during the lifetime of the communications equipment. These modifications need up-to-date international standards as well as new types of SIM cards that can be tiny, embedded elements to support IoT and consumer devices.

This chapter outlines the available and future alternatives for traditional and enhanced variants of smartcards. It also discusses the fact that SIM has no own Input/Output (I/O) capability and its impacts on the future solutions. Furthermore, it presents a technical description of contact cards with respective standards and current solutions, as well as contactless cards, form factors with current options and future extensions, and electrical and mechanical characteristics. This chapter also outlines the idea for the smartcard file structure and presents typical use cases of the cards as well as their future development.

Wireless Communications Security: Solutions for the Internet of Things, First Edition. Jyrki T. J. Penttinen.
© 2017 John Wiley & Sons, Ltd. Published 2017 by John Wiley & Sons, Ltd.

4.2   Role of Smartcards and SEs

Smartcards represent a type of SE hardware. Examples of solutions for providing SEs are the SIM, UICC, eSE and external micro Secure Digital (SD) cards. Smartcards can be categorized by their physical characteristics such as contact and contactless variants, and by their functionality such as memory cards, Central Processing Unit (CPU) or Multi Processing Unit (MPU) cards with embedded OS, multi-mode communications cards and hybrid cards. The cards may be used, e.g., as a transport access system's wireless component and contact chip, dual-interface cards and as multi-component cards like an integrated fingerprint reader. The most relevant cards for the wireless systems are the contact cards in the form of SIM/UICC, which may also support the contactless communications via connectivity from the chip PINs to the device's NFC chip and antenna. Furthermore, the M2M area has been the driver for the embedded SIM, which refers to the soldered SE with the same functionality as the removable SIM/UICC.

The smartcard concept is relatively old as such, having been invented in the 1970s, and commercial products introduced into the markets in the 1980s. In the early stage of the commercial phase, smartcards were utilized largely in prepaid telephony by applying a counter for the call credit utilization. Other examples of the early stage smartcards are vending machine and public transport payment cards, student ID cards (sometimes combined with the vending machine payment concept) and library cards.

The simple form of the smartcard is a memory card with embedded Electrically Erasable Read Only Memory (EEPROM) which is connected to address and security logics of the card, the latter having the input/output (I/O) signalling with the external world based on the clock signalling. The EEPROM of these cards typically has a very small storage size, low amount of write and erase events and about 10 ms delay in writing an individual cell or group of cells.

The microprocessor card is an advanced variant of smartcards as it includes a Central Processing Unit (CPU) which connects the card's internal EEPROM, ROM and Random Access Memory (RAM). The CPU also connects the external world via I/O and clock ports. In this type of card, the EEPROM stores data while the ROM contains the card's OS and the RAM module is meant for the working storage.

The early stage microprocessor cards had a small memory size for the EEPROM, RAM and ROM, and a slow and low-bit CPU, the capacity and performance increasing along with the development of the IT technologies. As an example, the SIM card of mobile communications was based on an 8-bit CPU and only a few kB of EEPROM in the early stage of GSM deployments, which was still sufficient to store small phonebook information. Nevertheless, as the SMS was deployed some years after the first-phase GSM networks, the card's internal memory no longer sufficed for storing the received messages until upgraded cards with a larger memory size were introduced to the market.

As the mobile communications system functionality is advancing with vast leaps, the current LTE MNOs typically require SIM size from 64 kB up to several hundreds of MB. There are SIM cards in the current markets exceeding 1 MB. Furthermore, the card's chip is nowadays typically based on a flash memory type that optimizes the memory allocation for OS and other required functionality, instead of dedicating the OS in a separate ROM block.

Along with the further development of smartcards in the operational environment, to support more memory size, advanced functionality such as flash memory and modern technologies to speed up the processor speed, they continue to be a useful and even indispensable anchor for providing the security platform in current and future mobile communications such as in the era of IoT. They also provide useful additional functionality, e.g., storing more information and secure domains for multiple stakeholders per single smartcard.

It can be expected that the increasing number of IoT devices will result in considerably more active intentions for exploiting the vulnerabilities of the mobile services and networks. Thus, it can be assumed that further modifications will be needed for smartcards to support IoT such as remote management of the SE, which in mobile devices refers to either removable or embedded SIM/UICC. The basic concept of the smartcard as such is a functional and secure platform for supporting current and new telecommunications and other mobile services, and for the further development of solutions like dynamic subscription management when switching operators, almost in real time, to download, activate, switch and delete the subscriptions. The technical base, including advanced chip technologies and smaller form factors for the embedded (soldered) HW elements, ensure the usability of the concept far into the future to support current and new types of tiny devices such as consumer wearables and ultra-small personal area network devices such as digested intestine monitoring equipment embedded within a medical capsule. The HW-based security of the tamper-resistant SIM/UICC can be assumed to guarantee better security and protection level compared to pure SW-based solutions such as HCE although both can be further enhanced. Furthermore, there are no obstacles to combine the SW and HW-based security solutions to jointly guarantee the highest security level combined with the flexibility of the cloud-based solutions such as token-based mobile payment.

The fact that SIM has no own I/O capability is both a limitation and a benefit as the direct manipulation of the contents is not possible, and the communication happens in a very controlled way via APDU messages and their acknowledgements. As an example, copying the stored keys or any other contents for their further observation is not directly possible in the way it is via the file manager of a typical computer OS.

Smartcards have been standardized internationally. The aim has been to ensure the interoperability at a general level although the final functionality depends on the adopted system. Some examples of highly interoperable environments are the NFC-based contactless bank cards as well as the contact SIM/UICC cards of 3GPP systems.

Smartcards can be divided into sub-categories to distinguish between contact cards, contactless cards and multi-component cards. Furthermore, the cards can be memory cards, CPU/MPU cards under both contact and contactless card types (or combination of these in the dual-interface card format) while the multi-component cards can contain many current and innovative new card types such as vault cards, dynamic token display cards and biometrics authentication cards based on, e.g., fingerprints or eye iris data. For the wireless communications, the contact card in the form of SIM/UICC is the most typical although along with the growing IoT/M2M devices as well as consumer devices such as wearables, the form factor of embedded UICC options is shrinking.

The major categories for the SEs are the UICC, eUICC and microSD. The *UICC* is the base for the 'traditional' MNO business model. Its base is still relevant even if its precedent variant, the first SIM cards, were deployed in GSM networks in 1991. The evolution path of the SIM/UICC include, e.g., Host Controller Interface (HCI), Single Wire Protocol (SWP) and NFC.

The *eUICC* is a feasible base for M2M environments, and increasingly also for the consumer markets along with the increasing importance of the wearables [6]. Nevertheless, the only internationally standardized eUICC is the MFF2 which might be somewhat voluminous for the smallest IoT equipment. The rest of the currently available variants are based on the chipset vendors' proprietary specifications (especially the size and PIN layout, as well as other electromechanical characteristics) unless the standardization community activates for defining the smaller eUICC. The benefit of the eUICC is that the SE is directly soldered into the device, but if the initial subscription is changed in the interoperable environment, it requires further standardization for the subscription management. The current options are discussed further in this book.

Finally, the *microSD* is an SE platform that is provided by a thirdparty, and it is independent from the MNOs. If such a removable element is supported by the device (as is the case typically in current smart device markets), it may be used for secure services such as mobile wallet and transport access.

The above-mentioned variants can be utilized in the complete ecosystem in various technical environments [19]. As some examples, the NFC-based payment solutions may rely on respective service providers, app developers, credit card issuers, MNOs, NFC-compliant merchants and TSM concept. That role division model typically includes certification schemes by the service provider, which need to be complied by the card vendors, OEMs and MNOs. The model has two sub-categories, basing on either a removable NFC-enabled UICC (with, e.g. an already obsolete SoftCard payment app) or an embedded NFC chip (with, e.g., a Google payment app). The certification model may be time-consuming. For that reason, the model has been challenging in some cases, as the SoftCard initiative showed, resulting in ramping down the consortium and respective mobile payment service while other mobile payment schemes appear on the market.

In addition to the payment environment, the UICCs and other forms of SEs can be utilized in the access and transit environments, basically replacing the physical keys with SW-based access. The access use case may contain roles for the access company via TSM and venue (like hotel) corporations. This model makes it unnecessary to rely on physical door keys or access cards as the same functionality can be provided via mobile devices with UICC or eUICC and a respective access app. The app can further provide physical access rights for many other locations such as home and office, depending on the included (and to be added further) rights.

Variations of this model can be used in the transit environment to replace the pass by including CA, TSM and payment provider. The involved app may contain passes in several transport systems such as subway, train, bus and ferry in a local and wider geographical areas as long as the apps support the systems.

4.3 Contact Cards

The physical and logical requirements of contact cards like SIM/UICC cards have been standardized and recommended in various standardization bodies and industry forums. The globally agreed principles ensure the interoperability of the cards and their applications, reader and mobile devices, networks, card issuers systems and services. The ISO/IEC 7816 standard set is the base for contact cards used in mobile communications [4,5]. The standard defines the physical card size, Form Factors (FFs), pin contact layout, electrical characteristics, I/O protocols that can be byte or block based and file structure.

Recently, there are more use cases valid in the mobile communications markets along with the M2M connectivity and dynamic management of subscriptions both in M2M and consumer environments. Thus, the interoperability needs to be extended to cover the events in subscription management like remote downloading of the subscription, and the activation, deactivation, changing and deletion of it.

The standardization organizations and industry forums relevant to wireless communications security are presented in Chapter 1 while the most important principles of the smartcard standards and recommendations are discussed in the following sections. They summarize the ISO/IEC 7816 sub-standards for the integrated cards.

4.3.1   ISO/IEC 7816-1

The latest version of this standard is ISO/IEC 7816-1:2011, which specifies the physical characteristics of ICCs with contacts of the ID-1 card type. It may include embossing, magnetic stripe and tactile identifier mark as specified in ISO/IEC 7811. The respective test methods are specified in ISO/IEC 10373-1. The standard applies to cards with a physical interface and electrical contacts.

The standard contains the requirements for the physical characteristics of ICCs, including exposure limits for X-rays, UV light, electromagnetic fields, static electrical fields and ambient temperature of the card. It also presents physical requirements for bending in order to ensure the functionality for the planned lifetime in a practical environment as the embedded chips need to maintain firmly their position in the plastic frame, and guarantee a certain stress level on the exposed connectors and the internal pins of the embedded silicon die.

The material of the card body is typically made of Acrylonitrile Butadiene Styrene (ABS) or Polyethylene Terepthalate (PET) which are both more than adequate for environmental compatibility compared to Polyvinyl Chloride (PVC). Nevertheless, the body can also be made of other materials such as compostable materials like cornstarch, hemp fibres, bamboo and cellulose. The treatment of these materials is not straightforward as they need to be treated in a special manner to meet international standards and norms, which is the reason for the popularity of the easier handled and environmentally compatible ABS, PET, Polyethylene Terepthalate Glygol (PETG) and Polycarbonate (PC), especially as the printing material (ink) is also ecologically acceptable.

The critical part of the SIM card is the plug-in module which contains the electrical chip module and metallic contact points. It needs to be physically strong enough when placed into the card reader tray of the device. Nevertheless, the rest of the card body may be of another type of material which complies even more with the environmental aspects. Also, the size of the tradition- ally utilized full card body (ID-1) can be reduced, from which one example is the half-SIM body (ID-1/2). Yet another solution is to provide a plug-in-only SIM card, which is not attached to the actual card body at all, the typically printed related information being in a separate leaflet.

In the commercial market, the material of the SIM card delivery can further be optimized by providing the accompanying packet with less material or more environmental compatible substances. Also, the environmental aspects can be enhanced by optimizing the SIM lifecycle so instead of changing the physical card during the lifetime, it is possible to update subscription- related information OTA methodologies which are described later in this book. Logically the recycling can also be optimized upon the end-of-lifecycle by recycling the utilized material, the respective providers ensuring the subscriber data is securely and confidentially deleted when the card is returned.

Table 4.1   The ISO/IEC 7816-2 ICC contacts

Contact	Use	Description
C1	Vcc	Operating voltage for the microprocessor. Originally, the supply voltage was 5 V, and the support for 3 and 1.8 V was added later. Nowadays, the support of 5 V is merely optional
C2	RST	Reset signal for microprocessor to initiate the reset sequence
C3	CLK	Clock for the card's microprocessor originated from the device. The clock rate dictates the operation speed and functions as a base for the communications to and from the microprocessor and external world
C4	RFU	Originally marked as 'reserved for future use'. Can be used for connectivity-oriented USB 2.0 interface (1/2) via 8-contact modules
C5	GND	Common ground
C6	Vpp	Programming voltage for the 1G ICC's EEPROM. In the later phase, used for NFC contact via standardized SWP/HCI
C7	I/O	Input/output, half-duplex serial data channel for reader and smartcard. This provides physical channel for exchanging APDU messages
C8	RFU	Originally marked as 'reserved for future use'. Can be used for connectivity-oriented USB 2.0 interface (2/2) via 8-contact modules

4.3.2   ISO/IEC 7816-2

The ISO/IEC 7816-2 standard dictates the number, function, dimensions and position of the contact areas of the electrical parts. The full-size ICC consists of eight electrical contact points which are named as C1...C8, not all of which are necessarily used by the respective embedded microprocessor chip. Table 4.1 summarizes the contact definitions of ISO 7816-2.

For the internal use of the card itself, the optional functionalities are Vpp, Vcc and CLK. The RST is utilized either as itself provided by the external device, or optionally as a combination with the card. Nevertheless, if internal reset is used, the Vcc is mandatory, which refers to the power supply input. Contacts C4 and C8 are defined separately in respective standards. The SIM/UICC card originally consisted of a full set of eight PIN connections. PIN 6, which supports SWP, forms an interface between the Contactless Frontend (CLF) and SIM/UICC. In practice, it is a contact-based, bit-oriented, full duplex protocol for contactless communication (like NFC) in such a way that the CLF is the master and the SIM/UICC is the slave. The CLF supplies the UICC with operating power, clock reference, data (via binary-state voltage and current levels of the single wire) and bus management signalling.

Along with the size reduction of the SIM/UICC cards, a variant for a 6-PIN layout was also introduced. Figure 4.1 shows the original and later pin layout for the 8-PIN variant, as well as the latest 6-PIN layout while Figure 4.2 summarizes the functions of the PINs.

4.3.3   ISO/IEC 7816-3

ISO 7816-3 is relevant for establishing communication with a smartcard and for writing I/O software from, e.g., microcontroller, serial and parallel ports as well as via the USB port. The basic idea of these communications is to send a signal to the card which returns an Answer

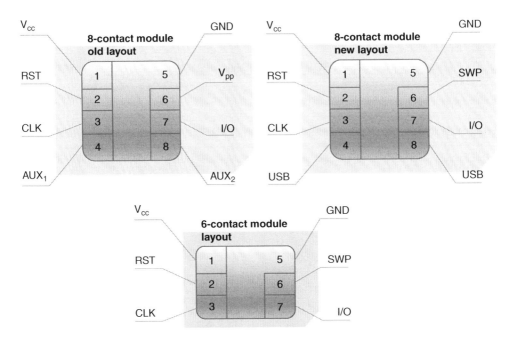

Figure 4.1 The physical connections of the UICC

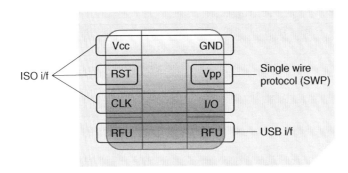

Figure 4.2 Physical interfaces of the 8-PIN UICC based on ISO, SWP and USB

to Reset (ATR). ISO 7816-3 contains definitions for the respective electrical signals, voltage and current values, operating procedure for ICCs, ATR in both synchronous and asynchronous modes, and Protocol Type Selection (PTS). The protocol type T=0 represents the original asynchronous half duplex *character-based* transmission protocol, while the T=1 protocol is a half-duplex asynchronous *block-based* transmission protocol. As Ref. [23] states, the terminal must support both T=0 and T=1 protocols, while the UICC may support either T=0 or T=1, or both.

4.3.4   ISO/IEC 7816-4

ISO/IEC 7816-4 defines the message contents, commands and responses between the connected interface device and the card itself. It also contains definitions for the structure and content of the bytes produced by the card for ATR, the file and data structure that is observed in the interface upon processing inter-industry commands for interchange, file and data access methods, secure messaging, and access to the algorithms processed by the card. It should be noted that ISO/IEC 7816-4 does not describe the respective algorithms.

4.3.5   ISO/IEC 7816-5

This sub-standard describes the registration of application providers. It defines the use of an application identifier to ascertain the presence of and perform the retrieval of an application in a card. The standard thus explains the procedure for granting unique application identifiers via the international registration. It also defines the respective authorities, the register related to the identifiers and the application providers.

4.3.6   ISO/IEC 7816-6

This sub-standard describes inter-industry data elements for interchange, and specifies the Data Elements (DEs) for inter-industry interchange of ICCs. The standard is valid for both contact cards and contactless cards. The standard describes the identifier, name, description, format, coding and layout of DEs, and defines the ways for retrieval of DEs from the card.

4.3.7   ISO/IEC 7816-7

This sub-standard describes inter-industry commands for the Structured Card Query Language (SCQL). As noted in Ref. [36], the commands for access are based on SQL functionality as defined in ISO 9075, and coded according to the principles of inter-industry commands as defined in ISO/IEC 7816-4. The database is a structured set of database objects referred to as Database File (DBF). Under a selected Dedicated File (DF), there cannot be more than one DBF which is accessible after selection of the respective DF. A database may also be directly attached to the Master File (MF). One typical example of the multi-application card is that under the MF, there may be several DFs. Under the selected DF, there can be a DBF at the same structural level with other internal elementary files and/or working elementary files. The respective DF and its contents underneath is referred to as an application with a database.

4.3.8   ISO/IEC 7816-8

This sub-standard defines inter-industry commands for security operations for contact and contactless ICCs which are used for cryptographic operations. The commands of this sub-standard are based on the ones defined by ISO/IEC 7816-4, with added new commands. The operations may be related to digital signatures, certificates and the management of asymmetric keys.

4.3.9   ISO/IEC 7816-9

This sub-standard defines industry commands for card and file management during its complete lifecycle. Typical use cases of these procedures include file creation and deletion, secure downloading of data into the card (including applets, keys and other code) and secure messaging.

4.3.10   ISO/IEC 7816-10

This sub-standard defines electronic signals and ATR signals for synchronous cards and terminals. The standard includes definitions for the power and signal structures.

4.3.11   ISO/IEC 7816-11

This sub-standard defines personal verification through biometric methods by describing the inter-industry commands and data objects that can be used for user verification. The actual commands are found in the ISO/IEC 7816-4 standard, and parts of the commands are found in ISO/IEC 19785-1. ISO/IEC 7816-11 also discusses the enrolment and verification, and emphasizes security issues.

4.3.12   ISO/IEC 7816-12

This sub-standard defines the USB electrical interface and operating procedures for cards with contacts (USB-ICC). The standard includes definitions for USB-ICC as an interface device, standard and class descriptors, bulk transfers and control transfers between USB-ICC and terminal, control transfers for version A and B protocols, interrupt transfers to indicate asynchronous events, and status and error conditions. The two protocols for control transfers are the protocol $T = 0$ (version A) or the transfer on APDU level (version B).

4.3.13   ISO/IEC 7816-13

This sub-standard defines commands for application management in a multi-application environment.

4.3.14   ISO/IEC 7816-15

This sub-standard is related to cryptographic information application, specifying a card application and a common syntax in the Abstract Syntax Notation One (ASN.1) format for the cryptographic information and sharing mechanisms. The standard also includes the card's storage of multiple instances of cryptographic information, use and retrieval of the cryptographic information, authentication mechanisms and cryptographic algorithms.

4.4   The SIM/UICC

4.4.1   Terminology

As for the terminology of this book, the SIM refers primarily to the HW of the original SIM of GSM. The original ETSI-based SIM is designed to work with 2G GSM mobile communications system devices, including further enhancements like GPRS, Enhanced Data Rates for Global Evolution (EDGE) and other GSM-specific solutions.

Nevertheless, the SIM has been upgraded for 3GPP systems at the 3G stage and beyond, including UMTS/HSPA, LTE and LTE-A. Along with the development of the 3GPP mobile communications systems beyond 2G, the original SIM HW has also been enhanced and is called the UICC in the 3GPP 3G systems. The UICC can support several radio access technologies as well as other wireless solutions, each being handled by a separate application on the UICC. The 2G and 3G telecommunications systems are managed via an application called USIM. In order to distinguish between 2G and 3G systems residing on the UICC, the application specifically supporting the GSM is called 'USIM with 2G context', or simply 'GSM', while the application supporting 3G (UMTS/HSPA) is called 'USIM with 3G context', or simply 'USIM'. Furthermore, the application supporting LTE as of 3GPP Release 8 refers to the LTE-specific files specified in the 3GPP TS 31.102. In all these 2G and 3G phases, the module is called USIM, but the LTE-specific functionality for the IMS voice calls is managed via a separate ISIM (IMS SIM) application.

In general, the original 2G SIM is compatible with later 3GPP generations, but is only able to provide services limited to 2G. Furthermore, in order to provide backwards compatibility, the 3G USIM can be utilized in 2G devices although the 3G-specific functionalities are not available in such setups if the device does not support them.

Not only 3GPP systems but also American CDMA networks (1xRT, EVDO, HRPD and eHRPD) can be supported by the UICC. The respective application is called Removable User Identity Module (RUIM) or CDMA SIM (CSIM) [38]. In this book, the terms SIM and USIM are used separately when emphasizing the 2G and 3G differences, respectively, and otherwise the general term UICC, or combined form SIM/UICC, are used for referring to the mobile communications subscription of all the 3GPP systems.

The physical aspects of the UICC are based on the ISO 7816 standard for contact cards. Furthermore, the ETSI TR 102 216 defines the UICC as a smartcard that conforms to the specifications written and maintained by the ETSI Smartcard Platform project. There is also the contactless standard ISO 14443 which is used as a basis for NFC.

4.4.2   Principle

The UICC is a security element which acts as a part of the complete security chain of the 3GPP mobile communications networks and in other supporting networks. It is a tamper-resistant HW-containing file and folder system for various security environments, complying with the Common Criteria Evaluation Assurance Level (CC EAL) 4. It can be considered sufficiently well protected for storing secure applets, which are applications that require high levels of security including the protection of application assets in confidentiality, integrity or availability at different security levels depending on the AP's security policy. Such a high level of security assurance is typically required for payment applications complying with CC EAL

4 with AVA_VAN.5 or with higher EAL class, Conditional Access (CA) mobile TV applications or digital signature applications. As an example, a Protection Profile, Secure Signature-Creation Device (PP SSCD) is required in Europe for qualified digital signature applications [7]. The PP defines the security requirements of an SSCD for the generation of Signature-Creation Data (SCD) and the creation of qualified electronic signatures. In this specific case, the assurance level for the PP is EAL 4-augmented [37]. More details about SSCD, including its lifecycle in development and operational phases, can be found in Ref. [37] Secure Applications (SAs) follow a CC evaluation and certification along with the certified (U)SIM [8].

The SIM/UICC is associated with its unique serial number (ICCID), IMSI as well as authentication and ciphering information. There are other related information fields such as TMSI associated by the network, a list of services the subscriber is entitled to and PIN1/PIN2 codes (primary and secondary) as well as a PUK code.

The SIM/UICC can be used as a base for offering services to the end-users via Java applets. Along with the development of content providing, there are increasing needs for securing the communications. Thus MNOs may want to further develop new value-added services based on UICC platforms for areas like mobile payment, electronic signature, mobile TV and mobile identity, which may require increased security levels of the traditional SIM/UICC, depending on the service provider's needs [9]. An example of a solution for such environments is TEE which is discussed more detailed later in this book. Higher level authorities may also be interested in ensuring sufficient protection for secure mobile services, e.g., the French Network and Information Security Agency has created certification reports that indicate the features of the proposed security targets, including the compliance for the EAL 4 or higher [10].

4.4.3   Key Standards

The SIM card was standardized by ETSI in TS 11.11, which defines the physical and logical functions of the card. Along with the transition of GSM and UMTS standardization work from ETSI to 3GPP, the SIM-related standardization was moved partially to 3GPP. As a result, 3GPP takes care of the further development of applications like GSM SIM (TS 51.011) and 3G USIM (TS 31.102) while ETSI continues developing the physical UICC.

While the 3GPP works basically on the USIM-related topics, some of the key ETSI standards for the UICC are the following:

- ETSI TS 102 221, physical and logical characteristics of UICC–terminal interface.
- ETSI TS 102 412, smartcard platform requirements, stage 1.
- ETSI TS 102 613. UICC–CLF) interface, part 1, for physical and data link layer characteristics.
- ETSI TS 102 600, characteristics of the USB interface in UICC–terminal interface.
- ETSI TS 102 484, secure channel between a UICC and an end-point terminal.
- ETSI TS 102 223, Card Application Toolkit (CAT).
- ETSI TS 131 102, application-specific details for applications residing on an ICC (USIM for 3G).

An important security-related topic for the UICC security is the compliance with the CC. It refers to standards denoting EALs from 1 through 7 as summarized in Chapter 1.

Especially for the SIM Application Toolkit (SAT), the initial specification of ETSI was TS 11.14. Nowadays, the enhanced SAT is defined jointly by ETSI and 3GPP, the key standards being ETSI TS 102 223, ETSI TS 102 241, ETSI TS 102 588 and ETSI TS 131 111. The SAT applications were originally based on proprietary APIs, but along with the introduction of the Java Card, it is possible to provide better interoperability of the applications.

4.4.4   Form Factors

The SIM card was introduced in mobile communications systems along with the first GSM deployments as of 1991. It is based on the contact card definitions of ISO/IEC 7816. The first SIM card with Form Factor 1 (1FF), as presented in Figure 4.3, was designed based on the size of a standard credit card to provide a convenient way for storing and fluent transportability of subscription data between all GSM devices. The main idea of this complete separation of the user equipment and user account is still applied in GSM and advanced 3GPP systems.

The initiation of the GSM SIM card era happened in 1991 when the smartcard provider Giesecke & Devrient provided the world's first subscription modules to a Finnish MNO, Radiolinja [2]. Soon after the first form factor (which is the same size as the ID-1 card body), the reduction of the GSM hand-held device's size resulted in the second SIM Form Factor (2FF), i.e., a mini card which is roughly a stamp-sized smartcard.

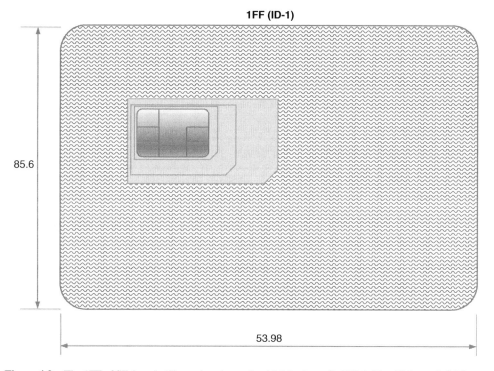

1FF (ID-1)

85.6

53.98

Figure 4.3   The 1FF of SIM cards (dimensions in mm), which is also called ID-1. The thickness is 0.76 mm. The ID-1 is used in practice only for delivering the plug-in units which are further snapped out from the card body when inserting them to mobile devices

The SIM card's plug-in part has further been resized as a result of the need for reduction of the physical size of the mobile devices. The third Form Factor (3FF), i.e., a micro card, was introduced in 2010 in the commercial markets. It was specified by ETSI in cooperation with the ETSI Project Smartcard Platform (EP SCP), 3GPP (for UTRAN/GERAN), 3GPP2 (for CDMA2000), ARIB, GSM Association (by GSMA SCaG and GSMNA), GlobalPlatform, Liberty Alliance and the OMA for optimizing the space for devices that benefit from reduced space. To date, the smallest consumer SIM with fourth Form Factor (4FF), or nano card, entered the commercial markets in 2012. Both 3FF and 4FF are backwards compatible with previous variants, so the contact area has been retained (for the active 6-PIN layout). In addition, the processing is executed with the same 5 MHz rate. The nano SIM is, though, slightly thinner than previous variants. Table 4.2 and Figure 4.4 summarize the key aspects of consumer SIM cards.

Typically, the plug-in unit, whether it is 2FF, 3FF, 4FF or combination of these, is attached to the ID-1 card frame. More specifically, the smaller card consists of the same PIN contact arrangement as for the full-size 1FF SIM card and is typically supplied within a full-size card carrier which supports the plug-in part via connecting pieces. This method is described in the ISO/IEC 7810 specification which defines the ID-1/000, and it provides a means for using the card either with a device (card reader) designed for a full-size card, or by physically detaching the plug-in part and inserting it to a device supporting that form factor. In addition to the form factor size definitions, the chip PIN contacts can have different shapes. The visual aspect is not only decorative but it dictates how the embedding of the card is done onto the frame. Ref. [12] describes the interface between the UICC and terminal (card reader), and Ref. [11] defines further the M2M UICC physical and logical characteristics. Figure 4.6 depicts the mechanical components of the smartcard.

The SIM plug-in modules have been typically delivered by operators attached individually in ID-1 frames. The challenge of this method is the increased diversity of user equipment and the variations in the supported SIM form factors. Typically, the devices only accept either plug-in SIM (2FF), micro-SIM (3FF), or nano-SIM (4FF) cards. This leads to the challenges of physical stock maintenance for offering this variable set of SIM. Typically, it is not clear for end-users which form factor each user equipment requires.

One solution to tackle the limited support of different SIM/UICC form factors is a physical adapter that is either 2FF or 3FF in size, and that can accommodate a smaller plug-in unit to guarantee compatibility for the SIM insertion. Another way is to cut the plug-in unit into

Table 4.2 Consumer-grade SIM FF

Form Factor	Name	Standard	Length (mm)	Width (mm)	Thickness (mm)
1FF	Full-size	ISO/IEC 7810:2003, ID-1	85.60	53.98	0.76
2FF	Mini/plugin	ISO/IEC 7810:2003, ID-1/000	25.0	15.0	0.76
3FF	Micro	ETSI TS 102 221, V.9.0.0	15.0	12.0	0.76
4FF	Nano	ETSI TS 102 221, V.11.0.0	12.3	8.8	0.67

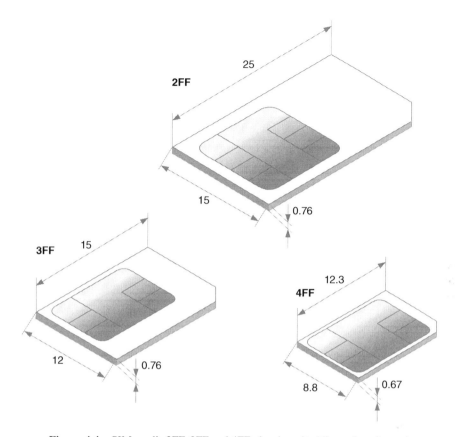

Figure 4.4 SIM card's 2FF, 3FF and 4FF plug-in units (dimensions in mm)

smaller 3FF or 4FF sizes as the actual electronics is within the surface of the visible metallic contacts. There are specially designed cutting tools available for this type of size adjustment, but it should be noted that such an operation may void the warranties of the card itself as well as the device the modified card is utilized with.

This situation has led to offering more sophisticated solutions like triple-SIM as presented in Figure 4.5 [1]. As can be interpreted from Figure 4.4, the nano-SIM is thinner than any other SIM variant. The triple-SIM solution takes this into account as the nano-SIM part is thinner at its rear side compared to the rest of the card surface. For MNOs, the benefits of the double- and triple-SIM solutions include logistical savings as a single card body for all the relevant form factors provides simpler inventory management. Also the Order Management (OM) process can be simplified at the point of sale, and there can be assumed to be less customer care calls as end-users do not need to solve the physical dimension incompatibility issues for inserting the SIM card into the device. This results in reduced customer care calls, and the overall end-user experience is enhanced. The end-users can also benefit from the fact that the new devices typically support smaller form factors; the changing of the old SIM card can thus be done by resizing the already obtained card to a smaller form factor.

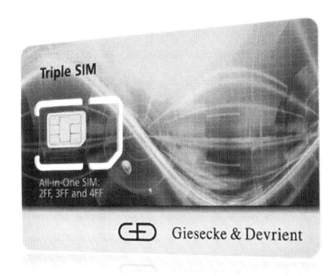

Figure 4.5   The plug-in units of 2FF or 3FF can be delivered within a single ID-1 card body. This eases the logistics and enhances user experience upon inserting the plug-in units into mobile devices. Photo reprinted by courtesy of Giesecke & Devrient

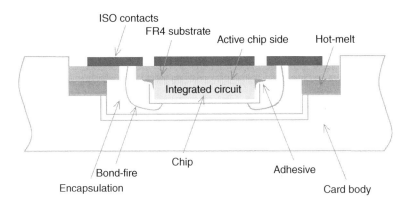

Figure 4.6   The physical building blocks of a smartcard. The ID-1 card body can be of plastics or recyclable materials, while the frame material of the plug-in needs to comply with typically stricter mechanical and environmental requirements making plastics the most feasible material

4.5   Contents of the SIM

4.5.1   UICC Building Blocks

4.5.1.1   Types of UICC Subscription Containers

The *SIM* was introduced into the markets along with the deployment of the GSM. The original SIM was designed to work in the ETSI/3GPP 2G environment. The physical SIM card is based on the ISO 7816 definitions which detail the UICC contact cards. Furthermore, the wireless variant based on NFC is defined in ISO 14443.

The *USIM* was introduced along with the 3G mobile system of ETSI/3GPP. USIM is backwards compatible with the ETSI/3GPP 2G system, although when the user is connected to the GSM network, the advanced properties of the USIM cannot be applied.

The *RUIM* is a smartcard variant designed for the CDMA networks, and standardized by 3GPP2. As is the case for 3GPP SIM and USIM of GSM and UMTS, the RUIM stores the user's subscription data including identity, phonebook addresses, feature-specific information, network settings and supplementary services for CDMA handsets. Typically, CDMA handsets include this data in-built to the device's HW (which means that CDMA handsets are personalized), but with the RUIM, the MNOs are able to implement smartcards and provide the advantage of subscription-related data portability as has been the case with GSM and UMTS. Furthermore, the RUIM provides the same SAT principle as has been offered via the SIM and USIM as a basis for value added services, as well as OTA management of data. When utilized within a multi-application card, other applications can also be stored in addition to the RUIM, like SIM and USIM, which provides roaming capabilities for the handsets supporting the respective parallel technologies. Please note that the CDMA-specific smartcard is also called CSIM (CDMA-SIM).

The *ISIM* is a recent addition to the previous set of applications. The ISIM is meant for the Voice over LTE. The ISIM thus refers to the application residing on UICC that provides access to the IMS.

The *TSIM* refers to the SIM in the TETRA system as defined in Ref. [39].

In general terms, the SIM, USIM, ISIM, TSIM and RUIM (CSIM) are referred to as *Subscription Containers* of the UICC. The UICCs are used for security procedures, data storage and applications. As part of the security, the UICC holds authentication and encryption keys for accessing networks and for securing OTA transactions. The data storage contains a set of information like service configuration parameters, phonebook, short messages, as well as service and emergency numbers. The card may also contain diverse services like Location Based Services (LBSs) and information services.

The concept of the card applications has developed vastly since the initial introduction of the SIM. The first phase of the development was based on proprietary applications which were fast to finalize but, at the same time, they lacked interoperability and they were challenging to maintain. At that time, the role of the SIM was to act in the lowest layer while the SAT managed the applications in the upper layer. Some examples of the early SAT applications were banking and location services. In a parallel fashion, there was also the Wireless Identity Module (WIM) residing on the UICC that took care of the WAP browsing services.

Later development of the open system application environment that is based on Java programming ensures better interoperability. This has had a positive impact on the time to market of applications. Furthermore, the application development no longer depends on the respective card manufacturer, thanks to the standardized API definitions as presented in the Java standards and 3GPP TS 03.19.

4.5.1.2 Profile

The card profile refers to the contents of the UICC, i.e., to the specification of the file system as well as to the general and individual data which is personalized onto the SIM, USIM and other subscription containers. In other words, the profile contains the UICC file system.

The profile can be visualized by a tree structure that presents the nodes as files and applications residing within the UICC. The root node is the profile itself. Under the root node, there is a set of information which indicates the card characteristics, the UICC file system including the master file, the UICC files and applications. The complete (U)SIM framework consists thus of the initialized module and profile, and the profile contains file system, applications and switches for the functionality.

The most relevant UICC profile-related specifications are the following. The 3GPP TS 31.102 lists Elementary Files (EFs) in the profile while the 3GPP TS 23.097 describes the multi-user profile. The 3GPP TS 22.097 describes the Multiple Subscriber Profile (MSP) service which allows the served subscriber to have several profiles, and to distinguish between telecommunications service requirements, e.g., for business and home. The subscriber data specific to the MSP is stored in the HLR and GSM Service Control Function (gsmSCF). The 3GPP TS 23.008 defines the data stored in the HLR. The HLR element contains all the common data, i.e., the data valid for all profiles, and some data specific to the default profile. The 3GPP TS 31.101 defines the four EFs that reside in the MF level. The EFs at the USIM Application Dedicated File (ADF) level contain service and network related information as defined in 3GPP TS 3.102.

For creating the (U)SIM profile, there are special SW personalization tools capable of producing system-related files (MF and EFs) and the actual personalization. The MF is created at the time that the telecommunications subsystem is installed. It should be noted that there also is a Configuration Elementary File (EFConf) which transparently stores the card configuration parameters. It is mandatory for the card functioning and must therefore not be deleted.

The minimum file system on the (U)SIM card must contain the following files: the MF which is created at the installation of the telecommunications sub-system, EF CONF, EF GSM_PIN_CFG, EF DIR, EF USIM_PIN_GBL, EF USIM_PIN_PUK, EF ARR, ADF USIM and EF USIM_PIN_LOC. In the data generation phase, these files must be created in the given order. The applications of the card (applets, or native applications) can be installed directly via the card commands by using the ISO interface or via OTA, e.g., by a RAM application.

4.5.1.3   SIM HW Blocks

The main HW blocks of the SIM card are CPU, RAM, ROM and EEPROM. The CPU may function in cooperation with the assisting Numerical Processing Unit (NPU). The RAM is meant for the operating memory while the ROM is meant for permanently stored contents like the OS. The EEPROM is a non-volatile memory block used for changing data such as a phonebook.

4.5.1.4   OS

Each SIM/UICC has an embedded card OS which manages the communications between the card and external world as well as takes care of the functions of the card such as cryptographic procedures. The card OS is SIM/UICC vendor- specific and it provides practical ways to differentiate between competing card providers. Nevertheless, the abstraction of the functions via Java Card provides a means to use interoperable Java applications at least to some extent.

4.5.2   The SIM Application Toolkit (SAT)

The SIM Application Toolkit (SAT) refers to a solution that enables the SIM to perform actions related to value-added services. The SAT typically initiates upon the powering up the SIM, and it provides a means to interact with the applications of the SIM. It is standardized by ETSI and 3GPP, and was originally designed for GSM in ETSI GSM 11.14 technical specification. Nowadays, there is also an SAT for the 3G USIM as defined in 3GPP 31.111 Release 4, named USIM Application Toolkit, or USAT.

The SAT/USAT contains commands within the SIM/UICC that can be executed apart from the respective mobile device or the mobile network. One of the use cases of the SAT/USAT is to display menus and collect user input by the SIM/UICC commands, thus facilitating interactivity between the SIM/UICC, network applications, access networks and the subscriber.

The key benefit of the SAT/USAT is the possibility to create a simple user interface that shows options of the commands to the end-user. Although by default not 'fancy-looking', this menu structure allows the utilization of respective services with various types of mobile devices including the basic and feature phones as well as the most complete smartphones and connected devices. The SAT/USAT is thus most useful in a low-level application environment.

Along with the initial Java Card concept, the platforms for SAT services started to emerge at the beginning of 2000. Ever since, the Java Card and SAT have been developed further as new 3GPP releases have been published while the interoperability aspects of the SAT and Java Cards are taken care of by the SIM Alliance. It can be estimated that the development will continue as new developments enter the mobile communications market. Some examples of these are services related to the Smartcard Web Server (SCWS) and NFC, which may interact with the (U)SIM in an advanced way. Since the early days of the SIM card, it has evolved from being merely a user authentication solution to a comprehensive services platform for operators, and provides a means for the execution of applications such as phonebook synchronization or SAT menus which enhance the discovery and usage of operator services.

As one step further in the SIM development, the SCWS represents a new generation of SIM application environments. The SCWS functions as a base for developing, executing and distributing content-rich applications from the (U)SIM, now based on web technologies. The SCWS provides an upgraded user interface which enhances the earlier, monographic SAT menus with more appealing visual aspects for the card applications. Furthermore, it is based on the HTTP 1.1 web server embedded in the (U)SIM which facilitates the development and distribution of SIM/UICC applications. The SCWS is thus an application residing in the smartcard which implements an HTTP 1.1 server defined by the IETF in RFC 2616. Thirdly, the SCWS enhances the connectivity with remote web servers which makes client/server applications adaptive to the evolution in network data rates [20].

The benefit of the SCWS is that any application developed using the SCWS technology can be used by the end-users using the SIM, independently of the device type. The SCWS provides communication with HTTP clients such as the handset browser, which at the same time works as a user interface for services and applications. Also, an administration server can manage SCWS-based applications and their content remotely, thus facilitating OTA SIM/UICC card customization, e.g., for updated applications and content.

The SCWS implements in a tamper-proof device security between the relevant nodes of its ecosystem. The solution includes security of data at the transport level with OMA standards based on PSK TLS and the implementation of GlobalPlatform standards benefiting the application-level security, which is relevant especially for operators and banks.

In practice, the SCWS content of the smartcard can be accessed via a URL, e.g., via an Internet browser which gives access to static resources like xHTML files, images, multimedia files and other formats managed by the HTTP client (browser). The URL can also be handled by a web application within the card so that the installed SCWS web applications may provide dynamic content. The application is triggered as soon as the Internet browser requests a URL mapped to a web application, resulting in the dynamic generation of the xHTML page.

The SCWS includes mechanisms to secure the HTTP connection in order to protect the environment. This can happen via HTTP-based user authentication mechanisms, or via the authentication and confidentiality by implementing HTTP over TLS (i.e., HTTPs). It is also possible to control access to the SCWS from device applications by OMA-defined Access Control Policy (ACP). As a result of these OMA mechanisms, the card is seen as a client, and the OTA platform acts as a server whereas the communication is based on web protocols (HTTPS) so that the security layer is represented by TLS in a pre-shared key mode.

The SCWS works with the classic ISO cards which communicate via Bearer Independent Protocol (BIP) TCP server mode with the handset, and the USB-IC cards which support high speed communication protocols and that communicate with the handset supporting either BIP TCP server mode or TCP/IP mode. The SCWS thus uses BIP over the ISO 7816 interface or TCP/IP stack over the USB UICC-Terminal interface [20].

4.5.3   Contents of the UICC

The UICC is a platform for a multitude of applications. It is specified by 3GPP TC SCP, and it consists of application-independent functions and features so that there is a clear separation between lower layers and applications. There is also a set of up to 20 logical channels for accessing the applications in a parallel manner as shown in Figure 4.7. The 'SIM' and 'USIM' are thus some of the many applications within the UICC, these ones providing telecommunications services whereas other applications may be related to, e.g., mobile payment.

4.6   Embedded SEs

4.6.1   Principle

The basic principle of the UICC is the transportability of the subscription data between the user's devices so that there is no need to update the HW of the device. Thus, the subscriber's mobile telephone number, among other data like phonebook, is possible to maintain with the new device by inserting the UICC card into it.

For the M2M environment, special conditions have emerged compared to the consumer UICC cards. The physical and logical characteristics of the M2M UICC cards are specified by ETSI Technical Committee Smartcard Platform (TC SCP) [11]. It includes, e.g., definitions of environmental classes which indicate the temperature ranges and other conditions under which the UICC is designed to work.

The evolution of the M2M environment has brought along the eUICC which is not physically accessible or replaceable by the user. Changing the subscription on such devices is challenging

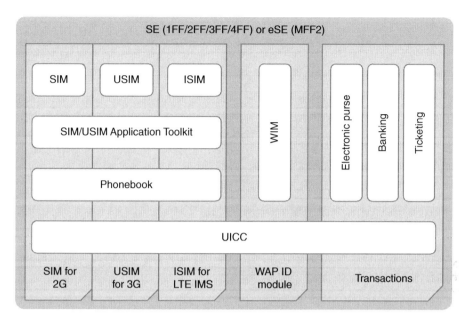

Figure 4.7 An example of the system level building blocks of a multi-application card based on the UICC. The applications may also include other subscription containers like RUIM for CDMA systems, and applets for many areas such as transit access and payments

if using 'traditional' ways, which has resulted in the need for new methods to securely and remotely provision the access credentials on the eUICCs, and furthermore, to manage the subscriptions if they are swapped between MNOs.

4.6.2 M2M Subscription Management

The ETSI specification TS 103 383 defines the requirements for the eUICC subscription management [34]. Release 13, published in October 2015, describes the remote management of the eUICC for the purposes of changing an MNO subscription without requiring physical removal and replacement of the UICC. It also presents several use cases such as provisioning of multiple eUICCs for M2M devices like utility meters, security cameras and telematics devices, provisioning of the eUICC for the first subscription with a newly connected device, and changing thesubscription for a device. Figure 4.8 depicts the logical aspects of the eUICC architecture and associated security credentials as interpreted from Ref. [34].

ETSI has documented use cases for the eUICC, e.g., for the M2M provisioning, end-user provisioning and network redundancy. Figure 4.9 presents potential use cases in this area.

The role of OEM, MNO and end-user for the eUICC activation has been under active discussion in ETSI, as well as in various other standardization and industry associations like SIMalliance, GSMA and GlobalPlatform during 2015. The current consensus indicates that the development of interoperable, international subscription management is highly important for both M2M and consumer environments which requires additional efforts to come to a final conclusion between the different standardizing entities.

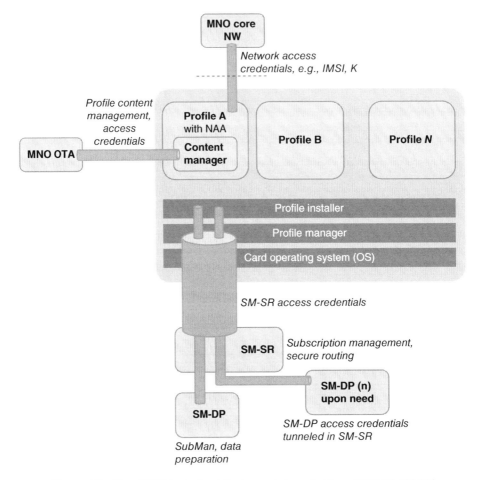

Figure 4.8   The eUICC logical architecture as interpreted from ETSI TS 103 383

As an example of the GSMA justifications of subscription management development, some common key statements for the benefits of the enhanced subscription management solution include accelerating the market growth of M2M and increasing operational efficiency for the M2M ecosystem, as well as enabling remote or OTA installation and management for operator profiles. Furthermore, the new solutions cut operational and physical logistics costs as there is no need to ship physically UICCs or change physical UICCs for an entire product lifetime. It also enables new business models, prevents market fragmentation by avoiding different, incompatible technical solutions, and drives economies of scale within the M2M industry. The GSMA embedded SIM has been developed to promote a common global remote provisioning architecture for the new era of M2M technology and to accelerate the M2M market [34]. The business environment of device types will thus change during the forthcoming years. According to the estimations of GSMA, the role of the mobile handset connections will soon lower from the 92% level in 2015 to 80%, while the role of M2M connected devices will grow considerably.

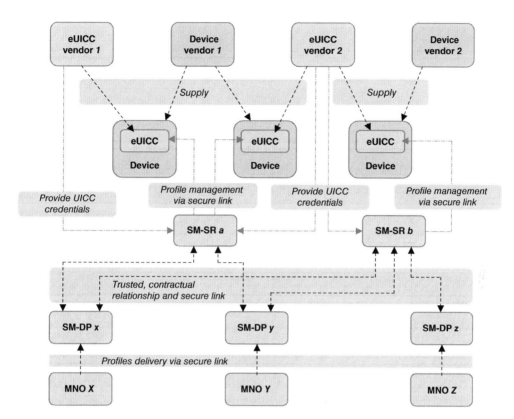

Figure 4.9   Some ETSI eUICC use cases for redundant subscription management

GSMA has identified various benefits for the eUICC. For many M2M applications, the use of traditional SIM cards is problematic because M2M modems are often inaccessible, making it difficult to insert or change SIM cards once the modem is deployed in the field. The use of the eUICC enabling remote operator profile provisioning overcomes these restrictions. The eUICC simplifies logistical processes as the UICC capability can be installed into the M2M modem at the point of manufacturing which can later be remotely provisioned with the operator profile. The eUICC removes the need for stock control and shipping of physical, pre-provisioned UICC cards. This operational flexibility is delivered with no compromise on security. Figure 4.10 presents the GSMA view to the eUICC architecture [33].

As can be seen from Figure 4.10, the subscription management changes the traditional linear embedded UICC lifecycle model which personalizes the UICC for a known MNO until the end of life of the UICC into the new model that is based on basic personalization. In the new model, after the distribution, the UICC can be personalized with the operational profile for a given MNO until the end of subscription, and the UICC can be changed for a new MNO by re-personalizing the card.

In the GSMA model, the Subscription Manager, Data Preparation (SM-DP) securely packages profiles that are ready to be provisioned on the eUICC, and installs the profiles onto the eUICC. The Subscription Manager, Secure Routing (SM-SR) securely transports the eUICC

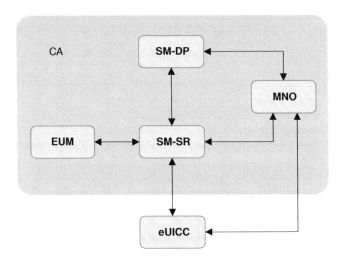

Figure 4.10   The embedded UICC architecture of GSMA as interpreted from Ref. [33]

profile management commands for loading, enabling, disabling and deletion of the profiles in eUICC. The role of eUICC manufacturer (EUM) is to provide the needed SM-SR functionality whereas the MNO infrastructure physically transfers the data onto the eUICC. The Certificate Authority (CA) is involved to ensure the trust chain in the information exchange.

The M2M and other connected IoT devices benefit the most from eUICC technology. Nevertheless, the importance of consumer applications can be expected to grow, the benefit being the enhanced user experience compared to the current switching and swapping processes.

In addition to the ETSI Ref. [34] about the topic, further information about the GSMA solution can be found in Ref. [33], which details the latest available technical description of the GSMA's remote provisioning architecture version 2.1 for the eUICC published in November 2015. Other entities currently driving for interoperable, international standards for the subscription management include SIMalliance and GlobalPlatform.

4.6.3   Personalization

The personalization of the SIM/UICC refers to the process of creating the file system of the card and loading the corresponding data such as USIM and other applications into the smartcard according to the profile. This data is placed into the memory of the chip which is typically flash memory. The customer profiles that are planned for a commercial environment are typically tested thoroughly by utilizing a set of laboratory and operational test cards among the stakeholders such as card vendors and operators. The profile configuration and format are vendor, operator and product dependent. Once tested by the stakeholders and validated by the operator, the commercial cards are produced via order management, which is an operator-dependent process. It may be based on automatized electrical or manual formats which detail the requested card parameters, including the graphical layout, form factor and delivery address and expected date.

4.6.4   M2M SIM Types

In practice, there are different M2M UICC types that each vendor provides, including the form factors utilized in consumer space and embedded UICCs. The industry and automotive grade UICCs are typically made to be more robust and durable compared to the consumer products to support long-term utilization. The aimed functional lifetime of the industry and automotive UICCs may be typically 10–17 years. The vendor-specific requirements are stricter for these variants due to harsher environmental conditions and use cases. As an example, it can be estimated that the UICCs in the automotive environment need to function for a number of years in a wide range of climate conditions.

The M2M UICC can take advantage of all the form factors. Typically, ruggedized 2FF as well as embedded MFF2 have been utilized in the M2M environment including the automotive industry. Along with the introduction of smaller equipment in the current and forthcoming IoT waves, the need for smaller SIM elements is obvious as well as the need for more dynamic management of the subscription data. The respective advances in subscription management is discussed in further chapters while the following paragraphs clarify the physical aspects of the embedded SEs utilized in M2M and consumer environments. The embedded UICC has been meant principally for M2M applications, but along with the introduction of small user devices such as wearables, it is increasingly useful for the consumer segment.

The surface-mounted eUICC does not differ as such from the removable UICC (e.g., 2FF, 3FF and 4FF) as all these variants provide the same electrical interface. The main differentiator is the size and volume, and the fact that it is soldered on the circuit board during the manufacturing process while the removable variants can be freely exchanged between mobile devices.

It could be argued that especially in the M2M environment, once the IoT device has been taken into use, there is barely any need to change the subscription information. This has been the case traditionally, but the high dynamics of the M2M communications is indicating a need for managing the subscription data, including changing the network operator or service provider, even more than once during the lifetime of the eUICC. An example of such an environment could be the automotive industry; once the initial subscription has been installed at the car manufacturing premises, the car could end up in different countries and under a different network operator's subscription. Furthermore, there might be a need to provide roaming and active communications, including changing the subscription, once the car is moved to another country. As the eUICC would not be replaced physically, this requires more advanced subscription management to guarantee updated subscription data downloading, switching, activation and deleting.

As for the electrical connections durability of the HW element, it can be assumed that in the M2M environment, which does not require changing the physical card/element, it is easier to ensure less wearing of the connector, which in turn improves the reliability of the functioning of the embedded element or permanently installed (and ruggedized) card.

At the moment, the Machine-to-Machine Form Factor 2 (MFF2) is the only internationally standardized embedded UICC form factor. It can also be called as a general term of the Machine Identity Module (MIM). Its form factor is referred to as DFN-8, SON-8 or VQFN-8, depending on the vendor environment. As for ETSI terminology, it is called MFF2, as it has been defined in the ETSI M2M UICC standard TS 102.671. The physical aspects of the MFF2 are standard in JEDEC Design Guide 4.8, and its size is $6.0 \times 5.0 \times$ less than 1.0 mm. There are other alternatives in the non-standard markets (complying vendor-dependent proprietary standards), some examples being Dual-Flat, No Leads (DFN), Ball Grid Array

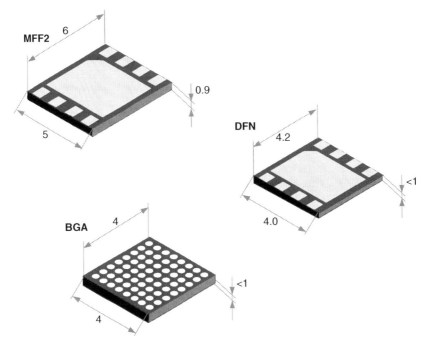

Figure 4.11 Some examples of the physically embedded SEs. At present, the MFF2 is the only standardized variant of embedded UICC. The smallest ones are typically based on wafer-level which can be very small in volume, such as the WLCSP which can measure, e.g., $2.7 \times 2.5 \times 0.4\,mm^3$, depending on each chip manufacturer's own specifications

(BGA) and wafer level solutions such as Wafer-Level re-distribution Chip-Scale Packaging (WLCSP). Figure 4.11 presents the aspects of some of the available options.

In addition to the physical element, whether it is a traditional removable form factor or an embedded element, there are also active investigations and discussions on SW-based UICC which sometimes is referred to as 'soft SIM'. The completely SW-based subscription such as the purest form of the HCE, compared to the HW-based SE such as UICC is discussed later. Also, the combination of the traditional UICC security and SW-based solution such as tokenization is feasible, as well as solutions 'in-between' such as HW-based security located apart from the UICC as in TEE.

For some practical solutions in the automotive field, the European Commission is basing the in-vehicle emergency call service, eCall, on the eUICC. In fact, new cars in the European Union need to comply with the eUICC-based eCall as of 2015 which ensures the instant connectivity between the car and the emergency services.

4.7   Other Card Types

4.7.1   Access Cards

A widespread use case for smartcards is access. Both contact cards and contactless cards are used for this purpose to serve as proof to open doors and access information systems. A growing business area is public transport, where the cards can facilitate the fluid entrance and exit of customers. More detailed descriptions of the most common solutions is presented in Chapter 5.

4.7.2 External SD Cards

In addition to SIM/UICC and embedded SE, the SE can also be provided by an external SD card. It basically refers to the SE which a service provider owns instead of an MNO. The microSD SE can be physically a HW solution with an integrated antenna for the wireless (NFC) environment and respective NFC capabilities built in, or a solution without antenna. For the microSD without an antenna but requiring external NFC capabilities, a separate antenna connection is required.

For a microSD, which is issued and owned by an external service provider such as a bank, there are basically no specific roles for MNO or OEM manufacturers. In fact, security, issuance and distribution of the SE is the responsibility of the service provider. This use case is feasible because a variety of mobile devices come with an integrated microSD slot. It should be noted, though, that the location and design of the slot does have an impact on NFC field strength and thus on the reading distance between the NFC antenna and respective reader if the antenna is built within the microSD. In order to cope with the requirements of the minimum NFC reading distance in these cases, the device manufacturers should ideally take this into account, although there are no guaranteed processes for aligning the external microSD card functionality with any of the devices supporting it as for NFC requirements.

4.8 Contactless Cards

4.8.1 ISO/IEC Standards

ISO/IEC 14443-1:2008 defines the physical characteristics of Proximity ICCs (PICCs), commonly known as proximity cards, or NFC cards. It is used in conjunction with other parts of ISO/IEC 14443. The set of ISO/IEC 14443 specifications have A, B and C variants, each indicating chip manufacturers. Parameter A refers to NXP (Philips) chips, C to Sony chips while B indicates the chip is from other manufacturers.

The proximity cards are based on some variant of RFID technology between the card and the reader so that the card is not inserted physically into the reader. The cards can be of simple read-only variants such as the ones meant for accessing buildings. These variants may have only limited memory. The RF of these variants may be based on, e.g., 125 MHz frequency band, or 860–960 MHz for the 2G Ultra High Frequency (UHF) cards.

The more complete contactless cards are capable of reading and writing. The communication link is on 13.56 MHz frequency as defined by ISO/IEC 14443 standard. An early use case of this category were transportation applications for real-time value consumption and reloading, which did require the highest security level as the wallet value was not considered particularly significant. These cards typically have protected memory types. In addition to the transport applications, they are also increasingly popular for storing retail value as the cards can provide faster transaction times yet maintain the transaction processing revenues.

4.8.2 NFC

In the modern contactless environment, the NFC-enabled cards are based on the ISO 14443 standard. They use the physical radio interface at 13.56 MHz frequency. The SWP is defined between the UICC and NFC chip, both residing in the user equipment. The SWP is an interface between the CLF and the UICC. It is, in fact, a contact-based protocol which is used for contactless communications.

The NFC is a short-range wireless communication technology which enables the exchange of data between two entities such as hand-held mobile devices and payment transaction readers. The technology is based on the high-frequency radio interface that provides functional connections within a distance of about 10 cm between the NFC-enabled devices, although the more precise requirement for the maximum distance depends on the entity, some payment certificates requiring possibly relaxation for the distance down to 4 cm. The maximum practical distance depends on the specifications as well as on the additional requirements which may be tighter for operators and credit card companies.

The NFC is based on the extension of the ISO/IEC 14443 proximity card standard referred to as RFID. Nevertheless, it should be noted that the NFC is not the same as RFID. Even if NFC and RFID do include common functionalities, the RFID is about small and economical tags that are readable within certain distances wirelessly. One of the examples of RFID is warehouse inventory tagging which provides instant information about the number of items, their types and characteristics that are found within that specific spot. Instead, NFC is about point-to-point two-way communications between NFC devices which can be physical, e.g., computers, cell phones, laptops and PDAs.

NFC combines the interfaces of a smartcard and a reader into a single device. The NFC device can communicate with both existing ISO/IEC 14443 smartcards and readers and other NFC devices, and is thereby compatible with the existing contactless infrastructure which is already in use for public transportation and payment. In practice, NFC is primarily used in mobile devices although there are various NFC-enabled devices on the market. Some of the typical use cases of NFC include the connecting of electronic devices according to the peer-to-peer data exchange concept, accessing digital content according to reader/writer concept and executing contactless transactions according to the card emulation concept. Figure 4.12 presents some high-level NFC use cases.

The use cases can be related, among a vast amount of the possibilities, to the following: enhancements to loyalty programmes (by tapping a loyalty card, e.g., in an airport), electrical format of offer coupons, content gathering and transferring, access card to physically closed locations, assets management, reporting and for making connections.

NFC development is taken care of by the NFC Forum. In addition to the specifications, it also follows up the implementation of the standards. The NFC Forum was established in 2004, and the number of participating members has grown significantly ever since. The mission of the NFC Forum is to promote the use of NFC technology by developing standards-based specifications that ensure interoperability among devices and services. The practical ways of work include encouraging the development of products using NFC Forum specifications, educating the global market about NFC technology and ensuring that products claiming NFC capabilities comply with NFC Forum specifications.

NFC readers and devices are basically always powered on when they function. They rely on a battery or an external power source. The readers need to be able to create an electromagnetic

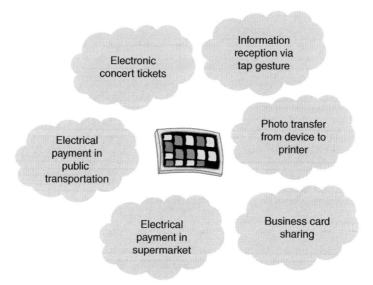

Figure 4.12 Typical use cases for NFC

field that is utilized for the radio transmission. The reader is active as soon as it produces an RF field. The tag normally has no power. It gets a sufficient amount of energy via the nearby RF field, and can thus respond via a carrier modulation which is one form of a passive communication. Since the tags are not able to activate radio channels, two tags cannot communicate with each other, unlike NFC devices.

NFC can be divided into the following features:

- *Tag reading* for telephone numbers, URLs and visit cards. Examples for tag readers include calling a taxi and reading bus stop time schedules.
- Easy *setup*, including Bluetooth headset and other accessory pairing by tapping the tag.
- *Sharing* of photo or other contents from a hand-held device to another phone or to a digital photo frame by touching it. As soon as the initiation has been done, the actual transfer of the contents happens via Bluetooth.
- *Payment and ticketing* can be applied in a similar manner as is the case for credit cards and digital bus tickets. The possibilities for the practical solutions are actually endless. It should be noted that payment requires a relatively complicated ecosystem with middleware SW and respective APIs, adaptation such as SWP as well as a HW platform such as NFC chip and antenna, UICCs, OTA, TSM, purchase readers and a set of wallet and payment applications and midlets by operators and credit card companies. In addition, the payment institutes and operators require certificates such as EMVCo, MasterCard Certification and Visa Certification. The NFC Forum has its own certificate for these purposes.

Chapter 5 discusses further NFC, and more information about the NFC technical releases of the SIMalliance can be found in Ref. [32]

4.9 Electromechanical Characteristics of Smartcards

4.9.1   HW Blocks

Figure 4.13 outlines the main HW blocks of the UICC. The main block is the CPU which is assisted by the NPU. The OS resides in ROM while the operating memory is formed by RAM. The EEPROM is a non-volatile memory block that is used for modifying data like phonebook [3].

Physically, there are either six or eight contacts on the UICC as summarized in Figure 4.1. Of these, not all are necessarily used for the mobile communications system. The underlying chip module consists of a die and housing. The combination of these forms the UICC.

4.9.2   Memory

The type of EEPROM dictates the writing rate and the maximum useful write cycles. Smartcard vendors personalize UICCs for different consumer markets like MNOs, industrial and automotive users. The useful lifetime of the SIM card may be around 2–3 years for the consumer space due to the relatively frequent change of the devices, whereas the industry and automotive grade UICCs may require a useful lifetime of 10–17 years.

The lifetime of the UICC is largely dictated by the maximum number of write cycles and by the tolerance for the temperature ranges and other environmental conditions. The UICC may also include additional functionalities like automatic error detection and recovery which works dynamically by reallocating the active writing areas and deactivating the non-functional areas of the chip. Furthermore, there may be value-added services provided by applets that are meant for monitoring the performance of the card during its lifetime and reporting possible problems in wearing of the memory, to indicate if the useful lifetime is being reached too soon. These solutions are especially useful for the automotive industry and other critical environments to indicate the need for a new SIM card beforehand as part of the next maintenance and thus avoiding extra customer visits.

The first SIM cards were limited in memory size. A set of 7 kB ROM, 3kB EEPROM and 128 byte RAM together with 8-bit CPU was a typical setup. This was sufficient in the initial phase of the GSM, but it clearly soon became a limiting factor for user experiences along with more demanding requirements such as the maximum number of stored contacts. Along with the development of CPU and memory technologies, smartcards have also evolved. Furthermore,

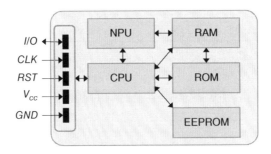

Figure 4.13   The block diagram of the UICC

the initial embedding of OS to ROM is not an optimal solution as it leaves unused memory space if the OS size does not coincide with the maximum available ROM size. More advanced variants are based on flash memory which is gradually replacing the older memory structures in current markets. Smartcard chips are also going towards single chip solutions with currently about 1–2 MB size, accompanied with some hundreds of kB of flash memory.

4.9.3 Environmental Classes

Ref. [11] details the environmental requirements for M2M UICCs. The main categories are indicated by the letters T (operational and storage temperature), M (moisture), H (humidity), C (corrosion), V (vibration), F (fretting corrosion), S (shock), R (data retention time) and U (minimum updates). Table 4.3 summarizes the environmental classes for the M2M UICCs, and Table 4.4 summarizes their required key values.

As can be seen from Table 4.4, there is a codification system developed for distinguishing the environmental performance capabilities of UICCs. It is based on a string of two characters which can be mapped onto the environmental property. These strings represent environmental property types accordingly. This two-character string represents the environmental performance in terms of the value range as seen in Table 4.4 in such a way that the first character stands for environmental property and the second indicates the level.

In practice, if the M2M UICC does not meet the level specified in Ref. [11] for a particular environmental property, it shall not be present in the string representing its environmental performance. There is no defined order for the environmental properties.

Table 4.3   The environmental classification; the main categories for M2M UICCs

Class	Description
T	M2M UICC performance for operational and storage temperature. Support of a given temperature range for an M2M UICC operation implies that the M2M UICC shall withstand (1) 500 temperature cycles within the full supported range; (2) 2 cycles per hour. Testing temperature cycling shall be in accordance with JESD22-A104 as presented in [13]
M	M2M UICC performance for moisture/reflow conditions that may be experienced during the manufacturing of the M2M communication module
H	M2M UICC performance in humid conditions
C	M2M UICCs shall be able to pass the salt atmosphere test according to JESD22-A107 [16] The conditions CA…CD refer to the duration of exposure to the salt atmosphere
V	M2M UICC performance in vibrating conditions
F	Fretting corrosion; M2M UICC performance when in a connector
S	M2M UICC susceptibility to shock
R	M2M UICC's ability to retain data over time. The data retention time property shall be able to fully operate with no loss of stored information over a required time as indicated in the class value, from the time of manufacture. Loss of information due to multiple erase/write cycles is excluded from this property
U	M2M UICC's expected minimum number of UPDATE commands, as specified in TS 102 221, supported for specified files, which are indicated as 'high' in the update activity field [12] The update of the specified number of file(s) per class value must happen without failure. Loss of information due to time factors is excluded from this property

Table 4.4   UICC environmental classes and required values

Class	Description	M2M UICC
TS	Operational and storage temperature, standard temperature range	−25 °C...+85 °C
TA	Operational and storage temperature, class A specific UICC environmental conditions	−40 °C...+85 °C
TB	Operational and storage temperature, class B specific UICC environmental conditions	−40 °C...+105 °C
TC	Operational and storage temperature, class C specific UICC environmental conditions	−40 °C...+125 °C
MA	Moisture/Reflow conditions, according to IPC/JEDEC J-STD-020D as presented in [14]	(1) Classification temperature (Tc) of 260 °C supporting Pb-free process; (2) Moisture Sensitivity Level 3; (3) Pb-free assembly reflow profile class
HA	Humidity, high	Shall support the high humidity condition as specified in TS 102 221 for UICC environmental conditions [15]
CA	Corrosion, condition A	Test condition A as specified in JESD22-A107 [16]
CB	Corrosion, condition B	Test condition B as specified in JESD22-A107 [16]
CC	Corrosion, condition C	Test condition C as specified in JESD22-A107 [16]
CD	Corrosion, condition D	Test condition D as specified in JESD22-A107 [16]
VA	Vibration, automotive	Shall be able to pass the variable frequency vibration tests according to JESD22-B103 [17]
FA	Fretting Corrosion	This item is for further study by ETSI
SA	Shock, automotive	Shall be able to pass the mechanical shock tests according to JESD22-B104 [18]
RA	Data Retention time, 10 years	Shall be able to fully operate with no loss of stored information over a 10-year period from the time of manufacture
RB	Data Retention time, 12 years	Shall be able to fully operate with no loss of stored information over a 12-year period from the time of manufacture
RC	Data Retention time, 15 years	Shall be able to fully operate with no loss of stored information over a 15-year period from the time of manufacture
UA	Minimum Updates, 100,000	Minimum of 100,000 UPDATE commands without failure
UB	Minimum Updates, 500,000	Minimum of 500,000 UPDATE commands without failure
UC	Minimum Updates, 1,000,000	Minimum of 1,000,000 UPDATE commands without failure

Ref. [23] sets the requirements for the consumer UICCs. It states that the standard temperature range for storage and full operational use shall be between −25 °C and +85 °C, but the support of specific UICC environmental conditions is optional for the UICC. Thus, the support of an extended temperature range is optional for the UICC. If supported, class A indicates the ambient temperature range of −40 °C to +85 °C while class B sets the limits to −40 °C to +105 °C, and class C to −40 °C to +125 °C. Furthermore, the extended humidity conditions are optional for the UICC. If the UICC supports the extended humidity, the limits are set based on the following environmental conditions for operation and storage in temperature of 85 °C, for the relative humidity from 90% to 95% while the duration is 1000 hours.

4.10 Smartcard SW

4.10.1 File Structure

ISO/IEC 7816 based cards have a standardized file structure. This applies to the SIM/UICC cards of mobile communications systems as well as to bank cards, transport cards and other environments that rely on the ISO/IEC 7816 smartcards.

After the initial SIM cards were utilized for the GSM networks, the 3G environment brought along extensions to the supported systems and functionalities of the SIM in form of the UICC. It contains different files, including the ones needed for the support for telecommunications. The UICC is a secure token that acts as part of the secure communications chain in 3GPP mobile networks.

The mobile telecommunications related files of the UICC are in effect applications that are able to support different radio access networks, and are referred to as Network Access Applications (NAAs). The UICC may thus contain one of various radio access networks according to the preferences of the MNO, i.e., the owner of the UICC. The GSM/GPRS SIM application manages the ETSI/3GPP 2G communications, i.e., it is the USIM in a 2G context, while the UMTS/HSPA can be managed via the USIM for a 3G context. For the LTE systems, the USIM includes LTE files that are specified by 3GPP in TS 31.102 as of Release 8. Figure 4.7 clarifies the idea of the NAAs. There may also be support for CDMA networks (1xRT, EVDO, HRDP, eHRDP etc.) via CSIM applications.

The UICC contains subscription data, including parameters that are stored in the file system of the UICC and used in authentication. The UICC also contains the algorithm for the radio interface encryption and decryption. This algorithm is also stored in the AuC of the 3GPP network, and is never exposed outside of the UICC or AuC. The UICC may also contain various applets that are nowadays based on Java. The benefit of Java is the interoperability and fluent adaptation of new applets by card manufacturers as well as by third parties.

The smartcard consists of an MF, DFs and EFs in a hierarchical way as shown in Figure 4.14. As an example, one of the dedicated files may be '3G' which contains related information for supporting the complete 3G services including the encryption keys. As Ref. [23] defines, an Application DF (ADF) is a particular DF that contains all the DFs and EFs of a single application while the general DF allows for a functional grouping of files. It can thus be the parent of a DF and/or EF, and the DF is referenced by a file identifier. Figure 4.15 depicts the principle as interpreted from Ref. [23].

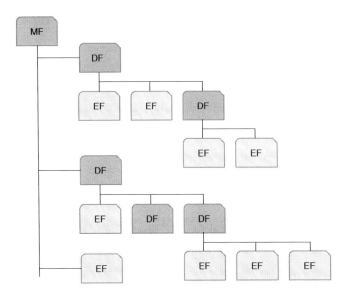

Figure 4.14   The overall principle of the file structure of the smartcard

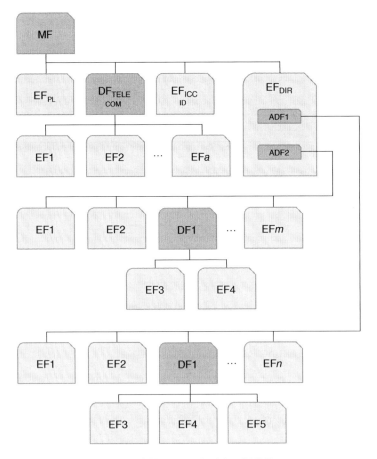

Figure 4.15   The principle of ADFs

Table 4.5   File types of smartcards

File type	Description
Transparent	Binary file
Linear Fixed	Has n records of fixed length
Linear Variable	Has n records of different lengths, yet the length of each one is fixed
Cyclic	This is a Linear Fixed file type so that the oldest records are overwritten
Execute	Special type of transparent file

Table 4.6   Some of the key commands of the SIM/UICC

Command	Principle
Create, Select, Delete File	General commands
Read, Write, Update Binary	General commands
Erase Binary	General commands
Read, Write, Update Record	General commands
Append Record	General commands
Increase, Decrease	General commands; applies to Cyclic File, changing the current position
Verify CHV	Access control-related command
Change CHV	Access control-related command
Unblock CHV	Access control-related command
Enable, Disable CHV	Access control-related command
Internal Authenticate	The card is authenticated to the terminal
External Authenticate	The terminal is authenticated to the card
Encrypt, Decrypt	Encryption-related commands
Sign Data, Verify Signature	Encryption-related commands
Initialize, Credit, Debit	Electronic purchasing

The length of the files are typically fixed size and they are addressed by a 16-bit file-ID (FID) which is typically divided into DF and EF. There is only one MF per smartcard, and its FID is standardized to 0x3F00. The standardized file types are listed in Table 4.5.

4.10.2   Card Commands

The smartcard commands are defined in the ISO standards. In addition, there may be provider-specific proprietary commands. Typically, the OS of the card is vendor-specific. The modern smartcard is based on Java so the respective Java applets are by default interoperable. Nevertheless, the applets that rely on OS-specific functions are typically vendor-specific, although it may be possible to use converters to adapt some applets between different card vendors.

Table 4.6 summarizes some of the standardized key commands. They include general commands as well as commands related to access control, authentication, encryption, electronic purse instructions and instructions specific for applications. The access control can be based on the PIN code or Chip Holder Verification (CHV), while the authentication and encryption are based on internal network functionality for challenge and response as described in Chapter 2.

It should be noted that ISO-7816 provides the commands, but the actual implementation is left open for vendor implementation.

4.10.3   Java Card

As a practical environment for the NFC UICC, Ref. [21] dictates that it needs to be compliant with the (U)SIM Java Card Platform Protection Profile. If the NFC UICC needs to be certified at EAL4+, it must use ISO/IEC 15408 CC against the (U)SIM Java Card Platform Protection Profile.

4.11   UICC Communications

The SIM/UICC communicates via the interface with the attached equipment, i.e., handset, laptop, mobile tablet device or other portable device with integrated cellular connectivity. It contains the same type of functionality as a smartcard reader, but substantially extends the messaging between the card and the device.

4.11.1   Card Communications

The UICC communications are based on secure channels which ensures the secure communication between the smartcard and external world (card reader and other entities wanting to communicate with the card). The respective secure channel protocols are the following, based on the definitions of ETSI and the GlobalPlatform [27,28].

- SCP01 refers to a deprecated symmetric key crypto-system [28].
- SCP02 refers to the symmetric secure channel protocol based on Triple-DES [28].
- SCP03 refers to the asymmetric secure channel protocol based on AES [30]. There is also a further modified variant of the SCP03, referred to as SCP03t [35].
- SCP10 refers to the asymmetric key crypto-system [28].
- SCP80 refers to the OTA remote secure channel protocol as defined by ETSI [24].
- SCP81 refers to the OTA remote secure channel based on SSL/TLS [29]. As ETSI TS 102 267 states, an application can open a connection to a remote server based on SCP81. The security for data exchange is provided by TLS. The HTTP protocol is used on top of TLS to provide encapsulation of the data. TCP/IP transport is provided by the BIP as defined in ETSI TS 102 223 or a direct IP connection as specified in ETSI TS 102 483 [29].

The overall secure communications principle for the UICC is described in Ref. [24,25,31. The communications between the smartcards such as the UICC and the external world takes place via the Card Content Management (CCM). It contains the loading, installation, personalization, extradition and deletion functions [26]. The communications are based on APDUs. There are two types of APDUs: Command APDU and Response APDU, as depicted in Figure 4.16.

The command APDU contains fields for Class of Instruction (CLA), which include category, SM and channel), instruction (INS), parameters 1 and 2 (P1, P2), length of the command data (Lc), command data and length of the expected response (Le).

Command APDU

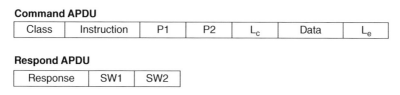

Class	Instruction	P1	P2	L_c	Data	L_e

Respond APDU

Response	SW1	SW2

Figure 4.16   The format of the Command and Response APDU

Table 4.7   An example of the SIM/UICC card response messages. The complete list can be found in ISO/IEC 7816-4 documentation

Response (SW1 I SW2)	Description
61 I xx	Response bytes still available; Command successfully executed; xx bytes of data available and can be requested using GET RESPONSE
62 I 81	State of non-volatile memory unchanged I Part of returned data may be corrupted
63 I 85	State of non-volatile memory changed I Secured transmission not supported
64 I 01	State of non-volatile memory unchanged I Command timeout
65 I 81	State of non-volatile memory changed I Memory failure
68 I 82	Functions in CLA not supported I Secure messaging not supported
69 I F0	Command not allowed I Permission denied
6A I F0	Wrong parameter(s) P1-P2 I Wrong parameter value
90 I 00	OK

The response APDU contains fields SW1 and SW2. Table 4.7 presents a snapshot of the responses from the card [22]. The response indicates the overall topic (SW1) and a more specific description of the response (SW2) for the normal and faulty behaviours. The APDU responses are used to ensure the correct updates of the card SW like the OS, and are useful in troubleshooting if misbehaviour is noted.

The card vendors and other stakeholders involved in the testing and troubleshooting of the SIM/UICC cards typically have a set of tools that can be utilized to monitor the card responses, as well as other communications between the card and reader. The APDUs are transmitted by Transmission Protocol Data Units (TPDUs) which are defined in ISO/IEC 7816-3. The most popular protocols in this domain are T=0 (byte-oriented protocol), T=1 (block-oriented protocol) and T=CL (contactless protocol).

4.11.2   Remote File Management

The SIM/UICC can be managed either locally by connecting a cable between reader and the device. Anther option is to manage the cards OTA. This item includes Remote Application Management (RAM) and Remote File Management (RFM). RAM and RFM, as defined in ETSI TS 102 226 and 3GPP TS 31.116, respectively, can be performed using either the CAT-TP mechanism or RAM/RFM over HTTPS mechanism, as defined in ETSI TS 102 225, and 3GPP TS 31.115 for PUSH SMS security and in ETSI TS 102 127.

As a practical example of the NFC UICC environment, Ref. [21] details respective requirements stating that the use of RAM/RFM over HTTPS shall be done as defined in ETSI TS 102

225 and 3GPP TS 31.115 for PUSH SMS security, as well as ETSI TS 102 226 and GlobalPlatform CS V2.2 Amendment B. The use of CAT-TP or RAM/RFM over HTTPS may differ between regions, which needs to be taken into account in the service provider interoperability considerations, so Ref. [21] also includes instructions for use by different MNOs, suggesting that at least one of the two transport protocols referenced in the document needs to be implemented by the MNOs. It should be noted that for small management operations, the SMS OTA is still allowed.

References

[1] Triple SIM. Three SIM cards in one: 2FF, 3FF, and 4FF. Whitepaper. Giesecke & Devrient, 2014. http://www.gi-de.com/gd_media/media/documents/brochures/mobile_security_2/cste_1/Triple_SIM.pdf (last accessed 7 December 2014).

[2] History of SIM cards. http://www.timetoast.com/timelines/cell-phones-by-whittney-williams-8th-period (last accessed 7 December 2014).

[3] Vedder, D. K. The UICC; The security platform for value addedservices. Sophia Antipolis, France. 4th ETSI security WS. 13-14 January 2009.

[4] ISO 7816 introduction. http://www.smartcardsupply.com/Content/Cards/7816standard.htm (accessed 24 January 2015).

[5] Smartcard basics. http://www.smartcardbasics.com/smart-card-standards.html (accessed 24 January 2015).

[6] Elaheh Vahidian. Evolution of the SIM to eSIM. Master's thesis. Department of Telematics, Norwegian University of Science and Technology, Trondheim. January 21, 2013. 90 p.

[7] SFR. (U)SIM Java Card Platform Protection Profile. Basic and SCWS Configurations. Evolutive Certification Scheme for (U)SIM cards. PU-2009-RT-79-2.0.2. June 17, 2010. 85 p.

[8] Trusted Labs, Guide de composition CC entre plateformes certifies et applications sensibles, CP-2007-RT-407-3.0.

[9] JP. Wary, M. Eznack, C. Loiseaux, R. Presty. Developing a new Protection Profile for (U)SIM UICC platforms. ICCC 2008, Korea, Jiju, September 2008. 18 p.

[10] Secrétariat général de la défense et de la sécurité nationale. Agence nationale de la sécurité des systèmes d'information. Certification Report ANSSI-CC-PP-2010/05. (U)SIM Java Card Platform Protection Profile/SCWS Configuration (ref.PU-2009-RT-79, version 2.0.2). 2009. 15 p.

[11] ETSI TS 102 671, V9.0.0, April 2010. Technical Specification, Smartcards; Machine to Machine UICC; Physical and logical characteristics, Release 9. 21 p.

[12] ETSI TS 102 221, V11.0.0, June 2012. Technical Specification; Smartcards; UICC-Terminal interface; Physical and logical characteristics, Release 11. 181 p.

[13] JEDEC JESD22-A104D. Temperature Cycling.

[14] IPC/JEDEC J-STD-020D.1. Moisture/reflow sensitivity classification for non-hermetic solid state surface mount devices.

[15] ETSI TS 102 221. Smartcards; UICC-terminal interface; Physical and logical characteristics.

[16] JEDEC JESD22-A107B. Salt atmosphere.

[17] JEDEC JESD22-B103B. Vibration, variable frequency.

[18] JEDEC JESD22-B104C. Mechanical Shock.

[19] Mobile/NFC Security Fundamentals Secure Elements 101. Smartcard Alliance Webinar, March 28, 2013. 42 p.

[20] SIMalliance. Smartcard Web Server; How to bring operators' applications and services to the mass market. February 2009, 18 p.

[21] GSM Association SGP.03.

[22] APDU response list. https://www.eftlab.com.au/index.php/site-map/knowledge-base/118-apdu-response-list (accessed 28 November 2015).

[23] ETSI TS 102 221 V8.2.0 (2009-06). Smartcards; UICC-Terminal interface; Physical and logical characteristics (Release 8). 174 p.

[24] ETSI TS 102 225 V9.0.0 (2010-04). Smartcards; Secured packet structure for UICC based applications (Release 9). 22 p.

[25] ETSI TS 102 224 V8.0.0 (2008-10) Technical Specification, Smartcards; Security mechanisms for UICC based Applications - Functional requirements (Release 8). 19 p.

[26] ETSI TS 102 226 V9.2.0 (2010-04) Technical Specification, Smartcards; Remote APDU structure for UICC based applications (Release 9). 43 p.

[27] Bharat Bhanjana. Secure Communication between Card and Server. International Journal of Scientific and Research Publications, Volume 5, Issue 8, August 2015 ISSN 2250-3153

[28] GlobalPlatform. GlobalPlatform Card Specication,version 2.2. March, 2006.

[29] Remote Application Management over HTTP. GlobalPlatform Specication, September 2006.

[30] GlobalPlatform Card Technology. Secure Channel Protocol 03. GlobalPlatform Public Release, September 2009.

[31] Smartcards; Secured Packet Structure for UICC based Applications (Release 6). ETSI TS 102 225 (V6.8.0), April 2006.

[32] SIMalliance, NFC releases, 25 December 2015. http://simalliance.org/nfc/nfc-technical-releases/ (accessed 25 December 2015).

[33] Remote Provisioning Architecture for Embedded UICC Technical Specification, Version 2.1. GSMA, 2 November 2015. 297 p.

[34] ETSI TS 103 383. Smartcards; Embedded UICC; Requirements Specification (Release 13), V. 13.0.0, October 2015. 29 p.

[35] Embedded UICC Protection Profile, Version 1.1. GSMA, 25 August 2015. 127 p.

[36] Phone book management with ISO 7816 part 7. 3GPP TSG-T3, Document T3-99167. Miami, June, 14th to 16th, 1999. 22 p.

[37] Protection Profile — Secure Signature-Creation Device, Type3. Prepared by ESIGN Workshop Expert Group F for CEN/ISSS. Version: 1.05, EAL 4+. 25 July 2001. 67 p.

[38] Proposed Correspondence to EP-SCP calling for establishment of a formal liaison to promote RUIM global harmonization. TSG correspondence of the 3GPP2, TSG-C, October 23, 2000. 3p.

[39] ETSI TS 100 812-2, V2.4.1. Terrestrial Trunked Radio (TETRA); Subscriber Identity Module to Mobile Equipment (TSIM-ME) interface; Part 2: Universal Integrated Circuit Card (UICC); Characteristics of the TSIM application. 3 August 2005.

5

Wireless Payment and Access Systems

5.1 Overview

This chapter outlines the current most relevant solutions for mobile payment and access control. As a base, the wireless security technologies detailed previously are discussed specifically from a payment and access point of view, followed by the description of NFC and other wireless techniques. Then, the EMVCo concept and other banking systems are discussed along with selected use cases such as e-commerce (Wallet solutions, Apple Pay), transport (MiFare, Cipurse) and access systems (access security classifications, solutions).

The mobile payment environment is developing strongly at the moment. Commercial markets are considering many mobile wallet solutions based on NFC, tokenization or HCE. Some examples of mobile payment can be found in Refs. [10,11]. Also, biometrics is becoming more relevant for providing additional security for payments, including fingerprint and eye iris based biometrics for mobile app authentication. The eye iris authentication discussed in this chapter is based on Eyeprint ID, which has been developed by US-based EyeVerify and AirWatch [9,12].

5.2 Wireless Connectivity as a Base for Payment and Access

Payment and access systems may rely technically on any known connectivity technology. One logical solution is NFC which is designed for providing very short distance two-way radio communications between devices such as a smartphone and NFC reader. Other typical solutions are based on RFID. Barcodes as such are the base for embedding product information to the object which can be mapped into printed price information by the cash register system as is the case in today's retail environment. The barcodes can also be produced dynamically as 'tokens' for approving subscription-related actions by printing them on paper and scanning, or by relying on the smartphone display to perform payments at the Point-of-Sale (POS).

Wireless Communications Security: Solutions for the Internet of Things, First Edition. Jyrki T. J. Penttinen.
© 2017 John Wiley & Sons, Ltd. Published 2017 by John Wiley & Sons, Ltd.

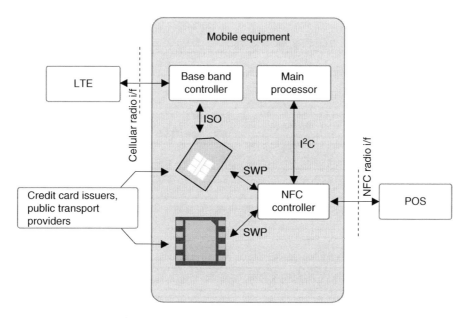

Figure 5.1   The development of mobile payment

As for highly fluent user experiences, NFC is one of the most logical solutions due to its ability for interactive communications and secure transactions. Figure 5.1 depicts the development of the mobile payment area based on NFC functionality as interpreted from Ref. [25].

As can be interpreted from Figure 5.1, the eSE can coexist with the multi-service UICC card within a single mobile device in such a way that the applications of the service providers may reside in both eSE and UICC. The UICC functions as a base for the MNO-managed businesses while the eSE provides additional businesses for the OEM customers. The NFC controller connects to the eSE and UICC via the SWP while it communicates with the main processor via the Inter-Integrated Circuit (I^2C). The following sections outline the commonly applied technologies and their utilization in payment and access as well as the basic technology behind these close-range solutions.

5.2.1   Barcodes

The barcodes include a range of 1D code systems which are adopted internationally in order to specify object classes for retail shops, industry component production, logistics and storage, to mention some examples. One might wonder why the barcode is relevant to the wireless environment; the answer is straightforward as it represents one-way near-field information transfer based on optical readers or imaging, and it also is possible to utilize via scanning via the smart device camera which typically is deployed in all consumer devices. Furthermore, the barcodes can be utilized as a basis for wireless transactions by generating a one-time code with a limited validity period, to be scanned with a consumer device for the payment transaction. As an example, the payment system security layer may be based on mobile transactions

Figure 5.2   An example of the QR code with embedded web link leading to further information about this *Wireless Security* book

in such a way that the credentials are stored in the cloud and present them in a face-to-face merchant environment by using Quick Read (QR) codes, barcodes or Bluetooth. This provides an easy method for low-value mobile payments as the credentials are not authenticated during the transaction even if the user's device is physically present at the POS [38].

A QR code is a practical manner to represent data which is related to the object to which it is attached. The barcode is designed for optical machine reading. There are many coding systems in the market that represent the identification of the object, such as the ISBN coding of books. Nowadays, a smart device (or any other device with embedded camera) is able to read the barcode and if it contains information about a web page, the device may enter the respective website. As another example, the IATA-standardized 2D format of barcodes is utilized in airport boarding via the Bar Coded Boarding Pass (BCBP), or 2D barcodes designed to be presented on a smart device's display for electronic boarding passes. Figure 5.2 shows an example of the general QR code which can be typically read with smart device applications (generated via Ref. [30]).

The original format of barcode presents the respective data by parallel lines with variable widths and spacing. The 2D-codes, instead, are based on a matrix presentation of the digitalized contents, defined by ISO/IEC [36]. The forms that the 2D barcode can take include rectangles, dots, hexagons and other geometric patterns which are placed in two dimensions, and a clear safety margin is added around the image. It also typically includes an embedded error correction technique, and the level of the correction can be selected according to the utilized system and need – the principle being that the better protected the image is, less useful data can be presented [37]. In strongly protected images, it is possible to print overlapping artwork on top of the QR code and the contents can still be read without issues. The barcodes have further been developed although the popularity of each advanced variant may not be as widely spread as the previously mentioned ones. The 3D codes are basically 2D variants with an additional parameter such as greyscale or colour palette included to provide an extra variable per pixel that makes it possible to embed more diverse information within the same area as 2D is able to do.

The barcodes are highly secure as they require a close distance for scanning. The potential security breaches of barcodes could be related to the embedded malicious code in the image which may execute a mobile device's functions without the user's confirmation or connect to malicious Internet pages which may cause further security issues. If the barcode is used for payment, e.g., via the reception of a promotional code that is presented at the POS upon purchase of the product, the security breach could theoretically be related to someone stealing the barcode and utilizing its value before the original user is able to do so. For that reason, such payments via the barcodes are most suitable in small-scale purchases such as coffee bars. Nevertheless, it can be argued that for the instantly generated and personalized payment tokens that the customer receives only upon physical presence at the POS, it is highly unlikely that the code would be too easy or even useful for others to exploit.

5.2.2 RFID

RFID is a feasible method for the payment and access. RFID is suitable for various other environments, too, such as asset tracking, race timing and inventory management. RFID forms part of the process for identifying items in a unique way via radio frequencies. It should be noted that NFC is a subset within RFID technology, working on 13.56 MHz frequency. NFC is a secure form of data exchange so that the NFC device may act as an NFC reader as well as an NFC tag, which makes it possible for the NFC device to communicate in a peer-to-peer mode [26].

The RFID tag can be active (providing largest coverage areas with their own power source) or passive (basing the power on induction from the reader for short distance). In addition to the tag, the RFID system includes the reader and the antenna as well as the accompanying system. The task of the reader is to send a request to the tag to which the tag responds via the antenna. The active RFID mode provides coverage up to about 100 m which is sufficient for, e.g., a toll road payment system. The range of the passive RFID tags is lower, typically up to about 25 m.

The passive RFID tag operates typically in the Low Frequency (LF) band of 125–134 kHz (with a range of around 10 cm), High Frequency (HF) band of 13.56 MHz (which is also the NFC frequency; providing a range of about 30 cm), and Ultra-High Frequency (UHF) band of 856–960 MHz (with a range up to around 100 m). Some of the relevant standards for the RFID proximity card and NFC include ISO/IEC 14443, ISO/IEC 18092 and FeliCa. The passive HF RFID tag is also compliant with the ISO/IEC 15693 standard.

The security aspects of proximity cards such as RFID are of high importance for protecting the contents delivery against copying and interception. For the shortest range proximity solutions, the distance as such may protect the transactions although it may be possible to locate an eavesdropping device close to the reader to capture the communications either by placing the respective antenna very near to the device, or by utilizing highly directive antennas.

The RFID and NFC tags are increasingly popular in product posters and signs for delivering compact, additional information like web links to the consumers. In practice, there is a wide selection of commercial, low-cost mobile RFID readers such as the ones shown in Ref. [22]. They can be paired with devices like smartphones and tablets, and the devices can be used for a variety of business applications. Along with the growth of the consumer smart device and tablet market, the RFID readers are also becoming more popular. The two main available categories are devices that are designed to be attached to mobile devices to convert them into RFID readers, and low-cost RFID readers for smart devices. These readers support passive UHF EPC Gen 2 protocol, and they can be used for business applications such as access control, authentication and verification, inventory management, logistics and transportation.

As the market develops, it can be more profitable for business users to obtain mobile RFID readers attached to consumer devices compared to conventional handheld RFID readers. Some examples of the enterprise products can be found in Ref. [23] which adds the next generation RFID with 1D and 2D barcode scanning to the mobile devices.

5.2.3   NFC

The components of a typical NFC-enabled mobile device include SE which can be SIM/UICC, eSE or microSD, as well as NFC controller, NFC chip, protocol stack and CLF. The NFC-based application such as Mobile Wallet is needed for payments, and a user interface application for the consumer interaction. The communication protocols and interfaces include ISO-7816, ISO-14443, SWP, Universal Asynchronous Receiver/Transmitter (UART), I²C, Serial Peripheral Interface (SPI), and NFC needs to be supported by the SE's operation system such as Java or vendor-specific OS, and the mobile device's OS such as Android, iOS, BlackBerry or Windows. Figure 5.3 depicts a typical NFC device architecture.

NFC functions at 13.56 MHz frequency which is the same for the HF-variant of RFID readers and tags. The NFC equipment can, in fact, work as both reader and tag making it a two-way technology. Nevertheless, the maximum reading distance is limited to only few centimetres. Typical NFC use cases are related to the information sharing either based on the low-speed NFC channel, or most typically, via opening a separate bearer such as a Bluetooth transport channel which is triggered by NFC tapping.

The focus of the NFC Forum is on standardizing the application domain. As an example, the data storage standard NFC Data Exchange Format (NDEF) and mapping for tag types are

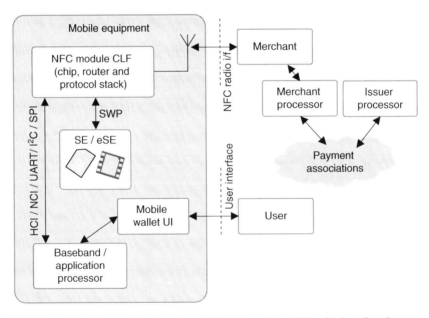

Figure 5.3   Example of the architecture of an NFC device. The NFC radio interface is connected to payment associations such as Visa, MasterCard, AmEx and Discover via the merchant processor

defined by the NFC Forum. It has been decided that the security should be handled on the application layer. The NFC Forum has a separate working group for security that focuses on identifying potential threats and attacks against NFC.

5.2.3.1 Architecture

An NFC Forum device needs to comply with the high level conformance requirements and it implements at least the mandatory parts of the NFC Forum protocol stack and operating modes. The mandatory NFC Forum operating modes are the NFC Forum peer mode and the reader/writer mode. The optional support means that NFC Forum devices may support the NFC Forum card emulation mode. Furthermore, the NFC Forum device can support optional parts of the stack as well as additional protocols and applications that are not defined by the NFC Forum.

An NFC Forum tag can be any contactless component that the NFC Forum device is capable of accessing, as defined by one of the Type X Tag operation specifications. NFC Forum tags are not required to support the complete specification for the NFC Forum protocol stack.

The NFC Forum protocol stack includes the protocols for the communication between NFC Forum devices, between NFC Forum devices and NFC Forum tags, between NFC Forum devices and technology-compatible contactless smartcards, and optionally between NFC Forum devices and existing reader/writer terminals. It does not make any assumptions about the implementation or the overall architecture of NFC Forum devices.

The NFC Forum protocol stack supports reader/writer, peer and card emulation modes. The NFC Forum reader/writer mode is capable of reading from and writing to NFC Forum tags. In addition, this mode allows the communication with compatible smartcards. The NFC Forum peer mode is meant to communicate with other NFC Forum devices, while the NFC Forum card emulation mode is optional and emulates the behaviour of a smartcard or tag. The communication with existing technology-compatible reader/writer terminals is possible in this mode.

In the NFC Forum reader/writer mode, the NFC Forum device has the capability to at least communicate with NFC Forum tags. The device can exchange data with NFC Forum tags by NFC Forum or third-party message formats. The device may also communicate with a variety of components like smartcards, memory cards and tags, provided they are compliant with some of the contactless technology types. The NFC Forum device supports the RF interface variants NFC-A, NFC-B and NFC-F.

In the NFC Forum peer mode, the NFC Forum device has the capability to communicate with another NFC Forum device. The service discovery protocol is the mechanism used to identify the common services supported by both NFC Forum devices.

The NFC Forum card emulation mode allows the NFC Forum device to act as a smartcard or tag in front of a conventional technology-compatible reader/writer. This mode includes the emulation of memory cards and tags, and the emulation of smartcards is intended mainly for portable devices that can be conveniently presented to reader/writers. Using this mode, existing technology-compatible terminal infrastructures (e.g., for payment and ticketing) can communicate with NFC Forum devices supporting NFC Forum card emulation mode.

Figure 5.4 depicts the NFC-defined architecture [7]. The technical architecture contains an initial set of mandatory tag formats based on ISO 14443 Type A, ISO 14443 Type B and

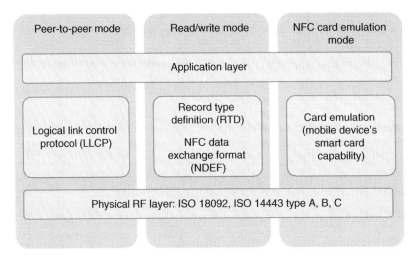

Figure 5.4   The NFC architecture as defined by the NFC Forum

Sony's FeliCa definitions. These include NDEF which specifies a common data format for NFC Forum devices and tags; NFC Record Type Definition (RTD) which specifies standard record types used in messages between NFC Forum devices and between NFC Forum devices and tags; Text RTD which is meant for records containing plain text; Uniform Resource Identifier (URI) RTD which is meant for records referring to an Internet resource; Smart Poster RTD which is meant for posters incorporating tags containing text, audio or other data.

5.2.3.2   Standardization

The NFC Forum coordinates the development of NFC. It was established in 2004, and the number of participating members has continued to grow. The mission of the NFC Forum is to promote the use of NFC technology by developing standards-based specifications that ensure interoperability among devices and services. The means can be to encourage the development of products using NFC Forum specifications, to educate the global market about NFC technology and to ensure that products claiming NFC capabilities comply with NFC Forum specifications [24].

5.2.3.3   NFC Use Cases

Payment is one of the high-level practical use cases for NFC transactions. The markets for NFC-based payments are only now forming and are highly volatile. One indication of this are the high expectations of SoftCard in the USA, which went bankrupt, and the rise of several alternative wallet solutions such as Apple Pay and Google Pay.

Transport is one of the most logical environments for NFC payment. Payment of train tickets, bus tickets and taxi rides can be done, e.g., with a mobile device with embedded NFC functionality on-the-go. The air flight environment also provides interesting cases that can be handled via NFC, like flight reservation and loyalty programme management, entering VIP lounges and boarding areas by utilizing the NFC-enabled device and luggage tracking can be combined easily with the same NFC concept. It is also logically possible to make payments beforehand in retail shops.

The NFC-enabled mobile device is suitable for ticketing as the tickets can be stored to the device in advance, and the device can be used for access to the transit areas and vehicles. At the same time, the user can investigate the time schedules and maps of public transportation by utilizing smart posters with the same device. As an extension to the traditional functions of the passenger, the user can also download special offers from the smart poster to the device, in order to get discounts. There have been practical deployments for ordering taxis via the tag. As an extension to this idea, the address of the user can be informed to the taxi driver's NFC-enabled device.

The *retail* environment represents another base which well suits NFC. The payment can be done with NFC mobile devices at contactless POS. The loyalty programmes and the utilization of coupons are straightforward. Furthermore, downloading of coupons and other special offers can be done directly from smart poster to NFC phone. Furthermore, the transfer of coupons to friends can be done fluently. In the other direction, the user can collect information about purchases by reading their product history. Touch tags can be used to collect shopping lists, with additional retailer offers. The NFC environment can combine various functions, like the collection of deposits from bottle recycling machines.

The *public sector* also benefits from NFC as it can be used to pay community services like car parks, with a record of the parking stall in the NFC phone. The NFC-enabled device can be used to access parking areas, buildings and offices by using NFC mobile device and contactless readers. In general, the NFC device can serve as ID card, visa and passport.

The *healthcare* environment with an NFC device can benefit from NFC payments by identifying the patient and showing the healthcare insurance information. The device can also contain the healthcare history of the patient, including access to graphical contents like x-ray images, possible illnesses that should be taken into account if the patient is found under severe health conditions and unable to communicate. NFC can also be utilized by the patient to access restricted areas in hospitals, and to show prescriptions in pharmacies that the doctor had ordered previously in paperless format. The doctor can see the history of purchased medicines and, if the device is connected to a larger mobile health management system, also the health values of the patient.

In addition to the payment solutions, NFC is also useful for many environments which do not include monetary transactions. Some examples are NFC tags that can inform summaries of public posters, or NFC tags located at the reception area of companies which can automatically call taxis by tapping it to a smartphone.

There are various NFC-enabled devices in the markets. Some of the typical use cases of NFC include connecting electronic devices according to the peer-to-peer data exchange concept, accessing digital content according to the reader/writer concept and performing contactless transactions according to the card emulation mode. The use cases can relate to the enhancements to the loyalty programmes, electrical format of offer coupons, content gathering and transferring, access card utilization to physically closed locations, assets management, reporting and create further connections such as Bluetooth to open an audio streaming channel between devices.

5.2.4   Secure Element

5.2.4.1   Principles

The secure payment via NFC can be based on an SE, which is a tamper-resistant device with an embedded microprocessor chip. The SE stores the applications to perform their secure execution, and keys to perform respective cryptography such as ciphering, authentication or signing for NFC services. The SE can store multiple applications to support the NFC services such as payment, offers and loyalty programmes. These SE applications are accessible by mobile applications through the baseband and accessible by contactless readers through the contactless interface. Physically, the supported SE can be the UICC and microSD card in their removable variants, or eSE which is integrated within the device's HW. In all these cases, the SE is based on a tamper-resistant chip.

The SE receives and accepts commands originated from the user's device. In the contact mode, this happens through the NFC controller for eSE, or through the ISO 7816 HW interface for the UICC. The commands can also be received in contactless mode from the external antenna based on the ISO 14443 interface. The features of the SE must be sufficiently compliant to cover the most common applications, including those that require fast cryptographic computations as is the case for secret and public keys. The respective memory requirements may depend on the possibility to provision new applications via OTA. The SE also needs to support Java Card specifications and implement Java Card API.

The protocols between the SE and the NFC chip (CLF) are the following: (1) the UICC uses SWP/HCI; (2) the eSE uses SWP/HCI, I²C, SPI, NFC-WI, and DCLB; and (3) The microSD uses SWP/HCI. As the SWP is the commonly available protocol for all these SE form factors, it is the recommended default solution for ensuring the widest interoperability of the devices. The upper layer HCI is used on top of the SWP although it would be possible to use it over any other physical protocol. The benefit of the HCI is that it allows standard interoperable layers to be leveraged for application development [27].

5.2.4.2   eSE

In the case of the eSE variant, the SE of NFC is integrated into the HW of the device. Its provisioning and management is typically based on the TSM concept. Google Nexus S, which uses NFC chips manufactured by NXP, is one of the commercial examples of this concept.

A key benefit of the eSE is that it provides a common architecture for content providers which does not depend on the mobile system. All the data is encrypted when it is stored and it also remains encrypted during the processing along the complete route the data is present.

Some of the drawbacks of eSE are, in turn, that it might be difficult to transfer applications to a new handset. Not all mobile device models support an integrated NFC chip, and for all the new device models, the payment applications must be re-tested, which may lead to delays in the device development. Furthermore, if the device requires physical maintenance, the SE is physically exposed. Even if this is a highly hypothetical case, and there is encryption involved, there could be fraudulent intentions against the SE.

5.2.4.3   UICC-based SE

The SIM card and its evolved variants have served as a base for mobile subscriber security-related procedures since the first deployment of GSM networks. Being a tamper-resistant HW element, the UICC provides a reliable means to authenticate and authorize subscribers, and is an important base for the secure billing. For that reason, it is one of the most logical bases for supporting mobile payments. In addition, the established procedures for OTA provisioning provides added value for the payment procedures.

Especially in the high-value payment environment, the process for payment certificates and the number of involved entities might be complicated because of the certification acceptance processes among operators, OEMs, payment service providers (banks) and payment applications developers. Nevertheless, the UICC-based SE with the NFC payment is typically preferred by many MNOs while it is controlled by the issuing party. As the UICC is already a mature and established technology, the solution complies with the security standards of financial institutions. Being a removable smartcard and interoperable with mobile devices, the UICC-based solution is also independent of the handsets, which provides faster development and deployment than embedded SE. Furthermore, it is possible to use OTA provisioning with this solution meaning that new secure payment applications can be downloaded remotely. Other benefits include the possibility to block the applications on the UICC by the operator if the device is stolen or lost. The UICC also supports multiple security compartments and thus a number of different cards can be taken into use according to the file structure defined by ISO 7816.

Some of the drawbacks of the solution include that it is based on the operator's processes and thus requires cooperation with the participating entities, which in turn may increase bureaucracy, overhead and delays, e.g., in the certification processes. In the case of various payment applications within a single UICC, it might not be straightforward to divide the responsibilities of the control and visibility of credit cards from separate banks. Also, the sharing of the costs between the operator and other parties, e.g., when the operator applies fees for transactions (revenue sharing vs. flat fee) is not necessarily straightforward to align.

5.2.4.4   Secure Digital Card

In addition to the embedded and UICC-based secure elements, the third solution is based on a combination of microSD card and NFC antenna that allows the handset to communicate with the contactless readers. The SE can thus be located to the microSD card (or smart microSD as the term is defined by the SD Association) while the handset takes care of the physical NFC functionality. This solution with the SE stored on the microSD card does not depend on the network operator or device manufacturer.

As an example, DeviceFidelity provides a microSD card SE. The company has partnered with Visa on its In2Pay microSD solution to offer NFC payment capabilities across the payWave platform of Visa. DeviceFidelity allows its microSD cards to be issued and personalized like traditional smartcards. It has partnered with Vivotech to add OTA provisioning capabilities to its In2Pay microSD product.

Some of the advantages of the SD-based SE solution include that it facilitates rapid application deployment and it functions with existing HW. It does not depend on MNOs or device manufacturers, and thus may look attractive for financial institutions as it allows the bank institute that issues the card, to own the SE.

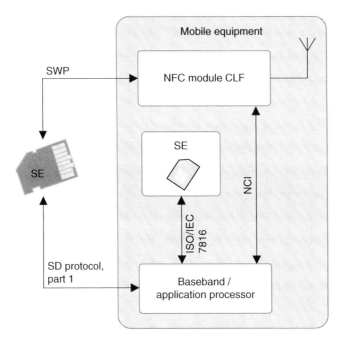

Figure 5.5 NFC device based on SE in microSD form and NFC chip residing within the device

There are two variants of the microSD card-based NFC solution [27]. The first integrates the NFC antenna and the NFC controller chip (CLF) into the device such as a smart device or wearable device, whereas the SE located in the microSD communicates with the device via the SWP. This variant provides the benefit of reusing the device's tested and approved interoperable NFC radio technology, and allows easy inserting of the microSD to the card slot of the device while the use of the SE is independent of the MNO. The drawback is that the selection of the devices supporting NFC and SWP – even if the microSD slot is getting more popular – is limited. Figure 5.5 depicts the principle as interpreted from Ref. [27]

The second variant of the microSD-based NFC device integrates the NFC antenna, chip and SE all integrated into the microSD. This standalone microSD card communicates with the baseband processor of the device via the SD protocol. Figure 5.6 depicts the principle as interpreted from Ref. [27]

5.2.5 Tokenization

The ideas beyond e-commerce that are based on the SE have resulted in cloud-based SW solutions. Nevertheless, the drawback of storing confidential credentials outside of the tamper-resistant UICC or other SE variant, e.g., on the SW of the device and/or on the cloud, creates new challenges for the security. In order not to expose the original payment card information via potential intrusion intentions, one of the solutions is to use tokenization. It refers to hiding the Personal Account Number (PAN) with time-limited equivalent data, or to a token that is used for payment transactions and that both payer's device and financial institute can map without others being able to interpret the original information.

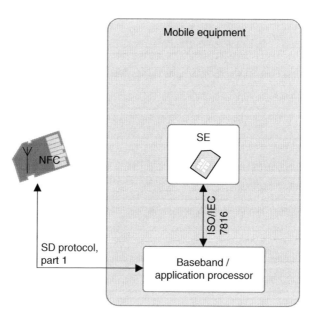

Figure 5.6 Device without NFC functionality can be used with microSD that is equipped with NFC antenna, NFC chip and SE

It can be argued that tokenization alone would not provide a sufficient security level for mobile payments. The EMVCo specification defines tokenization as an alternate PAN, but, in fact, tokenization does not refer to limited time use of data, it basically replaces the original card information for a longer term. This extended lifetime of the token in current commercial solutions might open doors for potential breaches [8].

The financial institute's card issuance is quite similar in the SE-based and in the HCE-based solutions, the key difference being that the SE environment is static while the HCE is dynamic. As the SE is tamper-resistant HW, the static approach is adequate for it. However, some of the key questions in the dynamic HCE environment are how the user's device can be authenticated in a reliable way and how the data is secured within the device. The integrity of the respective application also needs to be ensured, as well as the card data when it is transferred between device and mobile app. Thus, in order to increase the security level, the dynamic management of mobile-based card issuance needs further on-device security combined with management ensuring tokens can be transferred to the device in a secure fashion. One method for adding security in this dynamic environment is to evaluate the risk of all the transactions by transmitting additional data from the device such as phone number and device ID. Also the triggering of replenishment of account parameters can be applied, such as device thresholds and Limited Use Keys (LUKs).

The main components of the cloud-based payment solution are the tokenization system, digital issuance, on-device HCE client and app management. The role of the tokenization system generates and validates the tokens and serves as a safe token storage while the digital issuance system takes care of the cloud-based account for the card, the respective keys, card provisioning and replenishment. The task of the app management system is to manage the mobile endpoints and secure card data transfer to devices. Finally, the role of the HCE client is to protect the card-related data on the device.

It should be noted that the HCE and the SE, even with tokenization, are not mutually exclusive solutions but can be used jointly to create both flexible user experience and achieve a high level of security.

5.3 E-commerce

The markets for secure payment via mobile devices are expected to grow. Nevertheless, the current market looks somewhat fragmented, and the competition is fierce. This can be noted, e.g., from the high expectations of payment solution and certification provider SoftCard which has, regardless of the active support from the key MNOs in the USA, gone bankrupt.

The following sections presents an overview of some of the currently relevant solutions. It should be noted that the mobile payment area is highly volatile at the moment so the mentioned companies may equally increase or lower their role, and completely new players may appear in addition to the ones mentioned below.

5.3.1 EMV

The EMV Integrated Circuit Card Specifications for Payment Systems are global payment industry definitions that describe the requirements for interoperability between chip-based consumer payment applications and acceptance terminals in order to enable NFC payment transactions. These specifications are managed by EMVCo.

Named after the original organizations that created the specification, Europay, MasterCard and Visa, the EMV specifications were first published in 1996. According to the statistics of EMVCo, 1 billion active EMV chip cards were used for credit and debit payments in 2010, and 15.4 million EMV-acceptance terminals deployed around the world.

5.3.2 Google Wallet

The Google Wallet mobile application has been designed to store credit cards and offers of users on their phone hardware, thus it relies on eSE. When shopping at stores that accept Google Wallet, users are able to pay and redeem offers with the same device by tapping the phone at the NFC POS terminal [32].

Google has coordinated partnerships, e.g., with MasterCard, Citi, Sprint and First Data. Google Wallet's first version works with its Nexus S smartphone equipped with an NFC-embedded chip, combined with MasterCard PayPass terminals. Google Wallet has been gradually upgraded, and it also supports HCE as of the Android OS version 4.4.

As an example of a forerunner in this field, Google Wallet faced some issues in the early phase as can be interpreted, e.g., from Ref. [33]. According to the source, the PIN for the payment could be revealed by the subject-matter experts. As precautions, users were advised to refrain from rooting the phone, enable the lock screen, disable the USB debugging, enable full disk encryption and keep the handset SW up to date. It was noted later that the potential issue related solely to rooted devices.

5.3.3   Visa

In February 2012, Visa announced a mobile payment solution to compete with both Google Wallet and at that time the still existing SoftCard payment system. Visa's solution is based on the Visa-certified NFC-equipped smartphone that consumers can use to contact the company and activate the handset for mobile payments. In the solution, the device is linked securely with a user's bank account. This provides the mobile payments in those locations where Visa's payWave system is accepted. As is the case in the traditional secure provisioning of payment cards, Visa has extended the idea for mobile technology to securely provision mobile payment accounts OTA [34].

5.3.4   American Express

According to Ref. [35], American Express (AmEx) has a strategy in extending its proprietary payment network to online, mobile and NFC-based proximity payments space. AmEx, through its Serve platform similar to PayPal, aspires to be broader with a complete solution that integrates mobile payments, loyalty programmes and other social and connected services. AmEx has signed up Sprint and Verizon and its partnership with Payfone allows customers of both Sprint and Verizon to pay using their mobile phone number. The Serve digital wallet service is accepted by many merchants offering AmEx.

5.3.5   Square

According to Ref. [35], Square allows credit cards to be transacted via a mobile phone equipped with a Square reader. As a potential disruptor in the POS market, Square started off at the low end, creating its own market and moving up market to eventually dethrone traditional POS terminals vendors. Nevertheless, Square has no NFC presence so far.

5.3.6   Other Bank Initiatives

Some other bank initiatives are also active in the markets, as summarized in Ref. [35]. Concretely, Bank of America, Wells Fargo and Chase have formed a venture to enable P2P payments for their customers. ClearXChange allows customers to send money to each other without needing to open a separate ClearXChange account. These banks partnered with DeviceFidelity and Visa to use its In2Pay microSD solution to run NFC payment trials across Visa's payWave platform. These trials indicate that financial institutions are testing the mobile payment schemes with their own mobile payment applications.

5.3.7   Apple Pay

The Apple Pay concept was announced in 2014 to support in-app purchases and NFC payments at contactless POS terminals which consist of PayWave of Visa, PayPass of MasterCard, and ExpressPay of American Express, and it currently supports a wide variety of credit and debit card banks in the USA. It is an eSE-based mobile payment wallet service, and applicable for Apple devices of which some models support NFC. Apple Pay is a competitor for previous solutions that other retailers have deployed or planned for mobile payments services, like PayPal, Wal-Mart, Target, Google Wallet, and the obsolete SoftCard.

Apple Pay relies on single-use tokens generated by the eSE that replace the transfer of personal debit or credit card number information of the purchaser, combined with two-factor authentication which refers to fingerprint or button pressing (depending on the device type), as well as on the SE. The financial information of the SE is accessed upon the generation of a dynamic, randomized 16-digit number during the transaction. The SE is tamper-resistant and blocks in the case of physical attack intentions.

Apple Pay was initiated with the support of US-issued payments cards, then UK-issued payment cards and now one-factor authentication support is expanding in the international environment. Apple Pay does not relate to chip or PIN/EMV cards, but it is meant to replace credit cards, the device account number and a dynamic security code serving as bank card authorization [7].

5.3.8 Samsung Pay

Samsung Pay was announced in 2015, and is initiated in the US market. The service is supported by Samsung 6 and beyond. It only works on non-EMV terminals.

5.3.9 MCX

Merchant Customer Exchange (MCX) has been conducting trials of its CurrentC payment system.

5.3.10 Comparison of Wallet Solutions

As can be noted from the previous sections, the mobile payment environment is currently highly fragmented and competitive. Some solutions struggle, some have disappeared from the commercial markets while others have increase their market share.

The mobile payment can be based on NFC or other solutions like cellular connectivity and tokenization with cloud services. Payment can also be based on different variants of SE type, i.e., UICC, eUICC or microSD. Furthermore, there may be cloud-based solutions relying on, e.g., HCE. Also the tokenization can take place and be combined with many payment schemes with or without SE (the latter providing the highest security level).

In other words, there are many moving parts at the present, from which some work is based on standalone solutions and some on combined SE and SW solutions, while some combinations do not yet exist but are definitely possible (like UICC with tokenization). From trial and error, as the SoftCard initiative has demonstrated, it seems that the essential part of the game is fluent user experience as well as minimal complexity between the infrastructure and service providers. Figure 5.7 summarizes some of the involved items from which the participating entities should select the most optimal one – the task is far from easy.

The following arguments summarize the environment:

- SE provides a high level of security for the payment procedures. In the mobile payment environment, the question is who owns the SE as it also dictates who has overall control of the service.
- HCE removes the need for SE but it results in challenges in securely storing the payment information. Thus, HCE can be combined with other solutions giving a higher level of

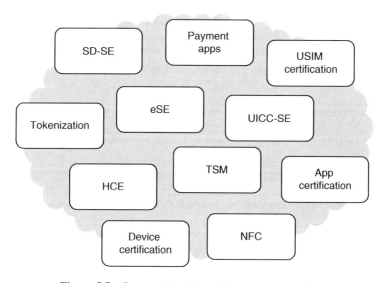

Figure 5.7 Some options for mobile payment solutions

security, such as tokenization and/or SE (whether UICC or SD based or embedded version). The hybrid model of HCE and SE would provide a highly secure environment.

- UICC-based SE is typically owned by an MNO who also owns the UICC and thus has control over the business conditions.
- eSE typically lowers the importance of the MNO in the complete picture. The eSE could be owned by the device manufacturer who integrates it to the device during the manufacturing process, although it could also be owned by the chipset manufacturer or the eUICC profile personalization provider. As an example of the current wireless payment environment, in the case of Apple Pay, the eSE is owned by Apple.
- There is also an alternative if NFC is not utilized, the Magnetic Secure Transmission (MST) transaction as Samsung calls it, referring to LoopPay. It basically emulates in a wireless way the magnetic stripe swipe procedure of a credit/debit card at the POS reader.

More information about the comparison of the currently most relevant mobile payment solutions can be found in Ref. [6].

5.4 Transport

There are various systems designed for the transport environment. The solutions are usually based on physical smartcard technology (microcontroller-based or memory cards) or RFID tags for the payment and access to vehicles [5]. The respective, most popular transportation ticketing standards at present are MiFare [1], CiPurse [2], Calypso [3] and FeliCa [4].

In general, as for the contactless transportation and ticketing systems, they have been based on highly localized solutions. This means that the systems have become fragmented, incompatible and only work normally at the city level without wider interoperability. Nevertheless, driven recently by governments, the trend indicates the favouring of more standardized solutions that would result in synergies between national transportation operators via the deployment of interoperable transportation network access cards. Therefore the credentials, once obtained via any operator, could be consumed within other participating networks as well as transport types. This is beneficial for defragmenting the national markets, but as Ref. [5] argues, it may have an impact on international common standardization. This can be seen from the different key standards which are discussed in the following sections.

5.4.1 MiFare

The GSMA NFC UICC requirements document [31] details the information for MiFare together with the mobile support. It specifies that the UICC may support MiFare implementation reachable through the MiFare Java Card Host Interface API. In this case, MiFare for the Mobile v2 application framework is required to manage it via OTA. For more information, MiFare for Mobile v2.1 specifications are available in Ref. [28]. Furthermore, the GlobalPlatform Secure Channel Protocol 03 shall be supported as presented in Ref. [29].

There are also competing solutions offered in the commercial markets which may indicate the fragmentation of the transportation ticketing and thus challenges in the wider interoperability. Nevertheless, according to Ref. [16], MiFare is expected to remain in pole position regardless of the advances of its competing alternatives, CiPurse and Calypso.

5.4.2 CiPurse

CiPurse is an alternative to MiFare defined by the Open Standard for Public Transport (OSPT) Alliance. CiPurse refers to an open ticketing standard that enables technology suppliers to develop and deliver interoperable transit fare collection solutions for cards, stickers, fobs, NFC mobile phones and other consumer devices. The cards provide a user-friendly way to pay for services such as taxi rides or healthcare. A related certification process ensures compatibility of CiPurse products from different suppliers.

Singapore's Land Transport Authority is one of the first parties that has used CiPurse adaptation [13]. Ref. [18] informs about the first CiPurse-compliant contactless cards that were issued in Brazil in 2013. Manufactured by Giesecke & Devrient, the cards are based on a CiPurse-compatible contactless security controller from Infineon Technologies. The CiPurse system has been designed to support flexible and contactless transport and ticketing systems from the start; it also allows the combination of identity and payment functions in mobile devices or multiple application cards. In this specific project, the local system integrator Rede Protege could move its applications to a higher security level and thus increased flexibility and performance using the existing reader infrastructure.

CiPurse cards allow the combination of multiple transport and ticketing applications as well as identity or payment functions on a single card. CiPurse also supports AES 128 encryption algorithm for fast and secure transactions.

5.4.3 Calypso

The GSMA NFC UICC document [31] defines that for Calypso support, the UICC will support both ISO 144443 type A and B specifications. If a Calypso-based OTA-downloaded application is present, it needs to comply with the Calypso 3.1 specification. The respective documents are Refs. [14,15].

5.4.4   FeliCa

FeliCa is a solution offered by Sony for mobile payment in different use cases. Like its competitors, it is based on an IC card that is shown over a reader/writer activating the data transmission to rewrite the data to the card. According to Ref. [17], the FeliCa card is suitable for high volumes of transactions and contains a security system. The FeliCa system has achieved ISO/IEC 15408 EAL4/EAL4+ security level, making it a suitable solution for protecting the card balance, e-money information and personal authentication against malicious attacks. FeliCa can be adapted into a wide variety of environments such as ticketing systems for public transportations, e-money and residence door keys.

5.5   Other Secure Systems

5.5.1   Mobile ID

When the customer utilizes the Internet and logs onto websites, there are typically many kinds of authentication data involved such as usernames and passwords which need to be typed several times. The user cannot ensure their personal data is safe via these methods. The aim of the Mobile ID solution is to simplify online authentication processes by only requiring the user to provide their details once. This data is stored on a protected server hosted by a service provider, making the mobile device the entry point to the web [19].

Mobile Connect is an initiative and standard of GSMA that aims to provide an interoperable and universal login service for everybody worldwide. It eases the global deployment of federated identity via Software-as-a-Service (SaaS) mode. Mobile ID also provides scalable identity management and authentication services for MNOs to act in a digital identity provider role and to set up the Mobile Connect services such as Identity Federation enabling Single Sign On service, onboarding portal services and end-user permission-based information-sharing service, and other services complying with level of assurance up to 4 [20].

5.5.2   Personal Identity Verification

Personal Identity Verification (PIV) refers to an ID chip card issued by a US federal agency. It is able to securely receive, store, recall and send information. The card encrypts data to provide secure communications between the services and the card, while using a common technical and administrative process. The encryption is based on the PKI which complies with federal policies, and is the accepted method of the Global Business Standard for Internet Security. The PKI also provides the procedure for digital signatures to ensure document authenticity. The PIV card thus encrypts data and verifies identity to ensure the confidentiality (only the cardholder can access the data), integrity (only the cardholder may modify the data), authenticity (the origin of the data is guaranteed) and non-repudiation (leaving no room for falsified data) [21].

5.5.3 Access Systems

Access systems can be based on contact or contactless smartcards as defined in the ISO/EIC 7816 and ISO/EIC 14443 standards. Other methods such as RFID can be applied. The respective card types can thus be utilized for *physically* opening doors and *logically* accessing data. There are no restrictions in basing the access to any other wireless means such as mobile communications. In fact, there have been practical solutions in the very early days of GSM for opening garage doors via a cellular call as the A-subscriber's MSISDN number indicates the legitimate user. SMS can also be utilized in accessing locations. However, these solutions may be vulnerable to security breaches as the A-number could be falsified.

As for the related systems and services, basically all the methods used in transport and described earlier in this chapter can be used in the secure access systems.

References

[1] Mifare home page, 5 September 2015. http://www.mifare.net/en/ (accessed 26 December 2015).

[2] CiPurse home page, 26 December 2015. http://www.osptalliance.org/the_standard (accessed 26 December 2015).

[3] Calypso home page, 26 December 2015. https://www.calypsonet-asso.org/ (accessed 26 December 2015).

[4] FeliCa home page, 26 December 2015. http://www.sony.net/Products/felica/about/ (accessed 26 December 2015).

[5] Summary of transportation ticketing systems, ABI Research, 5 September 2015. https://www.abiresearch.com/market-research/product/1013689-transportation-ticketing-standards-cipurse/ (accessed 5 September 2015).

[6] NFC mobile payments: An industry snapshot. Mobey Forum's HCE workgroup. Mobey Forum, May 26, 2015.

[7] Apple Pay explained. Cnet, 7 September 2015. http://www.cnet.com/news/everything-you-want-to-know-about-apple-pay/ (accessed 9 September 2015).

[8] Beyond tokenization. Ensuring secure mobile payments using dynamic issuance with on-device security and management. Sequent, 29 May 2015.

[9] EyeVerify. http://www.paymentscardsandmobile.com/first-internet-bank-uses-eyeprint-id-biometrics-for-app/ (accessed 23 November 2015).

[10] Samsung and Gemalto provide m-pay. http://www.mobileworldlive.com/money/news-money/samsung-gemalto-join-forces-for-m-pay-launch-in-europe/ (accessed 23 November 2015).

[11] Dual interface for EMV debit card. http://www.pymnts.com/news/2015/new-dual-interface-emv-debit-card-debuts/#.Ve5zpH28ohx (accessed 23 November 2015).

[12] EyeVerify. http://www.eyeverify.com (accessed 23 November 2015).

[13] CiPurse. http://www.nfcworld.com/2010/12/16/35479/ospt-alliance-debuts-cipurse-open-alternative-to-mifare/ (accessed 29 November 2015).

[14] Ref.060708 – CalypsoAppli 'Calypso Specification REV.3 – Portable Object Application' version 3.1, 10 March 2009.

[15] Ref.090316 – MU-CalypsoR3Amd1 'Calypso Specification REV.3 – Amendment 1 to Version 3.1, version 1.0, 1 June 2010.

[16] ABI Research, MiFare. https://www.abiresearch.com/market-research/product/1013689-transportation-ticketing-standards-cipurse/ (accessed 29 November 2015).

[17] FeliCa. http://www.sony.net/Products/felica/about/ (accessed 29 November 2015).

[18] CiPurse deployment; First CiPurse-compliant contactless cards issued in Brazil, 18 November 2013. http://www.finextra.com/news/announcement.aspx?pressreleaseid=52783 (accessed 29 November 2015).

[19] Giesecke & Devrient. Mobile ID. http://www.gi-de.com/en/products_and_solutions/solutions/mobile_authentication/mobile_operators_1/mobile_operators.jsp (accessed 29 November 2015).

[20] Gemalto. Mobile ID. http://www.gemalto.com/mobile/id-security/mobile-id (accessed 29 November 2015).

[21] PIV card. http://www.va.gov/PIVPROJECT/piv_card.asp (accessed 29 November 2015).

[22] RFID readers, examples. https://www.rfidjournal.com/purchase-access?type=Article&id=12470&r=%2Farticles%2Fview%3F12470 (accessed 29 November 2015).

[23] RFID devices in enterprise. https://www.zebra.com/us/en/products/rfid/rfid-handhelds.html (accessed 29 November 2015).

[24] NFC Forum. http://nfc-forum.org/about-us/the-nfc-forum/ (accessed 29 November 2015).

[25] Wolfgang Decker. Mobile Security – Securing Mobile Life. Giesecke & Devrient, January 2014.

[26] RFID and NFC comparison. http://blog.atlasrfidstore.com/near-field-communication-infographic/ (accessed 22 December 2015).

[27] NFC Secure Element Stepping Stones, version 1.0. SIMalliance, July 2013p.

[28] MIFARE mobile specifications. http://mifare4mobile.org, 22 December 2015 (accessed 22 December 2015).

[29] GlobalPlatform Secure Channel Protocol 03, Card Specification v2.2 – Amendment D version 1.1

[30] QR code generator. https://www.the-qrcode-generator.com/, 26 December 2015 (accessed 26 December 2015).

[31] SGP.03 NFC UICC Requirements Specification v6.0. GSMA, September 30, 2015.

[32] Description of Google Wallet. http://www.google.com/wallet/what-is-google-wallet.html (accessed 31 December 2015).

[33] Google Wallet vulnerable to 'brute-force' PIN attacks (update: affects rooted devices). http://www.engadget.com/2012/02/09/google-wallet-open-to-pin-attacks/ (link reviewed 27 June 2012).

[34] Announcement of Visa's mobile payment. http://www.bgr.com/2012/02/27/visa-announces-new-mobile-payment-solution/ (link reviewed 27 June 2012).

[35] Mobile Payments – A study of the emerging payments ecosystem and its inhabitants while building a business case. https://www.ftc.gov/sites/default/files/documents/public_comments/ftc-host-workshop-mobile-payments-and-their-impact-consumers-project-no.124808-561018-00013%C2%A0/561018-00013-82732.pdf (accessed 31 December 2015).

[36] Information Technology; Automatic identification and data capture techniques; Bar code symbology; QR code. ISO/IEC.

[37] QR code error coding. http://www.qrcode.com/en/about/error_correction.html (accessed 8 January 2016).

[38] Smart Card Talk. Quarterly newsletter, Smart Card Alliance, November 2014.

6

Wireless Security Platforms and Functionality

6.1 Overview

This chapter summarizes key aspects of wireless security platforms and their functionality. First, a summary of theories is presented along with justification why each specific security mechanism is utilized. It also discusses the impacts of not deploying them in the network, SIM and other stakeholders. The SEs have been described earlier in this book, but they form an integral part of the secure platforms – both HW- and SW-based options. The role of these are thus discussed further in this chapter.

The overall secure protocols for signalling, data transfer and SIM management via, e.g., SMS and the BIP, are discussed together with the role of the SIM, UICC and eUICC. This chapter presents typical OTA remote techniques for subscription management, including SIM OTA (for initiation of the subscription, subscription lifetime management, RFM, RAM, subscription management for physical users and the M2M environment), as well as the TEE, cloud and HCE. Tokenization as a basis for contactless payment is also addressed.

6.2 Forming the Base

The UICC is still one of the most secure means for authentication, authorization and encrypting the radio interface. Nevertheless, there are many available alternatives, each fitting into their optimal environment. These solutions include TEE, HCE together with the cloud concept, and tokenization, especially for mobile payments. The question is, how each stakeholder plays which part in the respective ecosystems.

The traditional way was based on the SIM card which is provisioned for mobile communications users by the MNO. Today, there is a growing need for equipment such as wearables that are based on the eSE which cannot be removed and added to other devices by the user, therefore there is a need to be able to change the MNO and service provider and utilize the very same HW component that previously informed the SIM.

Wireless Communications Security: Solutions for the Internet of Things, First Edition. Jyrki T. J. Penttinen.
© 2017 John Wiley & Sons, Ltd. Published 2017 by John Wiley & Sons, Ltd.

One way to tackle this challenge is to deploy advanced subscription management mechanisms which provide interoperability between the stakeholders, including SIM card providers, MNOs, OEMs and Original Device Manufacturers (ODMs). Numerous international entities are contributing to the common standards that would allow wide interoperability between the stakeholders. However, there are many other, alternative or supporting solutions that have been developed and deployed in commercial markets. These include the TEE, HCE with the cloud concept, and tokenization. Advanced mechanisms have also been developed for the trusted chain, including more thorough utilization of CAs.

6.2.1 Secure Service Platforms

The 'traditional' OTA platform offered the means for MNOs to provision new subscriptions and remotely manage the SIM card contents, e.g., updating applets. This concept is developing further to provide advanced functionalities. One example of such development is the Allynis Advanced Over-the-Air (AOTA) SaaS platform. Sprint uses this platform to facilitate LTE service activations and to manage the complexities of providing multi-band 4G LTE connectivity [1]. There are various other solutions in the market, including solutions based on the SmartTrust Delivery Platform [9].

6.2.2 SEs

SEs have an integral role in the most secure connections, provisioning, authentication and encryption of the communications. SEs have many different physical forms such as the traditional SIM/UICC with Form Factors of 2FF, 3FF and 4FF. In addition, there is a growing number of eSEs adapted from the M2M Form Factor MFF2, which so far is the only internationally standardized variant. It is expected that there will be a growing need for physically much smaller embedded elements along with the rapidly increasing number of IoT devices and wearables such as smart watches. More detailed information about SEs can be found in Chapter 4.

The SE that functions especially with NFC forms an important part of the NFC device. The SE in this case may be the UICC, eSE or external microSD. The environment also contains the NFC controller, NFC chip, and protocol stack for the communications between the NFC element and baseband processor of the device. NFC can be used as a very logical base for the payment platforms for mobile wallet solutions. This also requires fluent functioning and an intuitive user interface for the respective applications. The communication protocols and interfaces related to the NFC environment include ISO 7816, ISO 14443, SWP, UART, I²C and SPI.

The SE requires a respective OS which is typically capable of handling Java applets. This ensures the interactions between multiple systems of MNOs, TSMs, data preparation systems and POS transactions. As an example, GlobalPlatform, SIMalliance and ETSI are working on the SE interoperability for advanced platforms via international specifications.

The communications of the SEs take place via secure channels (between card and remote servers) and APDUs transmitted between the card and external world (card reader of the device). The benefit of modern SIM/UICC is that it can house multiple applications and security domains. Thus, it is possible to utilize various payment schemes, such as mobile wallets of different providers, as well as access and transit services and any other wireless solutions that rely on the SIM/UICC application/applet. The SIM/UICC security is useful for many platforms.

6.3 Remote Subscription Management

The most flexible method for updating a user's subscription-related data, such as SMS and MMS settings, is based on OTA methods, which can be called with the general term 'SIM OTA'. The SIM OTA also works for the initiation of the subscription, remote management of files and applications, and for the overall subscription lifetime management.

6.3.1 SIM as a Basis for OTA

Being a tamper-resistant HW element, the SIM/UICC functions well for the mobile communications environment. It provides user authentication, authorization, and ciphering for the users' connections. In addition to the basic functionalities, the SIM/UICC also works as a platform for adding applets that enrich the user experience. Nevertheless, the SIM/UICC also has some limitations such as the communications speed between the card and respective card reader (mobile equipment). For that reason, it might not be an optimal solution for environments where many fast transactions are needed, especially if the communications are based on the short message bearer. With more modern methods such as CAT/BIP, the communications speed of the SIM/UICC clearly increases.

The applications executed via the mobile device's OS provide fluent user experiences but they lack the highest security regardless of virus defences and other typical means for protecting the device. Thus, novel methods have been developed for coping with the security issues while providing fluent execution of the code. These methods include technologies such as TEE that relies on the processor HW, and HCE that may provide a completely HW-free environment. For the highest protection level with these new methods, the combination of the SIM/UICC is still valid.

The SIM/UICC is based on ISO/IEC 7816 smartcard definitions. Along with the development of the mobile system generations, the subscriber card has also evolved, supporting more advanced protection mechanisms, a better, more suitable Java Card for interoperable Java applets and additional services such as NFC. As the mobile communications' value-added services have been increased, the SIM/UICC has also been following the development and is capable of providing secure solutions, e.g., in automotive and banking environments.

As the user device types are evolving, one example is the growing popularity of wearables, the 'traditional' Form Factors of the cards face new challenges; the original Form Factors 2FF, 3FF and 4FF are simply not small enough to fit into miniature devices. This has resulted in the need for embedded Form Factors. There are also other variants that are still chip-set vendor-specific for the pin layout and size. The benefit of the embedded SIM/UICC is the considerable space savings, but new methods are needed for the OTA subscription management as they are no longer removable, unlike the previous 2FF, 3FF and 4FF.

The strong security of the SIM/UICC is based on the message types between the card/element and the reader device (mobile equipment). There is no direct I/O communication, instead the communication takes place via APDU messages as described in section 4.11.1. This makes the SIM/UICC comparable to a standalone 'computer' with its own memory elements, processors and messaging protocols with the external world. The main bottleneck of the SIM/UICC is related to the communications speed between the card reader and card. This might be a limitation if an embedded applet is utilized for real-time video contents display with very frequent key derivation.

The SIM/UICC, whether it is of traditional removable form or eSE, is resistant to security attacks, and is suitable for basic protection as well as any additional services the card can support. It is especially suitable for protecting payment transactions as the relevant information such as keys can be stored in a secure way.

Figures 6.1 and 6.2 depict examples of mobile payments based on NFC and the SIM/UICC or eSE. The mobile user, who has installed a payment application (typically referred to as a

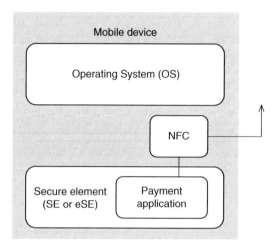

Figure 6.1 An example of the utilization of the UICC or eUICC as a part of the mobile payment service

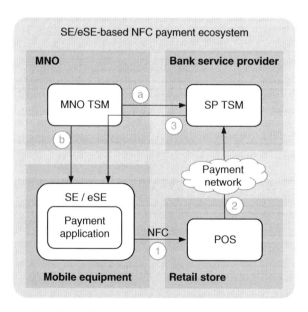

Figure 6.2 The NFC payment architecture based on the SE or eSE

'mobile wallet'), triggers the payment event (1) by tapping the wireless NFC reader located at the POS of, e.g., a retail store. The payment request (2) is sent to the payment service provider (bank) which has its own SP TSM. The TSM communicates with the secure mobile application (3) residing in the SE (SIM/UICC) of the MNO. In order for this ecosystem to be functional, the service needs to be first initiated, i.e., provisioned for the end-user by the bank and MNO (a, b). After the service has been initialized, the TSM of the SP is able to manage the funds of the payment app within the SE. In this mode, the MNO 'leases' SIM/UICC space for the bank's app, which is thus one of the 'tenants' of the card, now being able to communicate with the SIM/UICC in order to accept the payment transactions of the respective end-user.

6.3.2   TSM

TSM represents a 'trusted third party'. Its role is to bring the participating service providers together for the provisioning and lifecycle management of secure payment, access, transit and other transactions based on the credentials related to the respective app within and communications with the SE. Some commercial TSM providers include Giesecke & Devrient and Gemalto.

The TSM functionality includes the provision and deletion of services, key management, data preparation and post-issuance lifecycle management OTA. Different TSM models include TSM for the MNO (communications with the MNO's SE), and TSM for Service Providers (SPs). The overall model for the TSM ecosystem is presented in Chapter 3.

For the TSM deployment, there are different models, which come in simple, delegated and authorized mode (see Figure 6.3). In the *simple mode*, the MNO performs the card content management and TSM can monitor it. In the *delegated mode*, the card content management is delegated to a TSM via prior authorization. Finally, in the *authorized mode*, the card content management is fully delegated to a TSM.

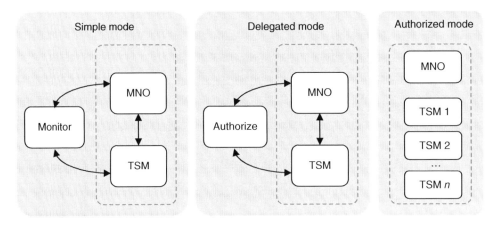

Figure 6.3   Examples of the TSM models

6.3.3   TEE

The TEE is located outside of the SIM/UICC but is still based on the HW within the user device. The idea of the TEE is to divide the processing functions of the mobile equipment into normal and secure domains, or normal and secure worlds. The Normal World (NWd) refers to the Rich Execution Environment (REE) while the Secure World (SWd) refers to the TEE. In addition to the TEE, the SWd has a set of containers that are isolated from the external world. They are similar to the secure domains of the NWd. The containers store trusted applications, or trustlets, which are similar to the applications of the NWd. Figure 6.4 depicts the joint architecture of the REE and TEE.

The OS of the NWd is called Rich OS, and it offers a variety of user experiences along with commercial OS variants such as Android and Windows and respective applications. The Rich OS is open for the third-party SW and, regardless of the versatile protection mechanisms, it is vulnerable to security risks such as viruses. The TEE, in turn, is a protected environment of the mobile device that is based on both SW and HW residing within the processor. The TEE protects against software security attacks that may happen via the NWd. The TEE participates in the access and isolates the trustlets from the NWd OS [3–5].

One example of the benefits of the TEE is the service provider's audio/video content display (such as a chargeable premium sports event) on the mobile device during a limited time period in such a way that the data stream cannot be replicated or copied to other devices or users. This contents delivery via TEE can be based on basically any bearer like PTP packet data or eMBMS.

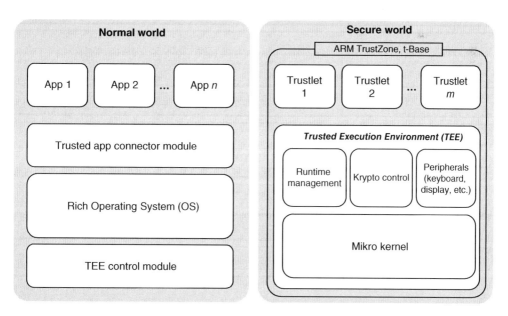

Figure 6.4   An example of the TEE architecture based on ARM TrustZone t-Base. The TEE is connected to the external world via communications protocols designed between the TEE and REE which provide the means for the safe execution of the trustlets

The TEE thus provides a protection mechanism for the apps execution as they are completely isolated in the processor chip level. The respective level of security is close to that provided by the SIM/UICC, extending the use cases for protecting the device functions such as display, keyboard and microphone. As a highly concrete example, an app that is executed via TEE can be protected in such a way that the keyboard strokes cannot be recorded by (potentially harmful) background app when typing confidential information such as PIN codes.

TEE is a protected part of the OS within the processor HW that can used independently from the architectures. Previously, there were equipment vendor-dependent TEE solutions but nowadays the TEE is standardized by the GlobalPlatform. Currently, the most popular practical solutions are based on Advanced Reduced instruction set computer Machine concept (ARM.) An example of a commercial TEE solution is Trustonic t-base which is located to the application processor. The trusted applications of the TEE (trustlet, or trusted applet) can be deployed and managed via a separate TEE–TSM element as long as the user's device supports TEE.

The TEE is applicable to many practical situations requiring data integrity, privacy and confidentiality. The current mobile communications environment has changed rapidly as a result of smart devices from the beginning of 2000. At the same time, the previously quite closed telecommunications services have been liberated and there is ever growing quantity of third-party apps. Also, the ports to the public data networks have opened doors for unpredictable security breaches. This has resulted in dangers such as installation of malicious software into the mobile devices that are capable of executing apps. The same protection mechanisms can be used in the mobile environment that have been applied in fixed Internet, but it is obvious that updating of, e.g. virus protection software, is not always able to respond early enough, especially in the case of zero-vulnerability cases.

The TEE is one of the solutions tackling these challenges. Optimal benefit is achieved via the commonly interoperable, standardized solutions, which means cooperation between the OEMs, MNOs and chip manufacturers. The standardized TEE provides a high level of protection in the MNO's services to the customers via the service providers, which in turn increases customer satisfaction. The contents providers can also benefit from the standardized TEE because it provides certified protection for products (such as video streaming) as well as maintenance independently from the utilized platforms, which in turn speeds up the entrance to market and unifies the user experience.

As a base for the TEE architecture, the following summarizes the key terminology:

- System on Chip (SoC) refers to the IC, CPU and any peripherals involved with TEE. The CPU in turn may contain apps and a baseband unit, i.e. modem.
- Silicon Provider (SiP) refers to the chip manufacturer. The task of SiP is to manufacture and provide SoC. Examples of SiP are Qualcomm, Intel, Marvell and CGT.
- Trusted Zone (TZ) refers to chip-set architecture of, e.g., ARM or any other TEE manufacturer.
- Normal World (NWd) refers to the area which contains Rich OS and related applications.
- Secure World (SWd) refers to isolated compartments, e.g., within ARM TrustZone-enabled SoC.
- Trusted Execution Environment (TEE) refers to the secure microkernel within SWd.
- Rich OS refers to the operating system working within NWd. Examples of Rich OS are Android and Linux.

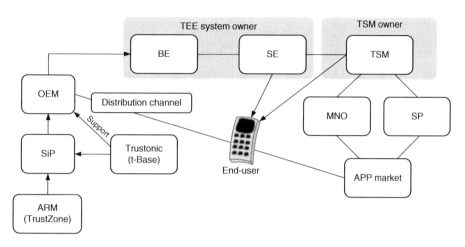

Figure 6.5   An example of the t-Base ecosystem

- Container refers to the secure area within SWd which is similar to the security domain.
- Trustlet is an executable unit, or trustlet applet, within Container which is used for secure operations.

Following the example of these terms specific to the t-base TEE ecosystem, the architecture contains elements as shown in Figure 6.5.

As can be observed from Figure 6.5, the heart of the TEE architecture of this example is a combination of Trustonic and ARM TrustZone support. ARM provides IP-based TEE HW TrustZone to the SiP, which in turn has integrated the t-base within the chip-set. The task of the OEM is to manufacture the TEE-capable device and ensure it is added with respective serial number to the t-base. The task of Back End (BE) is to store device-specific authentication keys such as K, AUTH and SOC while the Service Enabler (SE) manages the unlocking of the t-base Container to be utilized by the TSM. The TSM, in turn, provides key management services for the end-user device, and communicates with the and SP. The task of the MNO is to distribute t-base enabled devices for the customers while the role of the SP is to develop and make available TEE-based applications. The SP also authorizes the Container's (i.e., device-specific) activation by relying on apps developers. Finally, after this chain of roles, the end-user is able to consume secure apps that are stored within the secure world of the device.

In order to make TEE work, the SiP partners should cover important parts of the OEM devices while ARM provides the chip architecture and SiP embeds the TEE into the silicon level. When the equipment supports TEE, the task of OEM is to initialize the TEE with a key procedure for the BE to receive a serial number key table for its database. The task of the SE is to create the actual TEE with the end-user device by creating the SP container within it. The SP can be any party requiring trustful execution of the SP's apps (trustlets), such as those meant for secure banking services. As soon as the respective Container exists, the TSM provides lifecycle management for it in such a way that Trustonic provides the trustlet APIs.

The TSM works for post-loaded, secured applets so that the SP delivers secured apps for the TSM, which delivers them further to the device's secure world Containers via personalized OTA access, and which manages the lifecycle for operations and data to the respective devices.

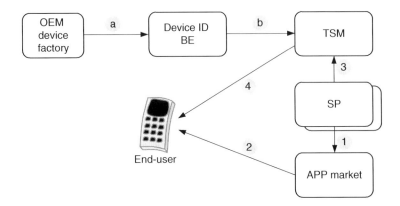

Figure 6.6   An example of the TEE secured application OTA lifecycle management

The TSM is thus a hosted service offering an adequate security level for the safe provisioning and execution of the apps in t-base trustlets.

When the user installs a new app, it triggers the trustlet process as summarized in Figure 6.6, via the TEE OTA lifecycle management of the secured apps.

Following the data flow of Figure 6.6, the initial phase of (a) makes the OEM device binding via Trustonic procedures, and delivers the key and device ID to the BE. Then, in stage (b), the key is delivered in a secured way to the TSM of the SP. The SP uploads a secure app to a typical app store (1). Now, as soon as the end-user downloads and starts to install the SP's app to his/her smart device (2), it triggers the delivery of the secured app to the TSM, which further installs the secured app to the TEE Container of the user's device via OTA (4). As soon as the installation is done, the end-user is able to start utilizing it.

In addition to the actual secure installation via OTA and utilization of the apps, TEE also provides useful functionalities adding value to the apps. Some of these functionalities include secure user interface and secure end-point.

This example is related to the ARM-based solution which represents about 95% penetration of the devices. Trustonic, on the other hand, is one of various TEE facilitators, having been initiated in 2013. At this moment, Trustonic is a joint effort of ARM and Gemalto while the former cooperating parties have been Nokia and Giesecke & Devrient, the latter providing the TEE TSM services (TSM owner). TEE product-wise, the TEE t-base 300 is a GlobalPlatform compliant solution. As for the commercial markets, Qualcomm, Samsung LSI and Broadcom have incorporated TEE in some chip-sets (application processors), and Samsung has released various devices equipped with TEE support. More information about TEE can be found in Ref. [2].

6.3.4   HCE and the Cloud

The basic idea of the cloud services is to distribute the mobile services away from the actual physical device so that the user may rely on the remote entity for executing the software and for storing contents. The term cloud refers to such remote functionalities supported by the network and its connectivity. Ideally, the benefit of the cloud service is the independency from the used device for utilizing any service and for storing data. Furthermore, once the execution of the code residing in the cloud has been initiated, it can be continued with another device, and the locally stored information can be transferred between the devices. In addition, cloud services can be used for a variety of environment, including mobile payment [6].

One of the practical solutions based on the cloud concept is HCE as depicted in Figure 6.7. HCE is especially suitable for small-scale payments in the format of a mobile wallet which are based on dynamically changing data objects, i.e., tokens, which serve for acknowledging the payment event. In practice, a token may be, e.g., a reference to the Personal Account Number (PAN) which is valid either for a single payment event, for a series of payments or during a certain, limited time period. Please note that the validity of such tokens may be set for several years. Regardless of the validity period, the idea of the token is to protect the original, permanent PAN of the user.

The token-based payment event requires a method for the digital issuance as well as for application management and an HCE element that resides either in the cloud or within the device. The HCE system takes care of the forming of the token, as well as its storage and integrity checking. The distribution system manages the virtual payment account residing in the cloud as well as the related keys, and offers business logics for the initiation of the virtual card. Application management, in turn, takes care of the secure transport of the information and utilization between each mobile device and HCE system while the SW of HCE client protects the data within the virtual card.

The challenge of the cloud services is that by default they are not based on an HW-based SE but on an SW which is either a separate application or is embedded into the SW or OS of the device. Thus, the security level of such a solution might be lower compared to, e.g., the SIM-based SE. Another challenge is related to the dynamic character of the token. If the token is selected on a one-time basis, for each new purchase the user requires a new token each time.

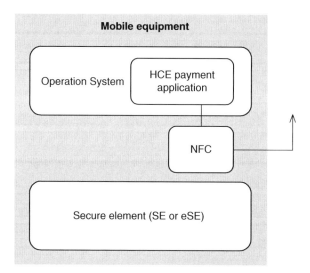

Figure 6.7   The payment application of the cloud service can be, in its basic form, within the SW-based OS located outside of the SE

If the user happens to be in an outage area of the mobile services, the new transaction would not be valid unless there is a locally pre-stored set of new tokens within the device.

In principle, the more tokens stored beforehand in the device increases the risk of potential cyber-attacks against that specific account. In other words, the sole token does not protect the payment transaction per se because if it is exposed, someone may be able to misuse it. The HCE services are thus critical for the protection against external intentions to capture and modify tokens. It is of utmost importance that HCE identifies correctly the device, end-user and the service. The virtual card stored into the device, as well as the related cryptographic functions, must be protected in such a way that they cannot be copied or changed physically from the device or via the communication link. HCE is thus especially suitable for small-scale one-time payments which do not include large risks even if the payment credentials were copied and misused.

The system architecture of the cloud-based payment service may be, e.g., as depicted in Figure 6.8. As can be noted, the role of the MNO can be removed completely due to the fact that the SIM card that MNO manages is not necessarily needed in this solution, nor are the TSM elements. The solutions thus simplify the payment service that is based on the SE. In the HCE service, a set of tokens are pre-loaded from the cloud server into the device for the acceptance of the forthcoming transactions. In the HCE solution, instead, a set of tokens (one or more) can be downloaded from the cloud server into the NFC equipment for the payment transactions. In addition to the principle shown in Figure 6.8, there can also be an SE based on the HW element such as the SIM/UICC or eSE. Commercial-wise, Android supports HCE as of the version 4.4 (KitKat).

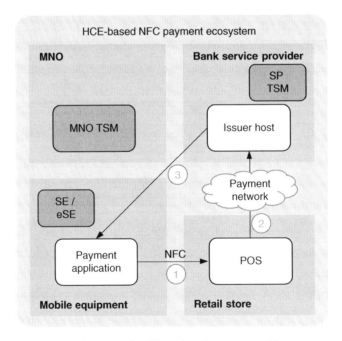

Figure 6.8 Example of HCE-based payment architecture

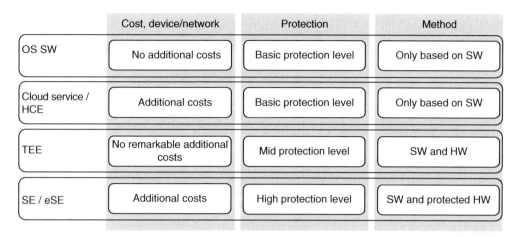

	Cost, device/network	Protection	Method
OS SW	No additional costs	Basic protection level	Only based on SW
Cloud service / HCE	Additional costs	Basic protection level	Only based on SW
TEE	No remarkable additional costs	Mid protection level	SW and HW
SE / eSE	Additional costs	High protection level	SW and protected HW

Figure 6.9   Comparison of selected protection mechanisms

6.3.5   Comparison

Each of the methods presented previously can be applied in their optimal environments. The SIM/UICC and its developed variants, permanently installed eSEs integrated into a microSD provide a good level of protection. Figure 6.9 and Table 6.1 summarize the usability of each solution.

6.4   Tokenization

6.4.1   PAN Protection

Tokenization refers to a concept that represents a high value with the help of something with low value, as is the case with casino chips. In wireless payment, tokenization is a functional solution for representing the PAN in such a way that the original PAN is hidden by replacing it with a token. EMVCo has produced the payment tokenization specification for the respective framework.

More specifically, EMVCo allows the replacement of the original PAN with a PAN-type of identification. In practice, the last four digits of the original PAN are typically maintained in order to ease the identification of the customer, e.g., for merchandize return and loyalty programmes. For the purchase transactions, the POS transfers the tokenized pseudo-PAN to the issuer. In order to recover the original PAN, the issuer relies on the tokenization system which maps the original and tokenized PANs. Only the trusted tokenization system is capable of mapping the PANs.

It is possible to use tokens with all the Cardholder Verification Methods (CVMs). Nevertheless, tokenization limits the environment in such a way that it can be focused, e.g., on only a single merchant. Also, it is possible to set a maximum expiration time for the tokens. Thanks to this limitation, together with the fact that the PAN cannot be reverse-engineered,

Table 6.1 Comparison of SE, TEE and HCE

	SE/eSE	TEE	HCE
Principle	SE can take the shape of SIM/UICC card, USB stick, microSD or permanently installed eSE. It protects well against cyber-attacks and intentions for modification of the contents SE/eSE can also work together with such solutions as TEE and HCE. Suitable in all mobile payment environments requiring a high security level, e.g., via TSM architecture. Nevertheless, the required ecosystem may be too heavy for low-value payment environments due to the complexity and economic benefits	TEE is an integrated area within the microprocessor of the mobile device that provides means for storing, processing and protecting data in a reliable way. The amount of the applications may be considerably higher than in the SE/eSE environment. Can also provide means for filtering the access to the applications residing in the SE/eSE	HCE provides a simplified ecosystem for payment services as the service provider can communicate directly with the payment application residing in the mobile device. Especially suitable for low-value, small-scale mobile payments. The protection level of HCE depends on the security of the OS and cloud service. The payment keys are protected by SW and the number of user keys and their validity time can be limited to offer additional protection
Pros	Especially suitable for NFC applications that are meant for high-value payments and environments requiring high security levels such as electronic signing via application installed into the SE	The protected and trusted code execution provides end-to-end security, access control and integrity assurance. TEE can also provide a trusted user interface for the delivery of the PIN which benefits (and is a requirement of) high-value payment solutions	The architecture and functionality of HCE combined with the cloud service are simpler compared to the SE-based payment services. As an example, a bank can offer a payment service without relying or depending on the MNO which, in turn, acts merely as a bit pipe. The amount and size of applications are practically unlimited due to the nature of the cloud principles
Cons	The drawback of the SE/eSE-based payment is the complexity of the ecosystem as it includes several stakeholders and certification schemes. The amount and size of the applications are limited due to the available memory and processing power	The availability of the devices supporting TEE is still limited	HCE is especially suitable for low-value mobile payments which are not critical for potential breaches or the reliability of connectivity to the network infrastructure. For better protection, there can be additional methods and solutions utilized on top of the HCE
Standard	ETSI, 3GPP and GlobalPlatform, based on ISO/IEC 7816 ja 14443 definitions	GlobalPlatform	Vendor specific. Visa and MasterCard have been worked on the specification

this ensures a good level of security for the transactions, making the Payment Card Industry Data Security Standard (PCI-DSS) risk on merchants minimal [7]. PCI-DSS is a proprietary information security standard for organizations that handle branded credit cards from the major card schemes including Visa, MasterCard, American Express, Discover and JCB. Private label cards – those which aren't part of a major card scheme – are not included in the scope of the PCI-DSS. In addition, the tokenization specifications include risk management measures which can be based on, e.g., account activity trends. Tokens need to have a special Bank Identification Number (BIN) range to ease the identification of the pseudo-PAN to indicate that it belongs to the contactless cloud payment realm.

6.4.2 HCE and Tokenization

Tokenization can be used basically with any payment method described previously. It is especially useful with HCE. HCE is based on the card data placed in the cloud. Visa and MasterCard have jointly worked on related HCE specifications. HCE offers an alternative approach for the SE-based procedures that are based on, e.g., the SIM/UICC. It also changes the business environment from the previously MNO-centric (as the network operator owns the SIM/UICC and its containers) into an MNO-independent direction. In the former, the processes for the payment solution certificates for devices and SIM/UICC cards as well as applications may be time consuming and limits the role of service providers. Instead, the HCE solution changes the environment towards service provider-centric as it reduces the number of stakeholders and eases the integration of the services. For that reason, financial institutes are evaluating the feasibility to move the card credentials from SEs into the cloud.

The first Android devices support HCE as of the OS 4.4 (KitKat), making it possible to move business and personal data into the cloud from the mobile device's SIM/UICC SE. However, the transfer of such sensitive data in an encrypted way for all transactions is not feasible as it would lower the security as well as user experience. HCE thus includes tokenization, together with other techniques such as limited use keys, account replenishment and risk assessment scores, to ensure security by balancing the user experience when transactions are executed.

More detailed information about tokenization can be found in Refs. [7,8] as well as from the EMVCo payment tokenization specification.

6.5 Other Solutions

6.5.1 Identity Solutions

The mobile equipment is a highly useful base for many services and solutions requiring high protection levels, for both individuals as well as businesses. One example of the new business environment can be see, e.g., via the wider adaptation of the mobile equipment and tablets with cellular connectivity to replace the old POS terminals, and converting them as mobile POS terminals (mPOS). Table 6.2 presents some typical identity-related solutions as interpreted from Ref. [9].

Table 6.2   Comparison of mobile security solutions

Solution:	ID on card	ID on device	mPOS-ID	Secure cloud storage	Secure cloud messenger
Method:	Secure computer logon	Secure mobile access	Device identification	Secure storage and sharing	Secure cloud messenger
Description:	Secure identities via authentication cards, several form factors	Secure identities on the mobile via capable SIM and NW service such as SmartTrust Licentio	Converting mobile device into mPOS terminal; TEE to protect the mPOS ID	User can securely upload and share data by using secured cloud service and capable SIM	User can securely send and receive messages via capable cloud service and SIM

6.5.2   Multi-operator Environment

The MNO partnering trends include the concept of *planned partnering*, which refers to the solution forming local partners as a part of the proposition to the OEM. The practical setup of this concept include MNO alliances formed and marketed to global and regional OEMs, and M2M service providers and MVNOs seeking to expand their service offering to current and new M2M customers. The other trend is called *dynamic partnering*, which refers to project-specific partnering. This means that large global enterprises and OEM customers select their preferred MNOs in different regions. As an example, the MNO as the issuer of the SE may retain ownership of the end-user through contractual agreements with its MNO partners.

The GSMA specification that was released in December 2013 for embedded SIM remote provisioning architecture is one of the feasible industry solutions as it provides a means for deploying full profile download and security domains. The solution may use HTTP OTA as the primary channel, and it allows the support of existing M2M UICC HW while providing SIM card multi-vendor interoperability.

References

[1] Press release, Allynis OTA platform. http://www.gemalto.com/press/Pages/Sprint-extends-relationship-with-Gemalto-to-manage-growing-LTE-deployments-across-the-U-S.aspx (accessed 22 December 2015).

[2] GlobalPlatform TEE guide. https://www.globalplatform.org/mediaguidetee.asp, 23 December 2015. (accessed 23 December 2015).

[3] GlobalPlatform. TEE Client API Specification v1.0.

[4] GlobalPlatform. TEE Systems Architecture v1.0.

[5] GlobalPlatform. TEE Internal API Specification v1.0.

[6] Use case example of HCE. http://www.gi-de.com/en/about_g_d/press/press_releases/global_press_release_35712.jsp (accessed 23 December 2015).

[7] Tokenization principle, Smart Card Alliance. http://www.smartcardalliance.org/wp-content/uploads/EMV-Tokenization-Encryption-WP-FINAL.pdf.

[8] Tokenization, Sequent. http://www.sequent.com/what-is-tokenization/ (accessed 23 December 2015).

[9] Wolfgang Decker. Smart! Discover the world of mobile security. Giesecke & Devrient, January 2012. https://www.gi-de.com/gd_media/media/en/documents/complementary_material/smart__newsletter/smart_1-2012.pdf (accessed 26 December 2015).

7

Mobile Subscription Management

7.1 Overview

This chapter presents technologies for managing mobile devices and subscriptions based on OTA methodologies. The principles, benefits and challenges of mobile subscription are outlined via the 'traditional' subscription management for post-paid and pre-paid customers. In addition to the initialization of the subscription which refers to the provisioning procedure, the subscription lifecycle management is explained, i.e., modifying, adding and terminating services.

In addition to the consumer space, the principles for M2M subscriptions are outlined. Followed by current solution descriptions, the near-future and most probable long-term solutions for both consumer and M2M subscription management are discussed with concrete examples from the standardization field. These novelty OTA platforms provide operators with a greater level of freedom via increased interoperability along with the development of M2M devices that are equipped with eSEs. These elements cannot be removed physically, however, there are increasing needs for changing the subscriptions between operators during the lifetime of such equipment. This chapter discusses the respective, ongoing standardization in various frontlines that aims to tackle the current limitations of the subscription management.

7.2 Subscription Management

7.2.1 Development

Upon purchasing a new mobile device, the user needs to establish a customer relationship with his/her preferred MNO. There are various ways to do this. A typical procedure is to establish the relationship at the same time as the actual contract is signed in the retail store or at the operator's own customer service point. There are two contract types: post-paid and pre-paid. In either case, the subscriber information needs to be established both in the network side and

Wireless Communications Security: Solutions for the Internet of Things, First Edition. Jyrki T. J. Penttinen.
© 2017 John Wiley & Sons, Ltd. Published 2017 by John Wiley & Sons, Ltd.

in the respective SIM/UICC. Also, in either case, the customer can typically select the device from the MNO's pre-selected (subvention) set, or use his/her own device (often referred to as 'Bring Your Own Device') by inserting the SIM/UICC into any general device.

The activation of the subscription at the POS happens in such a way that the customer selects the subscription type as well as the parameters for the supported functionalities (e.g., data quantity and rate, value-added services such as voice mail box), from different options and price ranges. When entering the customer's personal data like name and address, the sales personnel typically need to ensure the identity of the customer. As the national regulations tend to be increasingly stricter for pre-paid customer ID registering, the sales service may require a proof of identity and address. This step aims to prevent the activation of fraudulent subscription. The false user information would provide means to use the pre-paid or post-paid line anonymously (e.g., for criminal purposes) or overlook the generated bills.

After ensuring the correct identification of the customer, the customer service dedicates an MSISDN which the network (HLR/HSS) will associate to the subscribed services. The customer may also choose the preferred MSISDN from a certain set of available numbers. The line is then activated, that is, provisioned [16]. The procedure includes steps summarized in the following paragraphs by generalizing a commercial example of an OTA On-Demand Activation (ODA) as described in [17]. The OTA activation process allows the mobile operator to dynamically activate and provision the subscription and SIM/UICC as the subscriber goes active 'on the air' for the first time. Figure 7.1 presents the principle of ODM.

In the operational environment, prior to the provisioning of the new user services, the registered subscriptions that have not yet been activated cause license fees and other expenses for MNOs because of the reserved resources. This challenge is especially relevant for pre-paid

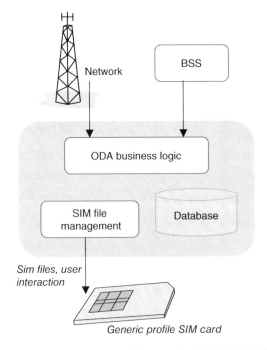

Figure 7.1   An example of ODA as described in Ref. [17]

SIMs that are physically stored in different POS locations to be purchased by new clients. With a large variety of offered SIM card types, the related logistics and storing costs of MNOs can be quite considerable.

There are solutions developed for these challenges such as the SmartTrust ODA of Giesecke & Devrient (G&D) which supports the activation, personalization and provisioning of the mobile subscription and the SIM card that is already distributed physically but not yet sold, and when the subscriber is about to activate the new subscription with new MSISDN information, or replace an old or lost SIM card with the same MSISDN. This solution is based on a single, generic card profile which is used for initialization and activation of the actual subscription, thus removing the need for reserving the subscription in the HLR/HSS and other platforms in advance when the physical card is not yet purchased. The benefit is the optimization of the respective capacity and licensing costs as they are only triggered upon the actual activation of the subscription including the customer-specific parameters, files, service set personalization, International Mobile Subscriber Identity (IMSI), MSISDN, priority lists for roaming, SIM-based value-added services, customer care entry in SIM-based phonebook and service provider name [17]. Furthermore, the end-users may be able to make selections via SIM-based menus for such items as the language and the phone number or subscription type during the activation process.

This service is an example of the solutions that integrate with the existing MNO infrastructure (such as HLR/HSS) so that the handling of sensitive information like the subscription's authentication keys is maintained safely within the same network. The service can also rely on the mobile network's signalling to trigger the activation, so there is no need for special adjustments in the respective interfaces as the existing network can be completely reused. Also, the solution may be integrated with the BSS of the MNO. Activation is possible from any compatible network in any country as the system is able to determine the country and network from where the activation takes place in order to select the proper operator profile within a cooperating operator group.

This type of SIM management platform product provides a complete solution for the SIM lifecycle OTA management by relying on the Remote File Management (RFM) and Remote Applet Management (RAM) as defined by ETSI specifications for Java Card based SIMs and by adopting vendor-specific OTA protocols from the participating SIM card suppliers.

7.2.2   Benefits and Challenges of Subscription Management

The subscription management can be done remotely via OTA methods, or in the simplest form, locally. Modern subscription management happens by relying on OTA methods due to the cost-efficiency while local subscription management (i.e., fixed line connectivity) can be assumed to be only relevant for devices that cannot be updated due to physical and economic reasons.

The subscription type selection is one of the essential tasks of the end-user. Depending on the need, the user may want to have only a basic subscription including voice, messaging and light packet data services, whereas more advanced users might need the highest available data rate. The task of the MNO is to optimize the available network resources by balancing the expected utilization of the capacity and charged fees. In some cases, the MNO may want to offer an unlimited data plan while others may control the utilization by setting capacity or data rate limitations on a monthly basis. Finding a good balance is not an easy task as it depends on competitive dynamics, so too unattractive subscriptions may increase the churn. The optimization of the churn is one of the priorities of the MNO, as well as the minimization of the customer care calls, as both are expensive to the MNO's operations.

As for the subscription lifecycle management, one of the most critical instances is the initial activation of the subscription. If something fails in this phase, it increases customer complaints and may lead to increased churn. The interpretation of the end-user (and POS personnel) may be that the service quality of the network is low, or that the mobile device or SIM card is of low quality. If the initialization fails when the customer is trying to activate the subscription remotely at home, it generates a customer care call which again is an added MNO expense and may result in churn if not solved promptly.

What then are the basic reasons for the failing activation, and what can be done to minimize these events? Some of the issues may be due to the incompatibility or failure of a component in the chain of the mobile device (OEM), SIM/UICC card (SIM vendor), service (provider), or network (MNO). The issue may also be due to the different interpretations of the standards or errors in the design. The aim of early testing is to capture these potential issues before the device enters market, including the problems that may lead to security holes.

When the issues are minimized and eliminated successfully, the HW then has a certain expected lifetime. It depends on the components of the device, such as the internal memory and any external memory surfaces – including the one found within the SIM/UICC card. The wearing of the memory happens gradually until part of it results in errors when writing to and reading from it. This type of HW behaviour can be assumed to be irrelevant fromr the security point of view as the ADPUs of the requests and responses between the SIM/UICC and the reader should prevent any problems – although the opening of a security hole may not be discarded completely due to the unpredicted responses in such a situation. To prevent incidents due to wearing, there are application-level solutions which monitor the memory such as the SIM Lifetime Monitor app as described in Ref. [20].

7.3 OTA Platforms

The following sections present the general architectures and functionalities of the SIM OTA platforms, and outline examples from the commercial field for OTA subscription management.

7.3.1 General

The user's subscription provisioning and management can be handled fluently by OTA methods during the lifetime of the active subscription. An alternative for the initialization and maintenance of the subscription is to connect the device via fixed wire with the operator's management server at the customer service, but this represents a special case which is not practical in the modern commercial environment. The SIM OTA refers thus to the management of the SIM/UICC lifecycle remotely via a cellular radio interface. After the first provisioning that initializes the new subscription, managing the SIM card is of utmost importance to the MNO in order to ensure a fluid experience while accessing and updating mobile services. The management is handled via the OTA platform.

The SIM OTA provides an adequate remote communications method for the management of the contents of the SIM/UICC card over the radio interface. The benefit of the SIM OTA for the MNO is the possibility to deploy easily new SIM services without requiring the customer to bring the device or SIM to a physical customer care location. It also provides the means to modify the contents of the SIM/UICC cards efficiently and economically.

Although the OTA procedure was initially related to the consumer space, management of the M2M environment is now becoming more important as a result of the current growth of IoT. For the respective OTA Device Management (DM), there are various commercial solutions available on the market. As an example, the SmartTrust Delivery Platform provides a base for G&D's OTA products. The SmartTrust AirOn enables remote management for the lifecycle of M2M SIM, related devices and value-added services such as activation, subscription management and deactivation. As another example, SmartAct is a SIM management solution compatible with other vendor cards. SmartAct provides MNOs the option to deploy NFC-based services for authentication, digital signatures and transactions. As a third example, the SmartTrust SmartàLaCarte functions as an enabler for mobile value-added services and operator-specific services for basic and feature phones. Since the solution includes the dynamic SIM Toolkit platform, it provides the end-users with a service menu on the device's display that allows users to call up applications that have been saved directly onto the SIM card via the OTA installation. Although the legacy and basic mobile device models are supported in such offerings, it is expected that smartphones will gradually become the universal service engine. Also, the smartphone architecture would become the default for multiple services and the coexistence of the SIM/UICC and eSE.

7.3.2   Provisioning Procedure

The initial phase of subscriber provisioning includes adding a new customer with new mobile equipment onto the system. The procedure contains the following steps for the activation of the new customer services in generic mobile communications networks: (1) The new subscriber purchases the mobile equipment at the POS of an authorized sales location of the MNO, and the sales personnel issues the respective new SIM/UICC. (2) The subscriber gets a new personal account, and the new mobile equipment and the UICC are linked to it via the POS terminal. (3) The UICC is inserted into the mobile equipment, and it is powered on. Upon the network access initialization procedures of the equipment, the real-time activation with respective information download is started from the OTA platform as depicted in Figure 7.2.

Another option for provisioning (i.e., activation) of the UICC is to perform it physically at the POS by relying on a card reader as depicted in Figure 7.3.

The OTA Provisioning Manager (OPM) provides services such as activation, deactivation, SIM swap, MSISDN change, feature change, push file, manage application, get subscriber information and roaming awareness. The OPM is integrated into the MNO's Billing System (BSS) interface. The OPM manages the administrative operations that are initiated in the BSS, such as the activation, SIM swap and subscription update.

7.3.3   SMS-based SIM OTA

The SIM OTA is based on the server and client architecture. The SIM/UICC represents the client while the MNO's back-end system – together with many other roles such as Business Support System (BSS) and customer care – contains the SIM OTA server. Typically, this back-end system

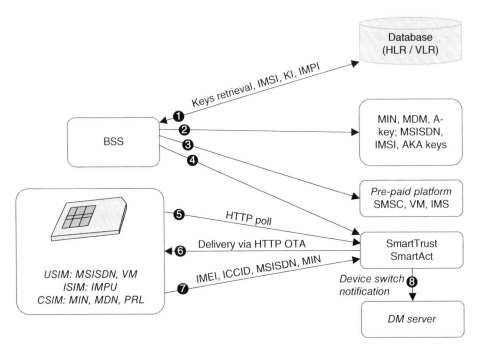

Figure 7.2   The high-level signalling flow of the real-time provisioning procedure as applied in the SmartTrust SmartAct solution

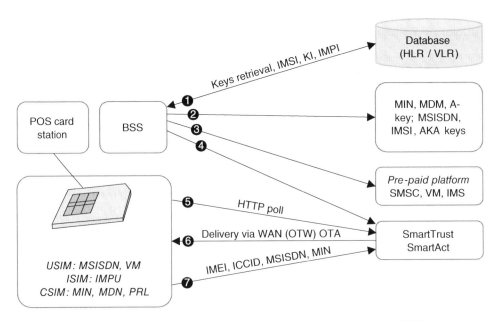

Figure 7.3   An example of the UICC activation, i.e., provisioning by utilizing a POS card reader

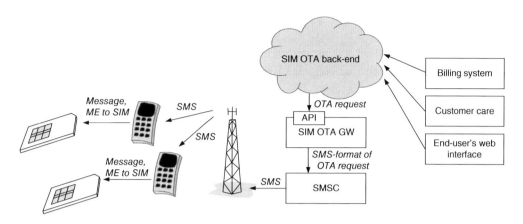

Figure 7.4 The principle of SIM OTA messaging

communicates with an OTA gateway by delivering service requests as presented in Figure 7.4. The gateway interfaces with the Short Message Service Centre (SMSC) by converting these service requests into Short Messages (SMs). Furthermore, the SMSC delivers these SMs to the SIM/UICC of a single device, or a set of SIM/UICCs of a respective larger group of devices.

The idea of the SIM OTA is to update and modify data which is stored in the SIM/UICC without the need to reissue it. Thus it is possible to download and use new services remotely instead of physically visiting an MNO's customer service or retail store.

In order to function, the SIM OTA needs access to the MNO's back-end system to create the requests, the SIM OTA gateway to convert the requests into SIM/UICC-specific format, the SMSC to forward the requests via the mobile communications network, the bearer (such as SMS) for request transport, the user device for receiving the request via the radio interface and for delivering it to the SIM/UICC card and, finally, the actual SIM/UICC for the execution of the request.

The SIM OTA back-end system may include a variety of entities like billing system, customer care and end-user's subscription self-management web interface. The service request may be related, e.g., to the activation, deactivation, loading, updating or modifying the contents. The request message also includes identification of the subscription and data for executing the service. The SIM OTA back-end then sends the request to the OTA gateway via a gateway API that indicates the respective SIM/UICC, which is identified with the help of a database that contains information about the cards, e.g., ICC Identification Number (ICCID), IMSI and MSISDN (Mobile Subscriber's ISDN number). Furthermore, the OTA gatewa contains information about the vendor-specific format which is needed in order to form the message to comply with the way each vendor's cards handle the requests.

The SIM OTA gateway forwards this vendor-specific formatted message to the SMSC by relying on the parameter set according to the ETSI specification GSM TS 03.48, issuing a sufficiently large set of short messages in order to complete the request. It is valuable to note that the SIM OTA gateway needs to manage the messages and ensure the integrity and security of their delivery. The SMSC then forwards the messages to the user device(s) via the SMS bearer within the mobile communications network according to the standardized SMS procedure.

As defined originally in the ETSI SMS service descriptions, each individual message can consist of a maximum of 160 alphanumeric characters, so the complete request may (and typically does) consist of several short messages that are compiled in the device side. If the device is not reachable upon the delivery of the messages, they are stored in the SMSC until the device activates the network next time, or until the maximum expiration time is reached. In order to support the SIM OTA functionality, the user equipment needs to comply at a minimum with ETSI phase 2+ GSM standards, and it specifically needs to support the SIM Toolkit functions. Finally, the SIM/UICC communicates with the user equipment via the standardized, secure interface, to deliver the vendor-specific contents to it.

7.3.4 HTTPS-based SIM OTA

The benefit of the SMS-based SIM OTA is that it is robust and straightforward to set up in any environment supporting short messaging. The drawback is the slow speed of the delivery of the commands due to the limitations in the interface between the mobile device and SIM card. An OTA update of the SIM profile may take several tens of kB with a multitude of separate messages which, in turn, might last about 10 minutes. Furthermore, not all the networks support 3GPP SMS format by default, as is the case for the CDMA 1xRTT.

The solution for this low transmission speed is to transfer the commands and contents via a packet data bearer which can be based on, e.g., GPRS or LTE data service while the protocol used on top can be the secured HTTPS. This solution may reduce the time for the update down to some tens of seconds instead of the minutes that the SMS method typically takes.

7.3.4.1 CAT

As the standard ETSI TS 102 223 [21] states, the Card Application Toolkit (CAT) is a set of generic commands and procedures for use by the ICC, irrespective of the access technology of the network. In this context, the UICC refers to an ICC which supports at least one application (NAA) in order to access a network. Furthermore, the ICC is considered as a platform based on either ETSI TS 102 221 [22] or ETSI TS 102 600 [23] for the 3G platform, or on ETSI TS 151 011 [24] for the 2G platform. As Ref. [21] further indicates, the NAA can be any of the ones summarized in Table 7.1.

Table 7.1 The options for the NAA as defined in Ref. [21]

NAA	Standard	Notes
USIM application	ETSI TS 131 102	Can reside only on a 3G platform
SIM application	ETSI TS 151 011	Can reside either on a 3G or a 2G platform
TSIM application	ETSI TS 100 812	Can reside only on a 3G platform
ISIM application	ETSI TS 131 103	Can reside only on a 3G platform
RUIM application	TIA/IS-820-A, 3GPP2 C.S0023-0	Can reside on a 2G platform. Alternatively, other applications residing on a 3G platform or a 2G platform

Source: Printed from [21] by courtesy of ETSI

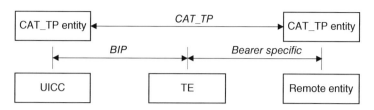

Figure 7.5 Data exchange as defined in ETSI TS 102 124

Ref. [21] also defines the interfaces to ensure manufacturer and operator independent interoperability between the ICC and a terminal. It also defines the commands, application protocol, mandatory requirements on the ICC and terminal for each procedure.

7.3.4.2 Bearer Independent Protocol

The OTA access to SIM/UICC cards has been based on the low-bandwidth SMS bearer for a long time. As a result of the updated standards, BIP provides considerably faster data transfer between the card, terminal equipment that supports BIP, and external world for accessing the card's contents.

An integral part of the modern HTTPS SIM OTA is in fact the BIP. It functions regardless of the underlying techniques in order to provide secured access OTA to the device and respective SIM/UICC, and it works as an alternative to the SMS-based accessing to the card. BIP is defined in ETSI TS 102 223 [21]. It allows the UICC's CAT application to establish a data channel with the terminal, and further through the terminal either to a remote network server or to a remote device in the PAN. BIP inherits the properties of the respective bearer and the network protocols and may be used on top of unreliable transport protocols such as UDP [18,19].

As described in ETSI TS 102 124, BIP provides the UICC a standardized manner to use Terminal Equipment (TE) bearers to communicate with remote entities in a WAN or PAN. The base for BIP is to exchange data first between the UICC and TE, and then between the TE and external server. Figure 7.5 depicts the principle.

Without CAT_TP, the CAT application is unable to know if the remote entity has received the data sent. Moreover, without CAT_TP, the remote entity may receivs data without transport information such as the emitter identity, packet numbering or transmission status. CAT_TP aims to provide the possibly missing transport functionalities.

More details about BIP, CAT and CAT_TP can be found in ETSI TS 102 124 [18].

7.3.5 Commercial Examples of SIM OTA Solutions

There are various commercial SIM/UICC vendors in global markets as well as respective SIM OTA management platforms. The basic functionality of the commercial solutions is similar for the lifecycle management, while the support of the more advanced set of features varies. Some of the globally recognized SIM/UICC vendors are Giesecke & Devrient (G&D), Gemalto, Oberthur, Schlumberger, DeLaRue and ST Microelectronics (STM).

As an example of the portfolio of G&D SmartTrust OTA products, they provide a platform for secure and reliable SIM card management by any vendor throughout the SIM lifecycle, from defining card specifications through pre-personalization to deactivation. The portfolio has been designed to support a variety of environments like NFC and secure mobile payment that are being deployed at a global level. The portfolio also includes device-aware products, integrated for SIM and mobile device management. Some examples of the portfolio are SmartTrust Delivery Platform, which is an OTA platform supporting advanced management of SIM cards and mobile devices, and SmartTrust AirOn, which is platform for secure OTA management of connected M2M SIMs, devices and their applications. More detailed solution descriptions can be found in [15]. The following sections generalize some of the commercially available solutions for presenting an overview of the possibilities of advanced OTA device management.

The OTA provisioning is typically capable of automatically downloading multiple types of settings to a variety of handset models. It may also support various download protocols such as OMA-CP and OMA-DM, as well as vendor-specific protocols of the device manufacturers. The solution may include operations for device and SIM personalization, configuration and reconfiguration, diagnostics and firmware management as well as device locking and erasing the contents in case the device is lost. In other words, the device management is designed to manage the whole lifecycle of the device.

The solution may also support event triggers for automated and manual operations, such as actions from the customer care and campaigns. Open API interfaces may be supported for the integration of third-party systems which may be based on, e.g., HTTP/SOAP. The solution can use a client installed on the device, and an IP-push mechanism to enable server-initiated device management actions over cellular networks and Wi-Fi hotspots. Furthermore, the solutions may include automated device detection functionality to ease the device management, which can be based on a Java applet installed on the SIM supporting Java Card and which makes the device detection independently from the network provider.

In general, along with numerous models on the market, the devices are provisioned according to the OS and provisioning protocol stack. The OTA provisioning system should ideally support all or near all of these devices and protocols in such a way that the provisioning with the respective settings happens automatically based on network and SIM triggers. The procedure can also be manual, based on customers' self-care provisioning for setting up the device, e.g., via a web user interface, short messaging or SIM-based menu of the device. The provisioning settings are typically related to GPRS, MMS, SIP, Wi-Fi, streaming and email settings. Logically, the device-dependent settings require respective terminal capability repository for storing the settings ad supported protocols per device, and for identifying the device in question, which means constant updating of the information by the provisioning system provider.

The following sections discuss the functionalities that may be included into the typical OTA provisioning system.

7.4   Evolved Subscription Management

Along with the increasing amount of advanced M2M devices, the related subscription management is improving. Not only are the machine-type communications but also the consumer environment going through major development, including the increasing popularity of wearable devices. This is leading to the need to manage the subscriber's credentials in a much more dynamic way, e.g., the user may want to change the primary communications device (typically a smartphone) to an alternative device like a jogging or health device which the user only wears for a short period of time during physical exercise, and using only limited functionalities of the same subscription. After that, the subscription could be changed to the original smartphone or into some other type of device. This example indicates the need for highly dynamic transferrable subscriptions between consumer devices.

The challenge is that the subscription management solutions described earlier in this book are not capable of adapting to such a dynamic environment, and they typically lack interoperability between different subscription management systems as well as SIM/UICC card types. Thus, there is a strong international standardization effort ongoing as of 2015 for providing enhanced subscription management systems which are detailed in the following sections.

7.4.1   GlobalPlatform

The card specifications of the GlobalPlatform can be found in Ref. [27], and the systems requirements in Refs. [28,29] summarize the new model for the customer-centric ecosystem for managing subscriptions.

7.4.2   SIMalliance

Ref. [30] outlines the role of SIMalliance in the development of new subscription management methods especially for the M2M environment, but the SIMalliance is ultimately useful for consumer space along with the development of wearables. The new specification allows MNOs to remotely load and manage subscriptions across deployed M2M and consumer devices in a standardized way. The respective eUICC Profile Package defines the interoperable format via technical specification v1.0.1 by describing a common coding process for subscription data to be built, remotely loaded and installed into any embedded UICC by any SIM vendor on behalf of any MNO.

Ref. [30] emphasizes the importance of standardized remote subscription management across eUICCs as it provides time and development efficiencies for MNOs and the wider remote provisioning ecosystem. Service providers can thus provision fleet or an installed base of devices fast and efficiently in a unified way across diverse terminals, MNO customer management systems and eUICCs. The benefit of the solution is simple application provisioning and lifecycle management combined with scalability and flexibility within the remote provisioning ecosystem.

As for the cross-functionality, the SIMalliance's specification is referenced in the GSMA Embedded SIM Specification within the latest Remote Provisioning Architecture for Embedded UICC Technical Specification v3.0, which defines a technical solution for the remote provisioning and management of the eUICC in M2M devices.

7.4.3   OMA

The OMA Device Management (DM) Working Group (WG) specifies protocols and mechanisms to define management of mobile devices, services access and software on connected devices. The OMA DM WG has operated since 2002 which has resulted in a suite of

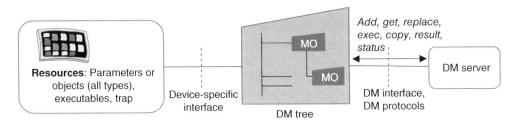

Figure 7.6   The OMA DM philosophy

specifications to provide simple, reliable and cost-effective ways to deploy new applications and services. The OMA also cooperates with other standardization bodies to avoid fragmentation and duplication of specifications.

The OMA DM technologies have been designed to manage converged and multi-mode devices on different networks, including devices that do not have a SIM card, as well as resource-constrained devices. The benefit of the OMA DM specification suite is thus the extensibility which is especially suitable for M2M communications. The OMA DM specifications define the protocols and the mechanisms allowing an OMA DM server to deliver configuration parameters to an OMA DM client by using a set of DM commands for a set of management procedures. These commands are executed within a defined and secure environment, referred to as the DM session.

The OMA DM client has been designed in such a way that it exposes the device data to the OMA DM server via the so-called DM tree, in the form of a hierarchical structure, as presented in Figure 7.6. It contains management objects, or sub-trees, which provide the functionality for the device management. The philosophy of the OMA DM is thus to manage device features via the DM tree which in turn virtualizes the device features and functionalities.

The management authority can remotely set parameters, perform terminal functionality troubleshooting, as well as install and upgrade SW via the OMA DM.

The device's applications are able to access the DM tree and interact with the management objects and the DM server via the interfaces specified by the OMA DM client framework API for receiving configurations and report data. According to the OMA DM smartcard specification, the DM server may be executed on the smartcard that is inserted in the device which optimizes network bandwidth and capacity.

OMA DM specification release 2.0 includes advances for reducing complexity and for providing better interoperability. The OMA DM specifications include management objects that implement various management functions such as *Firmware Update Management Object* (FUMO), *Software Management* (OMA DM SCOMO), *Diagnostics and Monitoring* (OMA DM DiagMon MO), *Connectivity* (OMA DM ConnMO), *Device Capabilities* (OMA DM DCMO), *Lock and Wipe* (OMA DM LAWMO), *Browser* (OMA DM BMO), *Virtualization* (OMA DM VirMO), *Management Policy* (OMA DM Management Policy MO) and *Gateway* functionality (OMA DM GwMO v1.0).

The OMA has designed remote management especially for the M2M environment via the OMA Lightweight M2M protocol, which focuses on constrained cellular and sensor network M2M devices. The OMA Lightweight M2M provides simple management interfaces based on feasible and available standards such as IETF (CoAP, DTLS; UDP and SMS binding). Figure 7.7 presents the principle.

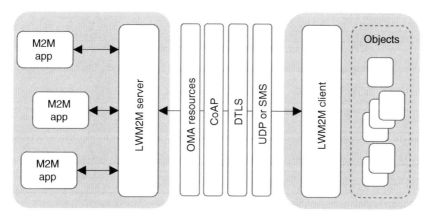

Figure 7.7   OMA Lightweight M2M architecture. The LWM2M communications between the client and the server is optimized via efficient payload, and is able to support interfaces for bootstrapping, registration, object/source access and reporting for very low-cost devices

7.4.4   GSMA

The GSMA is one of the entities establishing interoperable procedures for remote SIM provisioning for the M2M environment. The concept is integrally based on the embedded SIM/UICC, and the respective definitions are included in the GSMA Embedded SIM specification. It provides a single, de facto standard mechanism for the remote provisioning and management of M2M connections via OTA provisioning of an initial operator subscription, as well as the subsequent change of subscription between MNOs.

The M2M environment is one of the drivers for the development of interoperable subscription management because the respective SIM/UICC may be challenging to replace and change due to the restrictions of the physical access to the device, or if the element is permanently installed in the form of an eSE. The logical way to manage these devices is based on OTA methods for the provisioning of the SIM/UICC. Combined with the secure OTA channels, the method provides the same level of security as the removable SIM/UICC are capable of in the consumer market [26]. The GSMA Embedded SIM specification can be used for both removable and embedded SIM/UICC environments, which supports the development of the ever-growing markets of the IoT/M2M devices and the remote provisioning of operator credentials [1]. As an example, the automotive industry has spearheaded the introduction of remotely provisioned SIM/UICC elements in managing MNO subscriptions and to support the evolving services for cars [3]. Other examples of remote provisioning where GSMA is of use include utility and basically any other IoT device.

7.4.4.1   Embedded UICC

The remote provisioning architecture of GSMA for embedded UICC that focuses on the M2M environment is described in the GSMA documents in Refs. [2,6,8] and the test specification is in Ref. [7]. The definitions are based on current telecommunications standards like Global Platform (GP), which emphasizes the separation of roles and isolation of data. Nevertheless, there are also items in the GP that are not covered by GSMA, like the Issuer Security Domain (ISD).

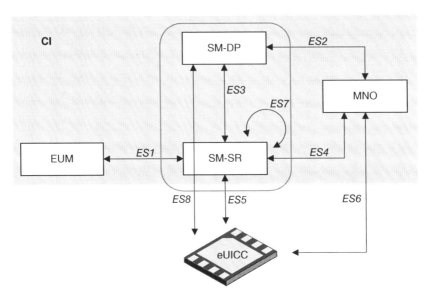

Figure 7.8 Remote eUICC provisioning architecture for M2M environment as defined by GSMA (version 2.1). *Source*: Printed with permission of the GSMA

The following section summarizes the GSMA remote provisioning architecture. More detailed information about the embedded SIM/UICC as seen by GSMA can be found in Refs [25,31]. The GSMA remote provisioning architecture of embedded UICC is shown in Figure 7.8 as interpreted from the version 2.1.

As can be seen in Figure 7.8, there are various roles and interfaces in the remote provisioning. EUM refers to the eUICC manufacturer which produces the physical SE. The MNO executes the profile order by communicating with the SM-DP (Subscription Manager, Data Preparer) and SM-SR (Subscription Manager, Secure Routing). SM-DP, SM-SR and MNO have a communication link to the secure element, eUICC. Furthermore, the integrity of the data is taken care of by the CI (Certificate Issuer).

The high level contents of the actual eUICC is shown in Figure 7.9 as interpreted from Ref. [31]. The respective contents are referred to as Security Domains (SDs) that provision platform and profile management use. In the GSMA remote eUICC provisioning architecture, each entity has a dedicated SD that consists of different privileges and settings.

Figure 7.10 shows further the structure of a single profile as interpreted from Ref. [31].

The mapping of the communication between the system and eUICC is presented in Figure 7.11.

Following the terminology of Figure 7.9 and Figure 7.11, the EUM installs and personalizes first the ISD-R (Issuer Security Domain Root) as an initial phase of the eUICC manufacturing. Technically, the ISD-R is associated with itself. ISD-P (Issuer Security Domain Profile) is a component that hosts a unique profile. In the GSMA solution, only one ISD-P is enabled on an eUICC [6]. The EID refers to the eUICC identifier related to the remote provisioning and remote management of the eUICC. It is comparable to the ICCID, but not equivalent. Thus, in systems that relate the ICCID and other identifications, such as IMEI, there is a need for mapping the new EID. It should be noted that there is no global database for this information

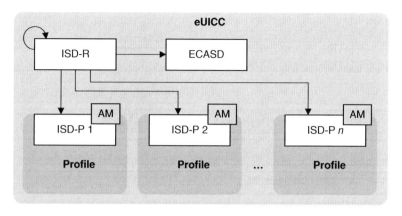

Figure 7.9   The contents of eUICC in GSMA remote provisioning systems. *Source*: Printed with permission of the GSMA

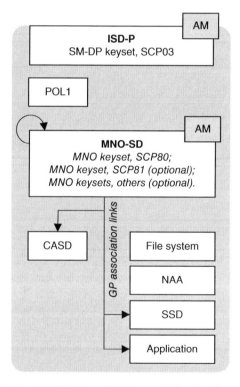

Figure 7.10   The contents of a GSMA profile. *Source*: Printed with permission of the GSMA

so operators need to rely on their own solutions. The (eUICC Controlling Authority Secure Domain (ECASD) is installed and personalized by the EUM during the eUICC manufacturing, assisted by the CI associated with the ISD-R. As soon as the eUICC is manufactured, the ECASD is set to lifecycle state 'personalized' as defined in the GlobalPlatform Card specification.

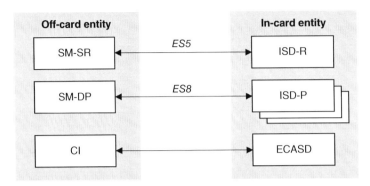

Figure 7.11 The mapping of the card entities with the provisioning system. *Source*: Printed with permission of the GSMA

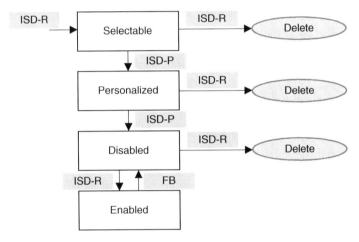

Figure 7.12 The ISD-P stages of GSMA remote provisioning eUICC. The transitions may be triggered by ISD-R or ISD-P itself. There also is a fall-back (FB) mechanism

The lifecycle states of the eUICC are presented in Figure 7.12. The lifecycle contains states called 'Selectable', 'Personalized', 'Disabled' and 'Enabled' in such a way that once the element is personalized, the transition can be switched between disabled and enabled for each ISD-P.

The further development of remote provisioning of the GSMA include V3 and V4 (GSMA+), of which the latter includes consumer use cases in addition to the M2M environment as presented in Figure 7.13. In fact, after the initial subscription management solutions, the importance of consumer use cases has been identified as highly relevant. This is due to the fact that wearables and other small devises are becoming more popular, and the respective changing of subscription information between the devices needs to be much more dynamic. The further evolution of GSMA subscription management is called Remote SIM Provisioning (RSP), and the respective SGP.21 architecture specification and SGP.22 technical specification set are divided into phases from which the phase 1 and 2 specifications have been available since the second half of 2016 for companion device and consumer use cases. They define further the architecture and add, e.g., Local Profile Assistant (LPA) into the eUICC. Figure 7.14

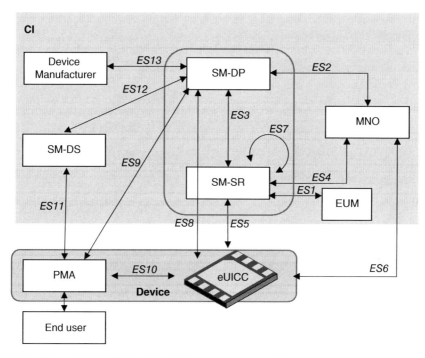

Figure 7.13   The evolved GSMA subscription management architecture (version 4) that includes the consumer environment

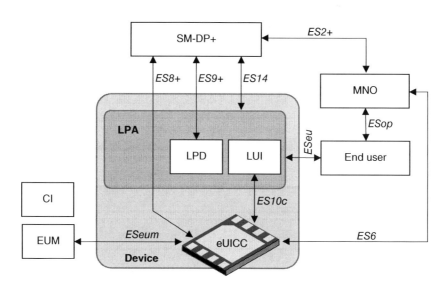

Figure 7.14   The GSMA RSP V1 architecture

summarizes the RSP V1 [33]. As the RSP architecture develops along with phases 2 and 3, the latest documentation of can be found in Ref. [32].

There have been several initiations related to this environment from which the GSMA has been one of the most active parties during 2015.

7.4.4.2   SAS

The subscription manager itself also has to comply with the security requirements which are detailed in GSMA SGP.07 both for SM-SR and SM-DP services. Compliance is needed for certification under the GSMA Security Accreditation Scheme (SAS) [4,5]. The SAS standard for subscription manager roles is explained in Ref. [13], and the SAS methodology for the roles is defined in Ref. [14].

7.4.4.3   Internet of Things

GSMA also considers the overall security related to IoT devices and apps for safe communication via the mobile network. According to GSMA, the IoT environment is dependent on the efficient and intelligent use of the mobile network. GSMA develops guidelines emphasizing efficient connectivity to ensure that the developers of IoT devices and apps have concrete information in order to follow a common approach and to create efficient, trusted and reliable IoT services that are scalable with the growing market.

The concrete means to make this documentation and information sharing happen is via cooperation between GSMA and the IoT ecosystem partners. In fact, the optimal functioning and connectivity of such a huge amount of IoT devices as is estimated for the forthcoming years in a scalable mobile network can be achieved efficiently only if all the stakeholders agree and follow a common approach [9,10].

According to the GSMA Connected Living programme, the M2M represents 10% of all mobile connections in the United States, driven by aggressive growth in the automotive and utilities sectors. In addition, the United States accounts for 19% of all the M2M connections worldwide [11,12].

References

[1] GSMA Connected Living, M2M remote provisioning. http://www.gsma.com/connectedliving/embedded-sim/ accessed 22 January 2015).
[2] GSMA Embedded SIM Remote Provisioning Architecture, Version 1.1, 17 December 2013.
[3] GSMA automotive SIM. http://www.gsma.com/connectedliving/mautomotive/sim/ (accessed 22 January 2015).
[4] GSMA SAS. http://www.gsma.com/newsroom/all-documents/sgp-07-gsma-sas-standard-for-subscription-manager-roles-v1-0/ (accessed 22 January 2015).
[5] GSMA SAS Standard for Subscription Manager Roles, Version 1.0. GSMA, 13 October 2014.
[6] GSMA Remote Provisioning Architecture for Embedded UICC, Technical Specification, Version 2.0, 13 October 2014.
[7] GSMA Remote Provisioning Architecture for Embedded UICC, Test Specification, Version 1.0, 13 October 2014.
[8] GSMA Embedded SIM Specification, Remote SIM Provisioning for M2M. Presentation, GSMA Connected Living, October 2014.
[9] GSMA IoT. http://www.gsma.com/connectedliving/iot-connection-efficiency/ (accessed 22 January 2015).

[10] GSMA IoT Device Connection Efficiency Guidelines, Version 1.0, 13 October 2014.

[11] GSMA Connected Living. http://www.gsma.com/connectedliving/gsma-driving-innov-connected-living/ (accessed 22 January 2015).

[12] GSMA Driving Innovation in Connected Living. The US flags the future of M2M. Presentation, October 2014.

[13] GSMA SAS Standard for Subscription Manager Roles, Version 1.0, 13 October 2014.

[14] GSMA SAS Methodology for Subscription Manager Roles, Version 1.0, 13 October 2014.

[15] Giesecke & Devrient SIM OTA, 19 September 2015. http://www.gi-de.com/usa/en/products_and_solutions/ products/sim_lifecycle_management/sim-ota-and-lifecyclemanagement.jsp (accessed 19 September 2015).

[16] Gemalto OTA, 19 September 2015. http://www.gemalto.com/techno/ota (accessed 19 September 2015).

[17] SmartTrust ODA (On-Demand Activation). Giesecke & Devrient, January 2011.

[18] ETSI TS 102 124 V6.1.0 (2004-12). Technical Specification, Smart Cards; Transport Protocol for UICC based Applications; Stage 1, (Release 6).

[19] ETSI TS 102 223, V8.2.0. (2009-01). Technical Specification, Smart Cards; Card Application Toolkit (CAT) (Release 8).

[20] SIM OTA lifetime monitoring. http://www.gi-de.com/en/products_and_solutions/products/sim_lifecycle_ management/airon/airon.jsp (accessed 1 December 2015).

[21] ETSI TS 102 223.

[22] ETSI TS 102 221.

[23] ETSI TS 102 600.

[24] ETSI TS 151 011.

[25] GSMA embedded SIM. http://www.gsma.com/connectedliving/embedded-sim/) (accessed 1 December 2015).

[26] Subscription manager security requirements of GSMA. GSMA SGP.07.

[27] Card specifications of GlobalPlatform, 25 December 2015. http://www.globalplatform.org/specificationscard. asp (accessed 25 December 2015).

[28] Systems specifications of GlobalPlatform, 25 December 2015. http://www.globalplatform.org/specificationssystems. asp (accessed 25 December 2015).

[29] A new model: the consumer-centric model and how it applies to the mobile ecosystem. White paper, GlobalPlatform, March 2012.

[30] SIMalliance. New SIMalliance Specification Supports Standardisation in M2M Deployments, 25 December 2015. http://simalliance.org/media/press-releases/new-simalliance-specification-supports-standardisation-in- m2m-deployments/

[31] GSMA Remote Provisioning Architecture for Embedded UICC. Technical Specification, Version 2.1, 2 November 2015.

[32] GSMA document area, 21 May 2016. http://www.gsma.com/newsroom/gsmadocuments

[33] GSMA SGP.21 – RSP Architecture. Version 1.0, 23 December 2015.

8

Security Risks in the Wireless Environment

8.1 Overview

This chapter outlines potential security risks in the wireless environment and presents concerns of the wireless security mechanisms. Some typical security holes are related to the network, user equipment, applications, communication links and signalling – although the role of human errors can never be underestimated in the overall security. This chapter also summarizes some key attack types such as eavesdropping, data altering, overloading and some less typical threats such as RF jammers and Electro-Magnetic Pulses (EMPs). Also specialized techniques like electron microscope and power consumption analysis are discussed. Finally, some important impacts of wireless security on the utilities and applications are presented, and key protection methods are discussed as guidelines for network operators, device manufacturers, smartcard providers, app developers and users.

It can be generalized that the single most important security threat in any environment is related to human error. Even in the most secure platforms and delivery systems, the weakest link may be a mistake with the complying of the security process [9,12]. There are many examples related to this topic, e.g., protected delivery of security keys between SIM card production and the MNO/customer of the card as reported in Ref. [26]. In this specific case, the external attacker was able to capture a part of the user-specific Ki keys because they were delivered using a lower protection level than the secure transport process required. According to the reference, this case fortunately only resulted in a relatively low amount of compromised GSM keys with repairable damage. Furthermore, in order to take advantage of the compromised set of keys, an attacker that has access to a copy of the keys would need to find the corresponding users to attempt to unscramble the radio interface communications of those specific users, making the task comparable to someone finding a copy of a set of door keys and then trying to find the corresponding door for unauthorized access.

Wireless Communications Security: Solutions for the Internet of Things, First Edition. Jyrki T. J. Penttinen.
© 2017 John Wiley & Sons, Ltd. Published 2017 by John Wiley & Sons, Ltd.

Not only the personnel of industry stakeholders but also the role of the end-user is crucial – it really does not matter how well the whole communication link is protected by the operators and security vendors if the subscriber reveals confidential information to other people.

The radio interface of the 3G and the LTE are more advanced and better protected compared to the 2G systems [1,3]. The new protection mechanisms have been updated from the GSM principles which makes the eavesdropping of the radio interface much more complicated in real life, even if there are theoretical ideas to compromise the encryption. Since the radio interface is secured against potential breaches, the hacking intentions may now concentrate on other layers such as smart device applications.

The aim of the data network stakeholders is to block any potential security hole. Nevertheless, even if the network is protected against all known and future data breach intentions technically, the role of human error remains essential in the end-to-end security chain. This is the trickiest part of all the security-related work, and the preventive mode in tasks such as upgrading the network needs to be applied efficiently so that the participating team can ensure that no back-doors or other security holes compromising the security are left open to the network. Additional security assurance processes may be needed in the overall network planning and optimization, as well as in the development of user equipment and applications.

As the M2M is developing currently with such giant steps, the security of respective devices such as utilities starts to be a very relevant aspect which means that the stakeholders need to take the security processes and protection mechanisms into account in all the related areas. This is highly challenging as there will be a variety of extremely cheap devices which may completely lack the in-built security, and very innovative new kinds of devices that may not have the latest protection built in [13]. Also, the equipment for compromising the IoT equipment, e.g. in automotive environment, may be inexpensive [6].

8.2 Wireless Attack Types

Some of the threats in the wireless environment include cyber-attacks, eavesdropping, data altering, overloading, destruction, radio jammers and other RF attacks, EMPs and highly specialized techniques focusing on the component level such as electron microscope and power consumption analysis. One of the typical use cases is the attack via Wi-Fi access points [10].

8.2.1 Cyber-attacks

The term 'cyber-attack' is very broad and may include all kinds of security breaches. There are thus many definitions for the topic such as the *Tallinn Manual on the International Law Applicable to Cyber Warfare* which describes one of the key statements for a cyber operation so that *it is reasonably expected to cause injury or death to persons or damage or destruction to objects* [27]. Furthermore, Ref. [27] states that the broad spectrum of cyber-attacks could include intrusion, surveillance, recording of data, espionage, extraction, destruction and manipulation of data, theft of intellectual property, control of devices and systems, kinetic effect through control of devices, destruction of devices, property and critical infrastructure, individual lethal effect and operations with national impact.

The cyber-attack can be categorized into three elements: *intelligence* (for gaining access, executing a cyber payload, and understanding the target environment); *cyber weapons* (typically target-specific and activated under specially planned conditions); and *calculated human decision*. The combination of these elements form a *cyber force*.

From the human point of view, there are two types of consequences from the cyber-attack: those that are nonviolent yet impact on the functions of the information society (so-called virtual destruction and disruption); and those that lead to physical death and destruction. As an example of a cyber-attack from 2010, the Stuxnet virus infected targeted nuclear centrifuges resulting in them accelerating and self-destructing. This case is a concrete example about the attacks that destruct information storages permanently. Ref. [27] summarizes the cyber-attack as *a deliberate projection of cyber force resulting in kinetic or non-kinetic consequences that threaten or otherwise destabilize national security; harm economic interests; create political or cultural instability; or hurt individuals, devices or systems.*

As NATO has reasoned, there are three dimensions to cyber threats: confidentiality, integrity and availability of data. Confidentiality refers to the aim of protecting sensitive data. Confidentiality breaches are assumed to be the most common form of cyber-attack. There are many examples about stolen secret data in the defence force domain, which impacts on the strategic side as information about plans and equipment is discovered, and facilitates the opposing side to upgrade their own equivalent solutions [30].

As the role of wireless systems is increasing, cyber-attacks can be expected to concentrate increasingly on the radio access networks such as mobile communications and wireless local area networks, as well as on the low-protection, cheap IoT devices that are the most attractive targets since they are typically innocent-looking, small objects that may open important security holes, not only in the home and company environment but also in highly strategic locations.

8.2.2   Radio Jammers and RF Attacks

The old method for destruction of radio communications is based on deliberate interference, e.g., in the form of radio jammers. A simple variant of this category is any radio transmission which is set to function on the same radio frequency as the targeted communications or functionality. The directive antennas can be used to focus the transmission and/or reception – the basic principle being that a sufficiently high interference level (I) compared to the useful carrier signal (C) lowers the 'readability' of the contents as the system-dependent C/I ratio is pushed under the functional limit. An example for the still functional C/I for the full-rate voice codec of the GSM is +9 dB while the widespread CDMA signal, functional under the noise level, is more tolerant against narrow-band interference sources.

Historically there have been practical cases about radio communications used for interference, both for destruction as well as for protection purposes. A fascinating example of the latter is from the era of the Continuation War in the city region of Vyborg where the Finns learned that the radio-controlled mines equipped with battery-powered receivers were the reason for mystical explosions in 1941. The reverse engineering of an unexploded sample indicated that the mines were designed to activate as soon as the accompanied receiver captured a three-note chord broadcast on the pre-defined radio frequency which resulted in three individually designed tuning forks to resonate and trigger the explosion. Once the experts resolved the principle of the mines, the Finnish national radio broadcasting company started to transmit spectrum-rich contents in the form of *Säkkijärvi polka* by Viljo Vesterinen at the same operational frequencies of the mines, aiming to deliberately interfere the activation procedure. This continuous transmitting of the RF interference functioned as a protection mechanism to prevent the resonating and further explosions, until the rest of the mines were found and deactivated [28,31].

At the other extreme, the interference can be based on a high power RF pulse that effectively destroys part of the electronics, the chips being highly vulnerable, and thus damaging the communication and other equipment. Modern chips may be especially sensitive to these EMP attacks [29]. The EMP can be very directive within a broad spectrum. In addition to the strong RF emission, the pulse can also be caused by a nuclear explosion close to the targeted electronic devices, the most severe damage happening above the area. Not only on Earth (e.g., via nuclear missiles) but also the EMP can take place, e.g., between satellite equipment causing wide damage over the wireless and fixed communications networks relying on the satellite component. Satellite communications as such may be encrypted depending on the environment, and the protection level depends on the strength of the respective algorithms [22].

RF and EMP threats are relevant not only to the highly specialized environments as described above but also the interferences may be present in daily wireless networks, often unintentionally. A typical example of this is the utilization of various devices in unlicensed frequency bands, the signalling and communication devices generating inter-system and intra-system load and thus incrementing the respective interference level.

8.2.3   Attacks against SEs

In addition to SW-based attacks, there are physical means that aim to reveal the contents and functions of the HW such as memory elements and SEs. The methods can be based on the analysis of the data responses and respective electrical behaviour at chip level by observing the behaviour of the equipment with an electron microscope, or by executing a power consumption analysis. These may expose some internal crypto-processor's internal results for injected input data which might help in revealing the embedded and protected data such as user keys. Respective protection mechanisms have been designed, e.g., for the SIM/UICC environment to minimize the exposure of the internal card's signalling to such intentions.

8.2.4   IP Breaches

There are plenty of 'traditional' IP network breaches like viruses and malware familiar from the fixed Internet but relevant also in the wireless environment. The security breaches and attack types are comparable to the overall Internet security vulnerabilities, but the interfaces, devices and interfaces are also related to the mobile communications – both in radio interface and in the core and transmission networks. One trend, along with the enhanced network and system security of the mobile communications networks, is that the attacks are focusing increasingly on the application level of the user devices. Thus, even if the wireless device's application delivery ecosystem includes security protection mechanisms with security assessment and in-depth testing, there have been examples of hidden security holes that have been utilized against the user equipment. The topic is complicated as the applications may request wide privileges to access user data even if the nature of the functionality of the applications might not benefit from those as such – it is thus largely up to the end-users to consider how much data is exposed to the apps.

8.2.5 UICC Module

Regardless of the relatively long age of the SIM card and its enhanced variants, it still serves as a functional base for the protection of mobile communications for voice calls, signalling and data transfer, as well as for respective data storage, authentication, authorization and other procedures used in the mobile networks. The additional benefit of the UICC is that it provides an existing platform for multiple applications that can be managed and used via the same single user interface which nowadays is typically a smart device or some other variant of advanced equipment.

The most concrete threat against the UICC is the physical HW attack in order to reveal the contents of the card, like user data and keys that are permanently stored on the card. Some examples of such intentions include attacks against network elements, core network interfaces and applications. Especially for the latter, along with the growth of smartphones and the popularity of applications, the intentions for malicious SW increases.

There are also electromagnetic methods as well as localized, embedded eavesdropping smartcard methods like the one discussed in Ref. [32] which presents a combined attack tampering principle with the APDU buffer. The secure channel concept provided by the GlobalPlatform ensures the confidentiality and integrity of the communications between the terminal and card through cryptographic mechanisms. Nevertheless, the described attack results in accessing the APDU buffer array based on the fact that the APDU buffer is not meant only for the communication channel between the card and the terminal. More specifically, Ref. [32] details that the specifically developed re-startable task works as a tool for breaking the secure channel by knowing the initialization of a secure channel which is based on APDU commands *INIT UPDATE* and *EXT AUTHENTICATE*, and specific *CLA* and *INS* bytes. Ref. [32] claims that the eavesdropping may be focused by detecting the beginning of the secure channel session based on these commands. The deciphering (ciphering) and MAC checking (computing) of an incoming (outgoing) APDU is operated thanks to a call to a known method of the secure channel interface. If this method is called with the APDU buffer array as parameter, the attacker owning the re-startable task is now able to both eavesdrop and corrupt the communication.

As can be reasoned from the above description, this method requires physical access to the smartcard. Furthermore, the exposure of IoT devices will be much wider for potential hacking attack intentions, the eUICC also provides a security benefit due to its fixed installation. As Ref. [8] states, the eUICC delivered to the end-user is embedded onto the device, and for this reason the end-user does not have a direct interface to it. Ref. [8] has also identified a variety of security risks against the eUICC and respective protection mechanisms. As a summary, Ref. [8] concludes that an off-card actor or on-card application may try to compromise the eUICC by trying to perform unauthorized profile management, e.g., by altering profile data before or after installation, or unauthorized platform management, e.g., by aiming to disable an enabled profile. Ref. [8] defines a protection profile which covers these threats by defining security domains (SDs) so that the data and capabilities associated to an SD are accessible only to its legitimate owner. There are also various other threats identified, including physical tampering and cloning intentions as well as potential network algorithm flaws, to which the protection profile provides adequate shielding. The protection profile also tackles the known physical attacks such as side-channel analysis to leak the protected keys and fault injection to alter the behaviour of the target evaluation. The protection profile includes security objectives for the underlying IC, which ensures protection against physical attacks. More details about the GlobalPlatform's identified threats and shielding can be found in Ref. [8].

8.3   Security Flaws on Mobile Networks

8.3.1   Potential Security Weaknesses of GSM

When the GSM system specifications were developed in the late 1980s and the first commercial networks were launched as early as 1991, the security threats were not seen as too significant. Thus, the enhanced security that GSM provided compared to the already existing fixed network was more than sufficient at that time [7].

Nevertheless, novelty attack methods have been developed during the course of time. Even if GSM authentication and authorization are still highly secure especially with the most updated radio interface encryption algorithm A5/4, there is a potential threat that someone may create a false BTS close to the GSM user [4]. If the network ID selected is identical to the one used by the user's home operator, and if the spoof BTS radio signal level is sufficiently high, the fraudulent BTS may capture the calls in the initial phase of the call in such a way that the encryption is forced off – as it is the network that decides the selection of the algorithm, the user may only see a symbol on the display indicating that the encoding is not utilized. If the user continues with the call initiation, the actual delivery of the contents may be relative straightforward by applying a proper relying solution for reaching out to the originally intended receiving party (B-subscriber) while the fraudulent BTS operator can eavesdrop the communications of that compromised connection.

The GSM provides a good level of security for the access network especially with modern encryption algorithms. Nevertheless, while the radio interface is encrypted, the communications path between the base station and the rest of the fixed network is not protected by default. The original assumption of the GSM system is that the end-to-end communication is at least as secure as the fixed telephony network (which is not protected as it is assumed to be isolated from external attacks). The radio encryption equipment is located at BTS sites, which means that the communication may be possible to eavesdrop, e.g., via unprotected radio links within the GSM network as long as the eavesdropper finds the link and can apply the respective protocols for revealing the contents. Also, the old-fashioned signalling system of the core network may be vulnerable exposing possibly authentication and encryption-related information.

The deployment of the first GPRS networks during the early 2000s was a significant step towards an all-IP concept that helped overcome issues of the 'old-fashioned' circuit-switched (CS) data service. The CS data communications were well protected because of the highly isolated environment and as the connectivity was based on point-to-point communications. Then, the GPRS exposed the network to the public Internet. For that reason, a totally new planning of prevention methods was needed against the security threats familiar from the fixed Internet. As an example, unlike in CS communications, firewalls between the GPRS core and outside world were now needed with new methods for analysing suspicious utilization. Not only the eavesdropping of the data connections but also other aspects were noted to be at least equally important to tackle, including protection against DoS attacks and monetary frauds. An example of a simple DoS method is sending a big amount of GPRS PDP Context Activation requests from the Internet to the GPRS network by using imaginary mobile subscriptions of the receiving parties. Prior to the PDP context activation, the GGSN requests information about the location of the B-subscriber to direct the connection to the correct SGSN. As the user does not exist, the signalling results in a failed connection. By repeating these false call requests, the Internet user might cause register signalling overload and prevent the delivery of other traffic. Thus, this type of activity needs to be analysed in real time by the GPRS operator at the GGSN level in order to block attempts prior to the extra signalling.

Other potential GMS security threats relate to the data integrity, which was not provided in the initial GSM networks. The altering of IMEI code might also be possible in practice, although the specifications aim to protect it. It is worth noting that the home network does not have the means to know about the principles the connected roaming networks have for their authentication and encryption. The protection for these potential security threat aspects has been enhanced in the UMTS specification.

8.3.1.1   Altering IMEI

The IMEI identifies of the HW of the mobile equipment. It is thus an integral part of the mobile device and it explicitly identifies the device and its type. Nevertheless, the IMEI does not have a direct connection with the subscriber numbering like the MSISDN or IMSI because the removable SIM cards store all the relevant subscriber information. If the mobile equipment authentication is activated in the network, the call is allowed only if the IMEI is noted to be permitted.

The IMEI code is utilized for the identification of fraudulent or stolen equipment via the EIR. The register contains white, grey and black lists. The white list contains the permitted equipment types while the grey list is used for provisionally allowed equipment. The black list contains the IMEI codes that are not allowed to be used in the network. If the equipment is found in the black list, no communication or signalling is permitted. The one exception is the emergency call which functions regardless of the black list contents.

The IMEI contains 15 digits and includes fields for the Type Approval Code (TAC) of six digits, the Final Approval Code (FAC) of two digits, the Serial Number (SNR) of six digits and the Spare Number (SP) of one digit (which is always set to 0 if the MS sends it). Furthermore, there is the IMEI Software Version (IMEISV) number, which contains the TAC, FAC and SNR fields as well as a two-digit Software Version Number (SVN). Because the IMEI code explicitly reveals the type and commercial mark of the handset, it can be used as a basis for statistics collection by different stakeholders such as MNOs and resellers.

The IMEI is stored in each mobile equipment as well as in the EIR of the home operator. The EIR has a signalling connection with the Mobile services Switching Centre (MSC) to provide the network with the means for verifying that the mobile equipment is legitimate and allowed to communicate in the respective network. The EIR may contain a list of, e.g., non-approved (faulty) model types, stolen or tracked devices.

The EIR is not mandatory in mobile networks. It merely provides additional means for the operators to manage the permissions of individual devices and device types in the network. As for stolen devices, the EIR may have an impact on lowering crime levels. The EIR is s only used sometimeat the national level, or it is not deployed at all in certain areas. So, even if the stolen device can be blocked in the home operator's country, the device may work in other networks by swapping the SIM card if there is no active exchange of the list information between operators.

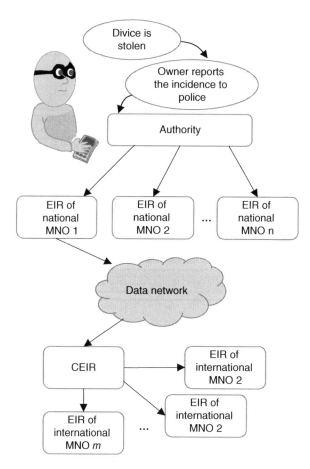

Figure 8.1   The principle of CEIR. Each of the connected operator-specific EIRs is synchronized upon the reporting of devices in their black lists

There also is an international version of the EIR called the Central EIR (CEIR). It collects the lists of operators connected to it via the internal packet network. It keeps the connected EIR elements updated and prevents the utilization of the stolen device in all these networks. Figure 8.1 depicts the procedure of the CEIR communications. The CEIR is maintained by the GSMA under the term IMEI Database to which the cooperating operators have access for updating and retrieving data.

The network may check the IMEI in the initial signalling as well as upon handovers and location area update procedures. The EIR is connected to the MSC via the F interface which is defined in ETSI 09.02. According to the 3GPP specifications, the IMEI should be protected against users' means for modifications. Nevertheless, there have been cases indicating that the IMEI can be altered. Furthermore, there may be devices on the market that do not comply with the standard requirement for the unaltered IMEI, instead, they might generate random IMEIs on a call-by-call basis.

Despite the good intentions, the CEIR does not connect all the operator, although some operators may have direct agreements to change the black lists. What is the impact of altering the IMEI? As the name indicates, it relates to the mobile device and is able to identify explicitly the individual equipment HW. In addition to the EIR and CEIR black list functions which prevent the use of devices with non-allowed IMEI, the operator is able to collect statistics about the IMEI distribution in the network in order to understand the share of different mobile device models as a function of region. If the IMEI of, say, a stolen device is altered, the user of such a device may try to use it regardless of the operator blocking the original IMEI. Naturally, if a certain operator does not have the EIR deployed, the device can be utilized in those areas regardless of the IMEI being on any other black lists. Nevertheless, the IMSI and respective MSISDN implicitly identify the user so there is no way of hiding the user's identity by altering the IMEI unless the IMSI/MSISDN have been obtained anonymously, e.g., by purchasing a pre-paid subscription without physical proof of identity or with an altered ID card.

8.3.1.2 Vulnerabilities of the Encryption

The GSM encryption for the originally strongly protected A5/1 has been noted to be vulnerable and it has been possible to compromise its shield. Ref. [17] presents a ciphertext-only cryptanalysis of the GSM-encrypted communication and active attacks on the GSM protocols. The reference describes a ciphertext-only attack on the least protected A5/2 which requires only few dozen milliseconds of encrypted off-the-air cellular conversation to find the applied key in less than a second on a personal computer. This attack may also be used as a basis for a ciphertext-only attack on the stronger A5/1, and attacks on the protocols of networks that use A5/1, A5/3 and GPRS encryption algorithms are indicated in Ref. [17]. The presented methods use the GSM protocol security flaws and can be applied if the mobile equipment supports the weak A5/2. The reference claims that the method does not require unrealistic information such as long known plaintext periods, or in fact any knowledge of the content of the conversation, but instead are practical and provide the means for decrypting conversations in real time or later.

8.3.1.3 Spoof GSM Base Station

The deployment of a spoof GSM BTS may be possible since the protocol stack of the GSM system is publicly available in the ETSI/3GPP specifications, as presented in Figure 8.2. It can thus be implemented as a simplified variant of the base station equipment by, e.g., creating a pure SW-based emulator that is combined with a low-power GMSK transceiver and portable antenna system. The emulator may contain only the minimum functionality for establishing the call as presented in Figure 8.3.

The BTS emulator can therefore create a local, small coverage area of a single cell which can produce higher received power levels for the mobile stations compared to the legal cells in that area. The GSM devices in the area receive the pilot signalling via the Broadcast Control Channel (BCCH) so the task is to send a copy of some of the legal network IDs (Mobile Country/Network Code, MCC/MNC) with the intention of making the GSM user device camp into that specific cell. As soon as the device has the fraudulent BCCH in the first position in the list of the strongest received cells, and when the user initiates the call, the device has no means of ensuring that the BTS is legitimate. So the device initiates the signalling by assuming

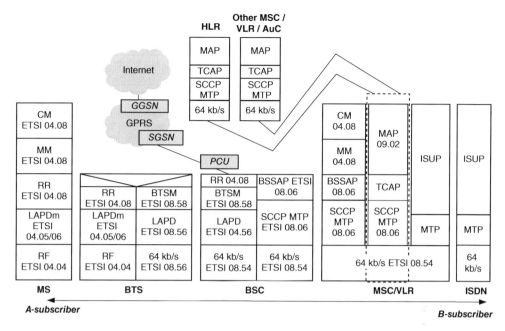

Figure 8.2 The original Phase 1 GSM system's protocol stack from the 1990s, added by the GPRS functionality of Release 97 from the early 2000s

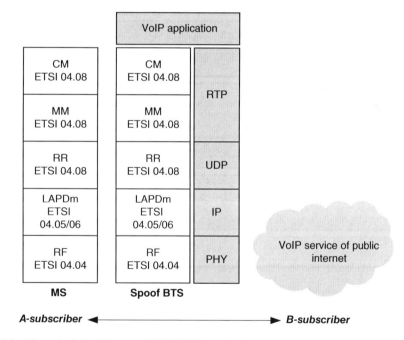

Figure 8.3 The principle of the spoof GSM BTS may be based on the minimum set of the radio interface protocol stack as well as the essential protocols in connectivity and mobility management layers. In this way, all the additional functionality like encryption, frequency hopping etc. can be eliminated from the connection while the interception and relaying of the clear-code call can be done, e.g., via a separate VoIP call

that the cell is either of the home operator or a legitimate roaming cell. As an additional trick to ease the camping, it is possible to block other BCCH frequencies via radio jammers.

This trick is based on the fact that the GSM BTS does not need to encrypt the radio interface, and basically all the standard mobile devices must therefore accept the initiation of the initiated call by utilizing the A5/0 if the BTS so dictates. In this way, the spoof BTS does not need to know the Ki key of the user. Instead, the spoof BTS can behave like a real base station, acting as a simple repeater for the voice calls by intercepting the B-number from the signalling of the A-subscriber which is the victim, and by dialling to the B-subscriber's number via any known means such as Wi-Fi VoIP or pre-paid mobile subscription that is not attached to the identity of the spoof BTS operator. The easiest way to further hide the spoof BTS is to bar the display of the calling number for the B-subscriber which then sees an unknown number calling. As soon as the B-subscriber answers, the spoof BTS is able to intercept all the communications. In addition, the spoof BTS is able to intercept the normally protected A-subscriber's IMSI from the initial signalling, which why these type of devices are referred to as *IMSI catchers*.

The emulated, low-power GSM BTS can fit into a small space as it does not require much processing power. By utilizing a directional antenna, the coverage area of such spoof BTS can be highly focused, and thus a targeted spying can be practiced, e.g., by directing the antenna towards the victim from the other side of the street – which causes additional challenges to find the illegal BTSs as the signal level is weak at street level.

According to the ETSI/3GPP specifications, the GSM device should be able to indicate the lack of an encryption algorithm. A more specific way for displaying it is left open for the device manufacturers. In practice, when only the A5/0 is used for active communications, the device might show, e.g., an open lock symbol – unfortunately the meaning of this symbol might not be too clear for the users. On the other hand, the symbol is not shown at all if the display functionality is deactivated from the internal SIM card settings. Especially in the smart device category, there are also non-compliant models that do not support the display in practice regardless of the SIM settings. Ref. [18] summarizes an example listing of such devices.

The spoof BTS is thus one of the most concrete security threats of the GSM networks as the users may not be able to interpret the warning signs of its presence [14]. Furthermore, it can be used in highly selective ways regarding its location and time to minimize its visibility. As an example, Ref. [2] indicates the probable utilization of such solutions in Oslo during 2014. That specific incidence might be related to the relaying of voice calls via an IMSI catcher.

Regardless of the packet-switched GPRS service's security being deeper within the GSM infrastructure, as the radio interface encryption algorithm is located at the SGSN element, it may be possible to emulate the whole GPRS protocol stacks in a similar fashion as for the spoof GSM BTS for the voice connections, so the relaying and illegal interception of the whole data traffic of the victim who initiates the PDP context might be equally possible [5].

Protection against such relaying threat is possible, e.g., via a separate, application layer protection mechanism like VPN for the GPRS connections or an application layer voice scrambling solution. In its basic form, sending short messages is trickier via such spoof GSM BTS. Also, if the spoof BTS manages to capture the call of the A-subscriber, it is not able to hand over the connection to legitimate base stations as the BSC of the legal network would not be aware of such a call – which means that when the user moves away from the spoof BTS's coverage area, the call breaks down. Nevertheless, as the emulated base station has more control over the RRM, which normally would be taken care of by the BSC, it means that the transmitting power level of the device can be commanded to be higher, up to the maximum specified. In GSM 900, is 2 W (33 dBm) while other variants support maximum power levels between 1–2 W (30–33 dBm). By increasing the power of the device as well as the base station itself, it is possible to extend the coverage area before the victim drifts too far away from the spoof BTS's field.

The spoof GSM BTS as described above functions only for capturing the communications that the subscriber initiates via the spoof BTS. The received call, instead, would be initiated via the legal radio network as it sends the initial signalling via the BCCH of all the base stations of the location area where the B-subscriber is registered, and forces the security algorithm to be switched on (except for the rare cases of a network not utilizing the encryption) in the one that is selected for the communications. Handover for the spoof BTS would no longer work as the BSC is now in control of the whole communications.

Even if the existence of the spoof GSM BTSs is assumingly not common, it can be speculated that the potential danger could increase near strategically important locations. In such places, a simple way to check the potential illegal routing is to send a short message by both parties on the active call to each other. If one of the subscribers is not able to receive the message, there might be a chance that one of the users is attached to a spoof GSM BTS as it is not able to deliver the messages at the same time as the legitimate network.

It is possible to reveal the potential spoof GSM BTSs by executing radio drive tests, and to correlate the BCCH frequencies found in the field with the known locations of the operator's own BTSs. Furthermore, the potential jammer transmitters blocking other legal BCCH frequencies can be located. If the user suspects the presence of such spoof BTSs, the local operator might be able to perform such tests.

8.3.1.4   Key Extraction via Spoof Base Station

A more advanced method compared to the basic spoof GSM BTS is the extraction of the user's permanent Ki key via the same man-in-the-middle attack described in Section 8.3.1.3, but with more sophisticated means. Such incidents have been reported, e.g., by the BBC [11].

The article speculates that this case is based on the solution referred to as 'Stingray'. According to publicly available information, such as Refs. [19,20], the Stingray could be an IMSI-catcher including both passive (digital analyser) and active (cell site simulator) capabilities. When it operates in the active mode, the device mimics the selected MNO cell with an intention to force nearby mobile devices to perform an attach procedure as soon as they initiate a call. The aforementioned references also indicate that the Stingray devices are available as handheld as well as vehicle-mounted devices.

What are the speculated technical details of the above-mentioned product? One educated guess may be that the user's key could be solved by using the spoof BTS to force the least protecting encryption scheme A5/2 for the call in order to solve the user's Ki. After that, as the Ki per individual user is the same for all the A5 encryption variants, once solved, it is possible to eavesdrop the normal calls even without the spoof base station by using the Ki of that specific user, whether the GSM network activates the A5/1, A5/2, A5/3, A5/4 or any other future variant of the encryption algorithm.

8.3.2 Potential Security Weaknesses of 3G

8.3.2.1 Jammers

Even if mutual authentication prevents the deployment of a 3G spoof base station, and the attack methods for the 3G encryption seem to be still merely theoretical, such as exposing potential weaknesses of outdated 3G algorithms or their settings, the existence of the overlapping 2G might still open a potential security threat for 3G users who are also able to use the GSM network [15,16]. One way to exploit this vulnerability is to place radio jammers in the field and prevent the 3G call initiation. Combined with a 2G spoof BTS, the initiating call can thus be forced directly to 2G mode by blocking the 3G radio frequencies locally, which functions as a workaround for bypassing the strong 3G encryption.

In order to prevent the calls from going to the 2G spoof BTS, the customer can enhance the protection level by forcing the confidential calls to be initiated via 3G/4G networks by setting the respective parameter locally on the device. The drawback of this method is that not all modern devices support such settings.

8.4 Protection Methods

The protection methods include various defence techniques by the operators, service providers and end-users. Up-to-date guidelines are constantly needed by network operators, device manufacturers, smartcard providers, financial institutes and app developers – and the end-users. As for the mobile communications area, the MNOs especially need proven security processes which are detailed in the following sections, taking the most up-to-date examples from the LTE domain.

The development of the security processes contains many items. The aim of all the security measures is to prevent in advance possible attacks by shielding the relevant mobile network interfaces and elements in such a way that outsiders have minimum possibilities to perform fraudulent activities. The enhanced phase of 3GPP LTE being one of the most relevant mobile communications systems at the moment, the following sections present key messages from the LTE environment, which can also be adapted largely to any earlier network technology [23].

8.4.1 LTE Security

The security design of LTE/SAE includes feature development according to the best knowledge about the current and future methods for the attacks together with their technical and business impacts on the network. For instance, security threats like a DoS attack can slow down, or in the worst case, paralyze a large part of the network and cause limited availability of services, which leads to loss of revenue and increases the chances of customer churn. One way of developing up-to-date measures against these security threats is to create a security process.

The first step in the security planning is to identify the security threats. Based on the security risk analysis of this phase, the LTE/SAE system is designed and updated accordingly in order to create all the possible counter-measures against the imaginable security risks. This leads to the list of security requirements, and to the specification of the security architecture layout at the system level.

The next step is to take into account the threats in the SW level, securing the safety of the code as much as possible in the SW development processes. The safety threats may be intentional, or they may be a result of accidentally left open backdoors etc. in the production of the code.

At the end of the security design process, a comprehensive security testing is needed, by taking into account the imaginable attack types in the normal operations of the network, as well as in the unstable conditions that are created either intentionally or accidentally.

This example of the security process is logically an iterative activity for the participating stakeholders. This means that as the technologies develop and new methods and ideas for attacks arise, they should be identified and taken into account at the earliest convenience so that the network protection can be updated accordingly to shield against the new threat types. As one part of the new security threat identification, a network fraud monitoring process should be implemented. This provides information about the possible new security threats to be taken into account in the security process.

In addition to the security process, it is recommended to execute security audits in the operator networks. This is an important task as the end-to-end chain of typical mobile networks tend to include a huge amount of different combinations of network elements with different versions and security levels. Both HW and SW can be audited in cooperation with the network vendors and operator. If any vulnerabilities are detected, the issues can be corrected by enhancing updated security threat counter-measures. Figure 8.4 summarizes key aspects of the security chain.

8.4.2   Network Attack Types in LTE/SAE

The LTE/SAE architecture has some special characteristics that should be taken into account in the enhanced security planning. The important fact is that LTE/SAE is based on a flat architecture, which means that all radio access protocols terminate to the eNB elements. Furthermore, the IP protocols are also visible at the eNB.

The challenges arise from the architectural realization of LTE/SAE because it is now possible to place eNB elements at more accessible locations exposing the HW to potential hacking intentions. Furthermore, the LTE/SAE network interworks with legacy and non-3GPP networks that might open unpredictable security holes even if their normal mode of operation does not expose special issues. There are also new business environments with networks whose trustfulness is not necessarily completely known. Comparing the purely IP-based LTE/SAE architecture with the previous 2G/3G principles, LTE/SAE requires extended authentication and key agreement methods to cope with modern IT attacks. This means that the key hierarchy as well as the interworking security are inevitably more complex. It also means that the eNB element has additional security functionality compared to the previous 2G base transceiver stations and 3G NB elements [24,25].

Identification of the potential network attack types in the LTE/SAE environment is one of the most essential preventive tasks. As the Home eNB concept basically means that the

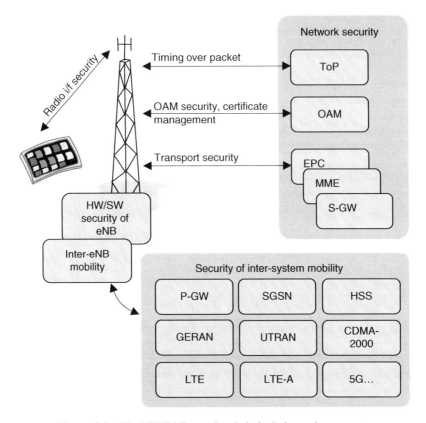

Figure 8.4 The LTE/SAE security chain includes various aspects

customer can try to access physically the HW and Sw of the element, it is one of the most potential avenues for fraudulent activities. Some possibilities are:

- Cloning of the Home eNB credentials.
- Physical attacks on the Home eNB, e.g., in the form of tampering.
- Configuration attacks on the Home eNB, e.g., fraudulent SW updates.
- Protocol attacks on the Home eNB, e.g., man-in-the-middle attacks.
- Attacks against the core network, e.g., DoS.
- Attacks against user data and identity privacy, e.g., by eavesdropping.
- Attacks against radio resources and management.

8.4.3 Preparation for the Attacks

A more detailed list of LTE/SAE security-related items to be taken into account in the security process include the following:

- Air-link security (U-plane and C-plane security). This includes the definition and description of the ciphering algorithm for the U-plane and C-plane, definition and description of the integrity protection algorithm for the C-plane, and description of the access stratum security signalling (including key distribution).

- Transport security. This item includes the definition and description of ciphering and integrity algorithms for the transport network, and description of the transport security signalling (including key distribution).
- Certificate management. This includes the definition of public key and key management concepts.
- Operations, Administration and Management (OAM) security (M-plane security). This includes the management plane security.
- Timing over Packet (ToP). This includes the synchronization plane security for IEEE v2 packets for frequency and time/phase synchronization.
- eNB requirement. This includes the definition of secure environment, requirement definition for eNB and secure key and file storage.
- Intra LTE and inter system mobility. This includes the definition of security aspects in handover cases (including key distribution).

It should be noted that different planes differentiate the traffic types, and this should be taken into account in the security planning. The planes in the LTE/SAE environment are: U-plane for the delivery of the user data; C-plane for the delivery of the control data; M-plane for the delivery of the management data; and S-plane for the frequency and time/phase synchronization information. Figures 8.5–8.8 identify the security-related aspects of these planes.

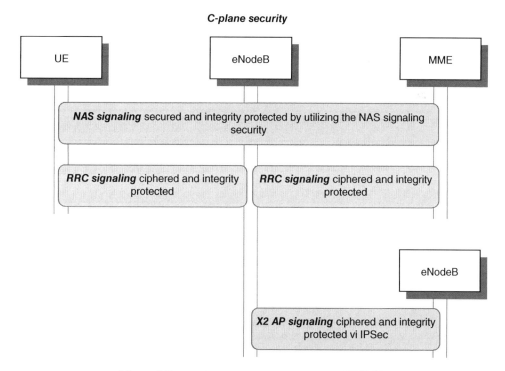

Figure 8.5 The C-plane security principle of LTE/SAE

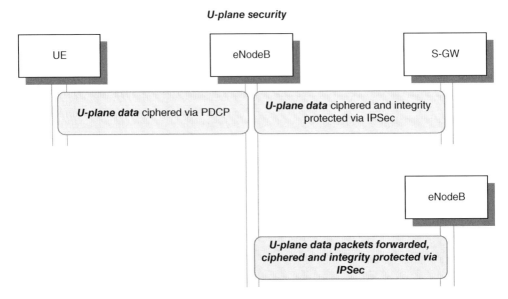

Figure 8.6   The U-plane security principle of LTE/SAE

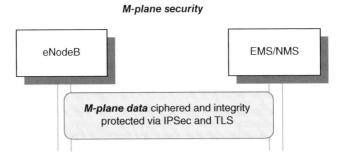

Figure 8.7   The M-plane security principle of LTE/SAE

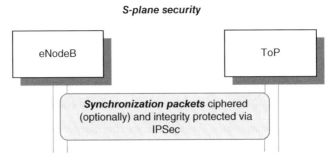

Figure 8.8   The S-plane security principle of LTE/SAE

As will be described in the next sections, IPSec is the 3GPP standardized solution for security on several LTE interfaces. These are *S1-MME* and *X2* control planes as well as *S1* and *X2* user planes. Security for the management plane is not standardized but the use of IPSec or transport security is one possible option. In addition, use of IPSec in combination with certificates makes it very difficult for any unauthorized person to gain access to the core network or eavesdrop the traffic between the eNBs and the core network. In this way integrity and confidentiality of data can be ensured.

An integral part of the preparation plan against security attacks include network security audits which can be an operator's own activity or an external managed service. The main aim of this activity is to check the adequate configuration of the network element settings, and ensure the potential security holes are identified and protected. Other techniques include network and security breach monitoring of the traffic, typically relying on DPI. It can be combined with information about historical behaviour and deviations to capacity, performance, traffic type, etc. Logically, the traditional methods such as virus protection are still applicable if provided as a real-time network service.

8.5   Errors in Equipment Manufacturing

One potential threat in the wireless environment is a result of the potential security holes that remain in the network equipment or user devices such as smart devices. A concrete challenge in this environment is typically tight time schedules in the equipment production which may sometimes cause reduced testing of the bugs in HW and SW. Special attention is thus needed in the testing. Furthermore, it is highly recommended for all the involved parties to improve the testing and optimize the issue 'hunting' as early as feasible. In this way, the OEMs can identify the issues – including security holes – in time. This early testing concept is thus essential in modern equipment manufacturing. The following sections outline the basis and idea of this concept, which is valid for security teams as well as other technical area representatives.

8.5.1   Equipment Ordering

As a general rule of thumb, equipment investment should not be done too early because of the warehouse storing costs, danger of the expiration of warranties, possible increasing of damaged units when stored, and potential existence of early version bugs which can jeopardize the security of the device. Nevertheless, if the investment is done too late, it might impact seriously on the deployment as the time schedules for the ordering might vary and it might not be possible to align the schedules with the actual deployment at the last moment. In practice, there are sometimes issues with component availability, logistics chain and the physical delivery of the equipment to the final sites.

Some operators may want to test the new equipment via special stress procedures which are sometimes only possible after the end-to-end chain is upgraded to an adequate level. This type of Inter-Operability Testing (IOT) is typically requested to ensure that the element is compatible with the operator's infrastructure. The potential problems require time to correct, with possible escalation efforts, which also should be taken into account in the time required for equipment order management.

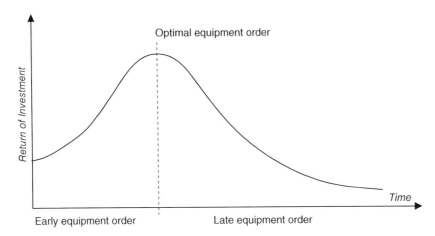

Figure 8.9   The correct timing for the equipment ordering has impact on the RoI

The correct timing is thus essential to ensure that the delivery chain is aligned with the LTE/ LTE-A deployment. This is one of the key optimization tasks of the operator apart from the actual network planning optimization efforts.

As Figure 8.9 indicates, there is an impact on the Return on Investment (RoI) related to the timing of the physical equipment ordering. If the equipment is ordered too early, the number of elements might not have the correct HW/SW level as the final deployment plan might change due to unexpected challenges, e.g., due to difficulties obtaining physical sites. If the equipment cannot be installed straight away, warehouse costs increase, and the equipment might be even become outdated and its warranty expire. Nevertheless, there should always be an optimal buffer for spare parts that assures the fast replacement of faulty modules. Another important aspect is the spare stock of the equipment. The size of the stock needs to be sufficiently complete for potential HW and outdated security protection problems of the equipment in the field.

On the other hand, if the equipment is ordered too late, there might be hard-to-estimate challenges in the production !ine schedules. If the line is saturated, the order might take a much longer time than average, which delays the new equipment installation. In addition to the HW readiness, the aligning of the SW level is also important. It does not make sense to order equipment with pre-loaded SW if the roadmap indicates that the SW needs to be soon upgraded.

For the overall equipment logistics, a good tool for optimal equipment ordering is an inventory management SW. It gives the possibility to track efficiently the logistics chain, and to assure the equipment arrives at the correct location on time.

8.5.2   Early Testing

Testing new network element functionalities early enough is one of the essential tasks of the operators and equipment providers for successful network deployment. It should be noted that this principle applies to all the network elements as well as the UE. In fact, the lifecycle and the introduction of new smartphone models to the markets is accelerating in such a fast way that there is a great potential to fail in the assurance of some performance targets or functionalities – including the assurance of an adequate security level. This is especially important in cases where the operator has ordered an exclusive model from the OEM. The functionality against the operator's own requirements is done by the OEM as deeply as the risk analysis indicates, but there may be remaining issues during the final acceptance testing of the operator, or even in the commercial phase after the launch of the device, e.g., due to the failure in quality processes of third parties. The late noting of the failures presents the worst-case scenario as it jeopardizes the public image of the operator and the OEM, and the corrective measures may lead to high expenses.

As the time schedules are typically very tight from the initial phase of equipment research, continuing into the development phase, and during the actual manufacturing, this means that the testing of previously correctly functioning items might be minimized in the faith that there would be no issues in new variants, either. This might lead to vast problems, as the new UE functionalities, the additional new RF bands, new MIMO and CA configurations, multi-mode operation, and chip-level architectural changes might cause hard-to-predict issues. As an example, the previously well-behaved GSM RF performance of the UE might suffer from additional LTE bands due to different filter adjustments, making the RF performance unacceptable in parts of the GSM RF bands. If the OEM does not perform (repeat) all the essential RF performance measurements in all the bands and systems, it means there will be unpredictable risks in the acceptance phase of the device – and in the worst case, it might result in non-optimal functioning of the device, which is only noted in the commercial phase of the device, in operational networks.

Along with the development of the digital cellular networks and devices since the 1990s, there are many examples about devices not working according to the specifications in certain situations, which was not recognized before the devices were utilized in the commercial markets. The operator has basically only a few options in these situations: (1) to accept a lower quality for the users of the faulty device which may even decrease the quality of the another user; (2) to modify the functionality of the network so that the faulty behaved device causes minimal problems; (3) to upgrade the device. Upgrading might be possible to execute via an OTA SW upgrade procedure, or in the most costly case, by collecting the faulty devices physically from the customers and upgrading them or replacing the devices completely. This is obviously so expensive that the business case for that device may turn out to be negative, especially in the case of devices with tight margins. The modification of the network functionality for supporting faulty devices is also sometimes utilized, if possible, to minimize replacement costs. The drawback of this solution is that the operator needs to ensure the adjusted network functionality in the longer run, e.g., when there are general SW upgrades or new releases. This obviously increases the complexity of the network management. So, early testing is the best solution for both operator and OEM in order to minimize later costs.

The introduction of LTE/LTE-A increases even more the probability of remaining faults in the network elements and user devices as a result of new, advanced technologies, functionalities and high performance requirements. Figure 8.10 presents an ideal high-level process for the new equipment manufacturing. This refers to any LTE/LTE-A network element as well as to user devices which contain new functionalities or enhanced performance, i.e., to equipment which has modifications or new HW and SW. In the real life, errors might only be revealed in the late phase of the testing, relatively close to the planned launch date of the equipment which, in the worst case, may lead to delays for market entry, as indicated in Figure 8.11.

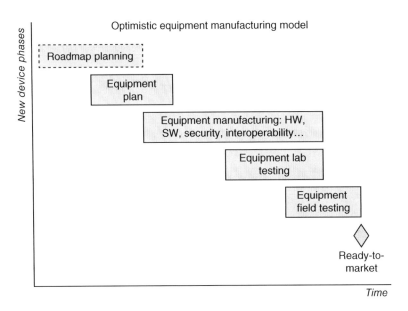

Figure 8.10   General principles of equipment manufacturing

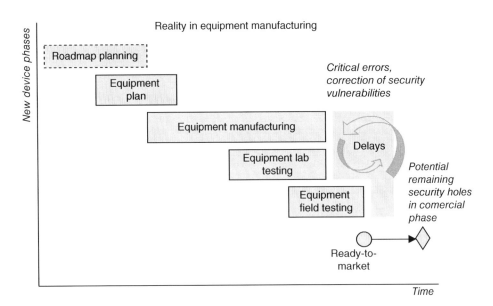

Figure 8.11   An example of a real-world scenario which sometimes may experience delays in commercial market entrance due to issues that are identified too late prior to launch

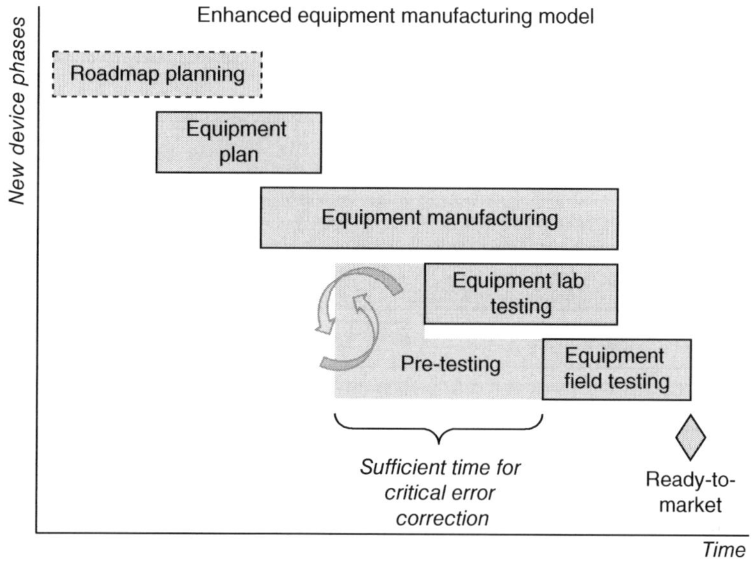

Figure 8.12   Issues resulting in delayed market entrance can be minimized via preliminary testing activities as soon as the equipment prototypes are ready

As much pre-testing as reasonably possible is recommended for the equipment manufacturing, prior to the deadline of the equipment manufacturing phase, yet balancing the expense. Figure 8.12 clarifies the early testing concept. This emphasis on early testing is due to the fact that new functionalities might have interfering effects on the equipment that was perhaps not an issue in earlier models.

Also, there might be some operator-specific demands and important requirements that need additional testing. If any issues related to such operator-specific functionalities are noted late in the field test phase, it might be much more challenging to get corrections on time. This is especially relevant for errors that are related to the operational system or chip-sets. In order to assure the capturing of potential problems, it is recommended to execute operator-specific field tests (trials, pilots, friendly operator's lab tests) during the development phase so that the fault management can react early enough for the requests for corrective actions internally and for the third parties. Corrections of equipment faults is typically done by opening a request for a fix, i.e., issue a ticket via an established process, as presented in Figure 8.13.

It should be noted that the principal threat in the device HW and embedded OEM SW is not because of faults in the trust chain – although this cannot be excluded completely, e.g., due to any untrustworthy individuals who might alter the device in the planning or manufacturing phase. It can be assumed that the potential security holes on devices – whether they are network elements like cellular base stations, Wi-Fi access points or core elements like routers, bridges, gateways etc. – are due to unintentional failures in the design or assembly line. Whatever the root cause for potential security vulnerabilities, the sufficiently thorough and early testing of the equipment for the functionality and performance but especially for the level of security are of the utmost importance to identify and correct any weaknesses. One important part for ensuring the HW functionality in the critical areas such as payment solutions require certification which

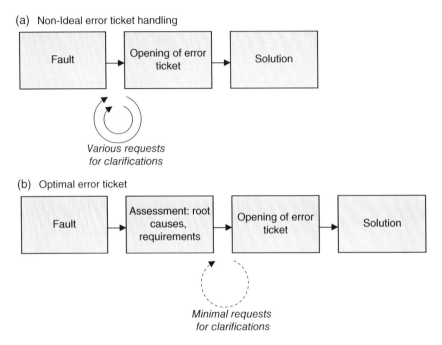

Figure 8.13 Process for the error ticket opening applicable to LTE/LTE-A UE and network elements. The optimal way is to assess deeply the background information prior to the error ticket opening in order to speed up corrections

is often time consuming but well worth the effort. As soon as the HW is approved and deployed, the tailored, periodic security audits are of great help in order to guarantee that the protection mechanisms as well as the potential, internal HW flaws are not exposed to the external world before proper correction takes place.

The same principles apply to accompanying components of the end-to-end chain, including the thorough security testing and audits of the expected functionality and interoperability of the SIM/UICC in the subscription management including OTA provisioning and advanced remote access methods.

8.6   Self-Organizing Network Techniques for Test and Measurement

8.6.1   Principle

Self-Organizing Network (SON) is a new technique being introduced within LTE/SAE as part of the next generation mobile broadband network technology, and is endorsed by the NGMN Alliance as a key requirement for future networks. The objective is to automate the configuration and optimization of base station parameters to maintain best performance and efficiency. In fact, the SON concept is not only highly useful in network planning and optimization, it also tackles potential security breaches by ensuring the correct setting of parameters.

Previously a drive test team would go out into the live network and take a 'snapshot' of the performance, then bring this back to the lab and analyse it to improve the settings. More snapshots provide more statistically significant data and hence a better base for the optimization, but this drive test based data acquisition process is expensive, difficult and not repeatable. In addition, this is a reactive method to solve problems after they have occurred, and it does not help to improve customer experience if the issue is already visible. A drive test is also heavily used at the start of a network deployment to measure cell coverage and set initial parameters for cell powers, frequencies etc. in order to control interference and maximize capacity.

SON should enable network operators to automate these processes using the measurements and data generated in the base station during normal operation. By reducing the need for specific drive test data, this technique should reduce operating costs. By using real-time data generated in the network, and reacting in real time at the network element level, this should enhance customer experience by responding more dynamically to changes and problems in the network much earlier so that users are less affected.

SON can simplify an operator's processes to install new cell sites, reducing the cost, time and complexity. SON gives an obvious benefit if deploying femto cells, as the operator is not strictly in control of the cell site and needs to rely on automated processes to correctly configure the cell into the network. In addition, the running costs of the site are reduced, as drive test optimization is reduced and site visits for fault investigation and repair can be reduced. All of this leads to OPEX savings for the network by using automated technology to replace manual operations. For the customers on the network, SON will lead to better customer satisfaction as coverage and QoS are driven and optimized by actual customer usage, and there should be reduced downtime or faulty cells. The OSS monitoring systems and SON should work together to automatically detect usage trends and failures and automatically take action in real time to correct errors.

SON is the top level description of the concept for more automated (or fully automated) control and management of networks, where the network operator has only to focus on policy control (admission control, subscribed services, billing etc.) and high level configuration/planning of the network. All low level implementation of network design and settings is made automatically by the network elements. The self-organizing philosophy can then be broken down into three generic areas relating to the actual deployment of the network, these are configuration (planning and preparation before the cell goes live), optimization (getting the best performance from the live cell) and healing (detection and repair of fault conditions and equipment failures). Each of these is further explained below.

8.6.2   Self-configuration

This is the first stage of network deployment, and covers the process of going from a 'need' (e.g., improve coverage, improve capacity, fill a hole in coverage) to having a cell site 'live' on the network and providing service. The stages involved here are roughly:

- Planning for location, capacity and coverage of the eNB.
- Setting of the eNB parameters (radio, transport, routing and neighbours).
- Installation, commissioning and testing.

The self-configuring network should allow the operator to focus on selecting location, capacity and coverage needs, and then SON should automatically set the eNB parameters to enable the site to operate correctly when powered on. This will in turn minimize the installation and commissioning process, and enable a simple 'final test' at the site to confirm that the new site is up and running. This will include the optimum setting for power levels, the choice of Cell ID, and correctly identifying the Cell ID of neighbour cells. The neighbour Cell ID must then be used by the eNB to negotiate with these neighbours using the Inter Cell Interference Control (ICIC) algorithms on the *S2* interface. This is critical to prevent interference where the coverage of two cells is overlapping.

8.6.3   Self-optimizing

Once a site is live and running, there are often optimization tasks to be made that are more of a 'routine maintenance' activity. As the geography of the area changes (e.g., buildings constructed or demolished), the radio spectrum changes (e.g., new cells added by the operator or by other operators, or other RF transmitters in the same area or at the same tower), then the neighbour cell lists, interference levels and handover parameters must be adjusted to ensure smooth coverage and handovers. Currently, the impact of such issues can be detected using an OSS monitoring solution, but the solution requires a team to go out into the field and make measurements to characterize the new environment and then go back to the office and determine optimum new settings. SON will automate this process by using the UE in the network to make the required measurements in the field, and the SON function will report them automatically back to the network. From these reports, new settings can be determined. This will remove the need for drive test teams to make such measurements. This concept can also be extended to managing QoS and load balancing by using quality reports to optimize the scheduling algorithms in the eNB.

8.6.4   Self-healing

When a site is fully operational and active then it is generating revenue and satisfying customers. If there are any problems with the site and it fails to provide a service or coverage then revenue and/or customers are lost, and so the site must be brought back up to full capacity as soon as possible. The third element of SON is to automatically detect when a cell has a fault (e.g., by monitoring both the built-in self-test, and also the neighbour cell reports made by the UE that are/should be detecting the cell). If the SON reports indicate a cell has a failure then there are two necessary actions: to indicate the nature of the fault so the appropriately equipped repair team can be sent to the site;, and then to re-route users to another cell if possible and to re-configure neighbour cells to provide coverage in this area while the repair is underway. After the repair, SON should also take care of the site re-start in a similar process to the site commissioning and testing. The self-healing functionality under the SON concept thus refers to the automatized fault management and correction.

8.6.5   Technical Issues and Impact on Network Planning

To deploy SON in a multi-vendor RAN environment requires standardization of parameters for reporting and decision making. The eNB will need to take the measurement reports from the UE and also from other eNBs and report them back into the Operations and Maintenance

(O&M) system to enable optimization and parameter setting. Where there is multiple vendor equipment involved then this must be in a standardized format so that the SON solution is not dependent on a particular vendor's implementation.

The equipment vendors who are implementing SON will need to develop new algorithms to set eNB parameters such as power levels, interference management (e.g., selection of sub-carriers) and handover thresholds. These algorithms will need to take into account the required input data (i.e., what is available from the network) and the required outcomes (including cooperation with neighbour cells).

Furthermore, as SON is also implemented into the core network (Evolved Packet Subsystem, EPS), there needs to be standards on the type and format of data sent into the core. Inside the core network, new algorithms will be required to measure and optimize the volume/type of traffic flowing taking into account the QoS and service type (e.g., voice, video, streaming, browsing). This is required to enable the operator to optimize the type and capacity of the core network, and adjust parameters such as IP routing e.g., in a Multiprotocol Label Switching (MPLS) network, traffic grooming, and so on.

8.6.6   Effects on Network Installation, Commissioning and Optimization

Equipment vendors have the opportunity to develop algorithms that link eNB configuration to customer experience, allowing fast adaptation to customer needs. Here the customer experience is exactly that which is measured by their UE in the network. The challenge is to link RF planning and customer 'quality of experience' closer together at a low level technical implementation. The benefit is that the network can adapt to meet the user needs in the cell without additional cost of optimization teams constantly being in the field. The network planners' simulation environment will now need to take into account the SON operation of the eNB when making simulations of capacity/coverage for the network. As the operator may not directly control/configure the eNB, the simulation environment will need to predict the behaviour of the network vendors' SON function in the network.

An operator/installer's site test must verify that all parameters are correctly set and working in line with the initial simulation and modelling. This will ensure that the expected coverage and performance is provided by the eNB. The SON function will then self-optimize the node to ensure that this performance is maintained during different operating conditions (e.g., traffic load, interference). This should reduce the amount of drive tests required for configuration and optimization (in theory reduced to zero), so that a drive test is only needed for fault finding (where SON is not able to self-heal the problem). We will see in the later section on live network testing, that a comprehensive suite of RF OTA tests can be made at the time of site installation, so that the SON can be correctly configured and verified. It is generally expected that SON will be able to reduce the level of drive tests needed to initially configure the network, but will not be able to replace the initial site commissioning/acceptance tests. So the preferred test strategy is to use the initial site tests to further strengthen the SON parameters setting.

The potential disadvantage of running SON in the network is that it requires the UE to make the measurements, and relies on enough data being available. The eNB is able to command the UE to make measurements and report them, but doing this on a regular basis will have an impact on the battery life of the UE. On current generation smartphones the battery life is already a limitation, so extra SON measurements should not significantly reduce the battery life further.

8.6.7 SON and Security

Although the SON concept was principally developed to ensure proper network functionality and performance, it could also be used as a feasible base for ensuring the security. Some ideas for the applied SON functionalities may include the automatized and repeatable security audits via controlled (isolated) stress testing in pre-commercial and commercial networks.

References

[1] I. Androulidakis, D. Pylarinos and G. Kandus. Ciphering indicator approaches and user awareness. Maejo International Journal of Science and Technology, 6(3):514–527, 2012.

[2] Aftenposten. The spoof GSM base stations revealed in Oslo, 16 December 2014. http://www.aftenposten.no/nyheter/iriks/Secret-surveillance-of-Norways-leaders-detected-7825278.html (accessed 4 July 2015).

[3] 3GPP TSG SA WG3 Security – SA3#25 S3-020557, 8–11 October 2002. http://www.3gpp.org/ftp/tsg_sa/wg3_security/tsgs3_25_munich/docs/pdf/S3-020557.pdf (accessed 4 July 2015).

[4] Wired. GSM spoof BTS demo, 31 July 2010. http://www.wired.com/2010/07/intercepting-cell-phone-calls (accessed 4 July 2015).

[5] Forbes. GPRS relay, 19 January 2011. http://www.forbes.com/sites/andygreenberg/2011/01/19/smartphone-data-vulnerable-to-base-station-spoof-trick/ (accessed 4 July 2015).

[6] Forbes. Information security of automotives, 8 April 2014. http://www.forbes.com/sites/andygreenberg/2014/04/08/darpa-funded-researchers-help-you-learn-to-hack-a-car-for-a-tenth-the-price (accessed 4 July 2015).

[7] J. Penttinen. The Telecommunications Handbook. John Wiley & Sons, Inc., Hoboken, NJ, 2015.GSMA. Embedded UICC Protection Profile, Version 1.0, 22 September 2014.

[8] Verizon Security Breach Report, 2015. http://www.verizonenterprise.com/DBIR/2015/ (accessed 19 April 2015).

[9] Wi-Fi security description by Wi-Fi Alliance, 2015. http://www.wi-fi.org/discover-wi-fi/security (accessed 13 June 2015).

[10] BBC. Mass snooping fake mobile towers 'uncovered in UK', 10 June 2015. http://www.bbc.com/news/business-33076527 (accessed 14 June 2015).

[11] 2016 Data Breach Investigation Report. Verizon, 2016.

[12] Intel Curie, 2015. https://iq.intel.com/tiny-brain-wearables-cute-button/ (accessed 15 June 2015).

[13] M. Green. A few thoughts on cryptographic engineering, 14 May 2013. http://blog.cryptographyengineering.com/2013/05/a-few-thoughts-on-cellular-encryption.html (accessed 4 July 2015).

[14] 3GPP TS 55.216 V6.2.0 (2003-09). Technical Specification, 3rd Generation Partnership Project; Technical Specification Group Services and System Aspects; 3G Security; Specification of the A5/3 Encryption Algorithms for GSM and ECSD, and the GEA3 Encryption Algorithm for GPRS; Document 1: A5/3 and GEA3 Specifications. (Release 6).

[15] M. Walker. On the security of 3GPP networks. Eurocrypt 2000.

[16] Elad Barkan, Eli Biham and Nathan Keller. Instant ciphertext-only cryptanalysis of GSM encrypted communication. Advances in Cryptology – CRYPTO 2003. Lecture Notes in Computer Science Volume 2729: 600–616, 2003.

[17] I. Androulidakis, D. Pylarinos and G. Kandus. Ciphering indicator approaches and user awareness. Maejo International Journal of Science and Technology, 6(3):514–527, 2012.

[18] Jen Valentino-Devries. Stingray phone tracker fuels constitutional clash. Wall Street Journal, 22 September 2011.

[19] WPG Harris. Harris Wireless Products Group catalog, page 4. 25 August 2008. https://www.documentcloud.org/documents/1282631-08-08-25-2008-harris-wireless-products-group.html (accessed 4 August 2015).

[20] L. Liang, S. Iyengar, H. Cruickshank and Z. Sun. Security for the flute over satellite networks. Proceedings, International Conference on Communications and Mobile Computing, Kunming, China, January 2009. Pp 485–491.

[21] M. Mahmoud, N. Larrieu, A. Pirovano. An aeronautical data link security overview. Proceedings, IEEE/IAII Digital Avionics Systems Conference, Orlando, USA, October 2009. Pp 4.A.4-1–4.A.4-14.

[22] 4G mobile broadband evolution. Release 10, Release 11 and beyond, HSPA+, SAE/LTE and LTE-Advanced. 4G Americas. October 2012.

[23] 3GPP TR 33.902, V4.0.0 (2001-09). Technical Report, Technical Specification Group Services and System Aspects; 3G Security; Formal Analysis of the 3G Authentication Protocol (Release 4).

[24] 3GPP TS 33.102, V2.0.0 (1999-04). Technical Specification Group (TSG) SA; 3G Security; Security Architecture, version 2.0.0.

[25] Example, SIM Ki key breach: http://www.gemalto.com/press/Pages/Gemalto-presents-the-findings-of-its-investigations-into-the-alleged-hacking-of-SIM-card-encryption-keys.aspx (accessed 12 April 2015).

[26] *Signal*. Incoming: What is a cyber attack? 1 January 2015. http://www.afcea.org/content/?q=incoming-what-cyber-attack (accessed 3 December 2015).

[27] Radio mines in World War II (in Finnish). https://fi.wikipedia.org/wiki/Viipurin_radiomiinat (accessed 3 December 2015).

[28] *Defense News*. Time To refocus on the EMP threat, 18 August 2015. http://www.defensenews.com/story/defense/commentary/2015/08/18/time-refocus-emp-threat/31915021/ (accessed 3 December 2015).

[29] Real Clear World. NATO tries to define cyber war. 20 October 2014. http://www.realclearworld.com/articles/2014/10/20/nato_tries_to_define_cyber_war_110755.html (accessed 3 December 2015).

[30] Suomen sotilas, historiikki Viipurin radiomiinoista. http://www.suomensotilas.fi/sakkijarven-polkka-eli-elsoa-jatkosodan-ajoilta/ (accessed 7 January 2016).

[31] Guillaume Barbu, Christophe Giraud andVincent Guerin. Embedded eavesdropping on Java Card.

[32] Dimitris Gritzalis Steven Furnell and Marianthi Theoharidou. 27th IFIP TC 11 Information Security and Privacy Conference, SEC 2012., Jun 2012, Heraklion, Greece. Springer, 376, pp.37–48, 2012, IFIP Advances in Information and Communication Technology.

9

Monitoring and Protection Techniques

9.1 Overview

This chapter discusses techniques for the protection of mobile communications, services, users and applications. 'Traditional' solutions such as firewalls are important means for the isolation processes, as well as techniques based on the IPSec gateway. The role and principles of monitoring techniques as well as subscription modules are also discussed, including remote estimation of HW deterioration as part of the preparation for possible faults that may even expose security weaknesses. Real-time network analysis and protection techniques are summarized such as deep packet investigation, virus protection and legal interception. Also a few words about location privacy and health-related security of wireless systems are presented.

The number of mobile communications users is growing at a global level, and as data utilization increases, the MNOs are facing a constant need to adapt to or further cooperate with the ISPs. At the same time, in addition to the 'traditional' MNO role of providing voice calls and data service via a sufficiently well-dimensioned bit pipe, the stakeholders have new challenges such as how to keep the performance and availability of the services sufficiently good, how to continue delivering new, rich and interesting user experiences and how to protect the networks, end-user contents, devices and applications against known and future security threats. If the MNOs manage to tackle these challenges, then they enhance customer happiness, retain churn and grow their businesses. In the current, highly competitive environment, these are elemental tasks for any MNO as well as any MVNO.

The enhanced functionality of smart devices and even less complex feature phones have brought new challenges as they function as a key base for the increasing demand for data consumption. So, operators need to satisfy the need for increased use of Internet services while ensuring proper protection against known and all kinds of new security risks in all the deployed network generations.

Wireless Communications Security: Solutions for the Internet of Things, First Edition. Jyrki T. J. Penttinen.
© 2017 John Wiley & Sons, Ltd. Published 2017 by John Wiley & Sons, Ltd.

The statistics clearly show that the smartphone apps have increased the network signalling which may have a negative impact on the overall network performance. In a certain sense, this is a positive challenge for the operators and other stakeholders as it increases revenue, but at the same time, the increased signalling load generated by the legitimate app users may be comparable to a non-malicious Distributed Denial of Service (DDoS) attack as Ref. [4] has noted. In addition, the new mobile devices and apps may bring along novelty vulnerabilities especially in the application layer. The overall trend for moving towards all-IP mobile network architectures, which the LTE/LTE-A deployments are spearheading, also opens potential security holes which may facilitate IP-based security attacks. This means that the MNOs need to constantly update their security processes and protection mechanisms.

9.2 Personal Devices

9.2.1 Wi-Fi Connectivity

Wireless devices, including laptops and smart devices which are equipped with Wi-Fi connectivity, are vulnerable to viruses and security breaches via public Wi-Fi hotspots. Reasonable care should be taken whenever the device is connected to publicly available Wi-Fi access points because the hotspot traffic may be monitored by hackers. For the most efficient protection, the safest way would be to completely avoid sensitive information transfer in public Wi-Fi areas as a hacker may have set up an innocent-looking but fraudulent Wi-Fi hotspot (with correct-looking ID) and route the user's data via the hacker's own computer which allows the hacker to monitor incoming and outgoing traffic. Also, hackers may monitor the legitimate signalling of a new user entering the hotspot upon the logon procedure, with the aim of capturing the credentials and computer's identity information, and logon later to the same services with the replicated set. As some services such as web email and social media may keep the original user logged on for a longer time period, the hacker may thus, once the overall access credentials are replicated, still try to access the original user's resources via the 'hijacked' Wi-Fi session.

The home Wi-Fi router is also a potential target for hackers. From the end-user's point of view, it is highly recommended to use a sufficiently complicated password to access the equipment, and a separate, equally complicated password to access the access point router within the home Wi-Fi radio coverage. Additional safety for lowering the risk is simply to switch off the router HW when not in use.

9.2.2 Firewalls

A firewall is one of the most commonly used protection mechanisms to limit the access to and from data networks. The firewall can be deployed within the data network infrastructure as a standalone component, or it can be integrated into other network elements such as the GGSN of the GPRS core. The firewall is very typical as a form of application installed in laptops and other wireless and wired devices.

The firewall can also be embedded into the SIM/UICC card. In this context, the application firewall refers to the functions of the eUICC Runtime Environment that restricts the capability of applications to access or modify data belonging to other applications. The Java Card System Firewall is an example of such an application firewall as stated in Ref. [7].

9.3 IP Core Protection Techniques

9.3.1 General Principles

'Traditional' isolation techniques are based on firewalls. As is the case in the fixed Internet, mobile communications networks have also had firewalls since the deployment of packet data services. The core of the mobile communications network consists of elements designed for exchange and routing functionalities. If the elements or interfaces are exposed to unauthorized parties via the public IP networks, the hackers may be able to intercept the traffic and interfere with the communications, modify the contents or block the services. A firewall therefore provides a straightforward and feasible protection method.

The GPRS initiated the all-IP concept for the GSM system. The packet system architecture already contained an initial firewall by default since the first ETSI Release 97 in the form of the GGSN. In addition to the firewall in the *Gi* interface between the GPRS core and Internet (or other packet data network), the GPRS also includes other interfaces such as the *Gp* towards roaming partners (GPRS Roaming Exchange, GRX) which might be equally compromised if not protected accordingly. For that reason, the security of IP-based mobile transmission needs additional methods for packet filtering to protect the network against spoofing attacks and billing alterations. There also needs to be Network Address Translation (NAT) to hide the internal GPRS network's user IP addresses (typically in the address space of 10.x.x.x) from the external IP address space. Additional traffic analysis methods are useful for discovering malicious signalling patterns such as intentions to bombard the HLR with false requests to overload the GPRS (and at the same time, the circuit-switched part of the GSM network).

The situation gets more complicated, though, as the previously relatively closed and protected 2G and 3G core networks are now in transition to full-IP networks as a result of the deployment of the LTE and LTE-Advanced systems. As a potential new risk, non-authorized people could access unencrypted user traffic or network control signalling traffic by utilizing IP-methods whereas the previously more isolated core network prevented such intentions by the architecture itself.

There are also potential dangers in the internal communications links. An example of this threat is roaming. Even if the roaming partners are trusted by default, there may be malicious attack intentions if hacker enters the international networks and aims to penetrate the partnering networks via such interfaces. With the public Internet and other untrusted external networks being increasingly involved with the routing, the security risks can be assumed to increase. As e radio interface protection has developed, the LTE and its evolved variants will need to be protected with special emphasis on the core side as it is based purely on the IP connectivity. A priority of MNOs is thus to ensure the protection of the mobile system interfaces, including the *Gi* towards Internet, *S1* for the radio access and *Gp* towards roaming partners [4].

Ref. [4] proposes a holistic approach via the use of a single security platform that integrates the inspection of the interfaces combined with other security applications such as IPS, VPN tunnelling, secure NAT, antivirus, anti-bot and web security. The benefits of such centralized security solution is the unified policy of the functions and a single point of monitoring and reporting which eases the management of the complete setup.

9.3.2 LTE Packet Core Protection

The LTE core network needs to be protected via the *Gi* and *SGi* interfaces by shielding against any Internet attack. One of the challenges is that modern smart device apps include highly advanced functionalities which are often based on the complex utilization of IP addresses. The MNOs therefore need to handle the growing amount of such IP addresses in a scalable way and still be able to identify a single user's devices, e.g., while not confusing the complex communications with DoS attacks. The public IP address and the translation between private and public domains can be handled by Carrier-Grade NAT (CGN) in the *Gi* and *SGi* interfaces. The migration from the IPv4 to the IPv6 is an essential part of this development which needs to be taken into account in the strategies for supporting both variants during the transition period [4].

The CGN can handle the excess of occasional signalling load which is a result of all kinds of activities in the Internet community, including random port scans, typical in the current environment. In addition to the general signalling load, there can be intentional focused signalling overloading to selected targets within the MNO's radio and core infrastructure blocking the traffic or, for the mobile devices of selected end-users, jamming the radio access.

9.3.2.1 *Gi/SGi* Interface

The idea of the CGN is to hide the core service and device IP addresses behind the *Gi/SGi* interface so that they are not visible in the public Internet. This method protects the services and devices against targeted DoS attacks. Furthermore, it protects against the potential 'hijacking' of the IP address of the device which could otherwise lead to charging attacks. Figure 9.1 depicts an example of the CGN firewall deployment as interpreted from Ref. [4].

Continuing with the solution presented in Ref. [4], the CGN can be used for securing the *Gi* interface in a stateful NAT firewall mode which is optimized for Internet traffic as for the voice service session-based applications and protocols. Stateful firewall refers to element tracking the operating state and network connection characteristics by traversing traffic and is thus able to distinguish legitimate packets from the bit stream. The risks of these cases are related to the threats of sharing IP addresses which may open doors for overbilling attacks against subscribers and carriers.

Ref. [4] advices that the NAT firewall, as depicted in Figure 9.2, ideally works as a single, scalable gateway. It can be managed by a single IP address and enables a single-console security and policy management functions thus easing the tasks and providing efficient traffic balancing. This is especially suitable for chassis-based solutions with multiple gateway modules stacked on top of one another. The NAT performance and throughput are essential factors along with increasing network traffic and number of devices.

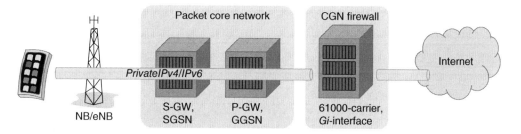

Figure 9.1 An example of CGN firewall deployment based on Check Point

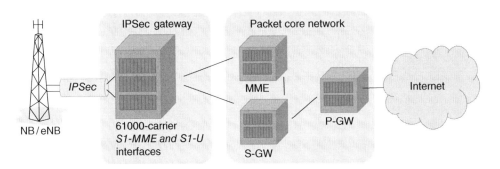

Figure 9.2   An example of Check Point deployment in an IPSec gateway mode, delivering the S1-MME signalling (SCTP) and S1-U traffic (GTP-U over UDP)

Ref. [4] further emphasizes the need for intelligence of the NAT firewall so that it may identify additional data sessions typical in billing attacks. The firewall needs to detect when the initiating party exits the session and force to terminate that specific session. Furthermore, the NAT firewall needs to deliver Deep Packet Inspection (DPI) with additional security functions such as IPS, antivirus, URL filtering, application control and anti-bot in order to protect the mobile network infrastructure and to prevent the network from being used for launching DoS attacks.

9.3.2.2   *S1* Interface

The radio interface of the LTE/LTE-A system is simplified from the previous 2G and 3G architectures as there is no need for a separate radio network controller. The LTE radio network has a flat architecture model. This opens potential security holes to the LTE/LTE-A radio interface as the eNB can also reside on the customer sites in the form of home eNBs, or small cells. The benefit of such a solution is the easy deployment of the coverage areas by customers, making the adaptation of the LTE/LTE-A services as easy as it is for the wireless home routers accessing the Internet. As these LTE sites are now located in public places like homes, hot-spot locations of small businesses and the office environment, they may be vulnerable to tampering intentions. In the worst case, an unprotected home eNB may open access for attacks towards the MME regardless of the HW shielding intentions, which means that the elements need to be protected with extra care. One feasible solution for shielding is based on the IPSec standard and security gateway.

The security gateway needs to support authentication between the LTE eNB and the packet core network in order to block intentions for establishing unauthorized access from the radio network. The interoperability with third-party PKI solutions is thus essential to enable certificate authentication for the eNB control plane and user plane. Ref. [4] emphasizes the importance of security gateway support for the Encapsulating Security Payload (ESP) and Internet Key Exchange (IKEv2) to ensure traffic confidentiality and integrity with AES, SHA-1 or Triple-DES encryption algorithms, to protect against eavesdropping and data tampering intentions on both control and user planes. Furthermore, the gateway needs to support Stream Control

Transmission Protocol (SCTP) deep packet inspection for the *S1-MME* control plane in order to provide protection against intentions to inject false traffic into applications. The protection needs to be scalable in order to provide carrier-grade IPSec throughput and performance and to minimize any network latency.

9.3.3   Protection against Roaming Threats

9.3.3.1   *Gp/S8* Interface

The protection of the *Gp/S8* interface shields the packet core network against malicious intentions regarding roaming, i.e., when users are connected to services via other MNO networks. The interconnection of the MNO networks is based on the *Gp/S8* interface which allows access to the GRX network. The GRX is in practice a centralized hub for connecting roaming users instead of each MNO relying on direct connections with each other. Due to overlapping radio access technologies, the inter-network roaming traffic from the LTE packet core to the 2G/3G packet core and vice versa needs to be supported on the *Gp/S8* interface by the MNOs in a secure way even if untrusted networks are involved.

According to Ref. [4],a typical security threat related to the *Gp/S8* interface is a DoS attack against service availability in the form of bandwidth saturation, data flooding, spoofing or cache poisoning. In addition, the *Gp/S8* interface may be vulnerable to overbilling attacks if the mobile station is capable of hijacking the IP address of a legitimate mobile station and starting data download without the original user's awareness.

9.3.3.2   *Gp/GRX* Interface

There are various security requirements for protecting the *Gp/GRX* interface in the roaming cases. Ref. [4] informs that the most important task is the proper protection of GRX networks against DoS attacks between MNOs' networks. The DoS attacks are possible if there is a way to insert IP packets into the GRX network domain from another IP network domain. Thus, the security gateway needs to have means for supporting deep and stateful packet inspection of the key protocols on the *Gp* interface, which are GTP, SCTP and Diameter.

The GTP delivers mobile data services, and thus understanding the GTP traffic eases the enforcing of roaming agreements based on carrier identity-based policies and provides protection against DoS, DDoS and overbilling attacks. The SCTP is located on the mobile network IP transport layer. Understanding the SCTP flow eases the protection against DoS attacks based on corrupt packets, and gives protection against unauthorized network access. Diameter is a signalling protocol for authorization, authentication, charging and QoS. A deeper understanding of Diameter data flow provides the means to protect the data against potential interception on untrusted, public IP transport networks between service providers. Figures 9.3 and 9.4 present practical options for protection.

Along with the heavily growing LTE/LTE-A traffic, the SCTP and Diameter traffic also increases. Ref. [4] emphasizes the importance of being able to inspect this traffic in order to provide a way to protect against possible malicious intentions like Data Exposer attacks from the packet core using unauthorized GTP or Diameter commands.

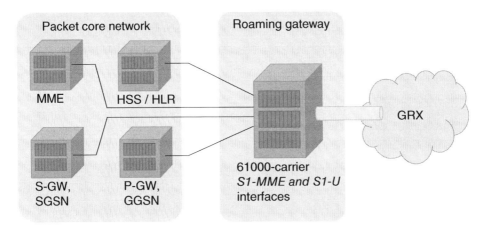

Figure 9.3   An example of Check Point acting as a roaming gateway

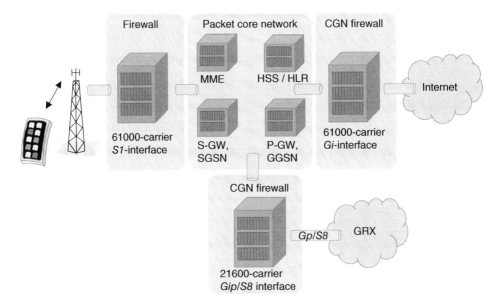

Figure 9.4   An example of Check Point protecting roaming networks

9.4   HW Fault and Performance Monitoring

It can be generalized that the wireless security, in addition to the actual encryption and communications protection mechanisms, also covers the assurance of the network elements and interfaces. This is logical because the non-ideals in any point of the network, whether this is due to the low performance of faulty SW/HW, may open doors for potential security attacks. Thus, performance monitoring and fault management of the networks can be considered an integral part of the security assurance.

The monitoring techniques include the means to monitor selected network elements and interfaces as well as the status of the user devices. The latter can typically be investigated remotely by the operator, and depending on the supported functionalities, it may be possible to retrieve information about the device HW, SW and the status of the SIM/UICC module.

9.4.1   Network Monitoring

There are various types of network monitoring methods and systems available on the market, which can be categorized into surveillance and security monitoring. Network (and computer) surveillance refers to the monitoring of the connected device's activity and data stored in its memory, or data that is transferred over computer networks such as the Internet. Computer security, in turn, which can be considered as a synonym of cyber-security and IT security, refers to the protection of information systems against theft or damage to the HW, SW and the information stored on them, as well as against disruption or misdirection of the services they provide [1].

Network monitoring as it is understood in the operational network management environment refers to performance monitoring and fault management. Both these areas may reveal potential security threats if the respective trend data in signalling or user data traffic starts to deviate significantly per cell or area. Network element vendors typically have integrated or additional solutions for such monitoring, and also external monitoring tools can be deployed for more personalized and focused analysis.

9.4.2   Protection against DoS/DDoS

As stated in Ref. [5], DoS can be defined as a temporary reduction in system performance, a system crash requiring manual restart or a major crash with permanent loss of data. From the early days of computers up to the major breakthrough of consumer markets for the graphical web and personal computers, DoS was not considered an important topic. Nevertheless, as the services are increasingly dependent on an electronic format with connectivity via the public Internet, the importance of DoS, and more its more powerful form, DDoS, have been causing major news as the essential functions of our daily life have been disturbed, such as banking and communications. The protection of wireless systems against Dos and DDoS attacks is thus a 'daily routine' for the MNOs.

9.4.3   Memory Wearing

As the memory modules of the devices – both user and network equipment – are made of physical HW, they have certain useful lifetime before they start to fail. This is especially important in devices that frequently read and write in memory blocks such as the SIM/UICC. The gradual wearing of the physical memory surface may cause unpredictable issues and, in the worst although extremely rare case, may open security holes. For that reason, it is important to ensure the correct functioning of the equipment during its planned use.

The UICC has a useful lifetime which is typically long enough to support the practical lifetime of the mobile device, or until the user changes the UICC for a more modern one. The degrading is principally a consequence of the number of read and write cycles in the memory. Some UICC types support, e.g., about 1 million cycles before error occurrence starts to exceed the design criteria whereas some less robust models may support a considerably lower amount of cycles. Nevertheless, there may be occasions when the UICC starts to fail unexpectedly, especially, if there is some active signalling app installed in the card that uses memory more often than dimensioned by default.

There are monitoring solutions that the MNO can deploy to follow the technical functioning of the subscription module. One example is the SIM lifetime monitoring tool (chip health monitor) of which an example can be found in Ref. [6]. These solutions may include the remote estimate of the HW wear and statistics for apps consuming the UICC. There is a certain expected lifetime for different categories of UICC in such a way that as the number of filed read/write cycles starts to increase gradually, the UICC will manage memory utilization by reallocating the memory blocks. So, the respective monitoring of the UICC wearing remotely reveals potential issues before the problem has advanced too far. In these cases, the operator can invite the customer to make a replacement before the UICC fails.

9.5   Security Analysis

Security analysis of mobile communications networks is an increasingly important part of an MNO's routine tasks. The aim of the analysis is to ensure that all goes fine with the traffic, and that no malicious intentions – whether accidentally or on purpose – can take place. The security thus refers principally to software and IT network related threats that may jeopardize the operator's or user's legitimate communications or expenses. The following sections describe security methods based on post-processing and real-time surveillance.

9.5.1   Post-processing

The histograms, i.e., historical comparison data, can be stored as part of the network statistics collection, and used as a basis for deeper posterior analysis for understanding deviations on the users' or groups' communications patterns. The monitoring, whether it happens in real time or later, may reveal potential malicious intentions to exploit the network, applications or devices. Thus, the respective monitoring tools may use this historical data as an important part in the complete security threat shielding process.

9.5.2   Real-time Security Analysis

9.5.2.1   Traffic Analysis

The regular network traffic tends to establish into certain repeatedly similar levels according to certain daily patterns (office hours and free time), weekly (weekday and weekend), seasonally (summer holidays) etc. If the long-term patterns in the traffic flow deviate suddenly without an obvious reason such as a football match at local stadium, it may indicate malicious intentions of attackers. The network attacks may be, e.g., DDoS efforts or intentions to brute-force certain element credentials found within the network infrastructure.

As the mobile communications networks are increasingly IP-based systems, there are many elements involved in the end-to-end traffic similar or equivalent to the Internet, such as routers and bridges, as well as the respective controlling and monitoring systems. The division can be done between Network Monitoring Systems (NMSs), Intrusion Detection Systems (IDSs) and Intrusion Prevention Systems (IPSs). The two latter ones are designed to detect security breaches and to prevent unauthorized activity while NMS indicates the performance of the network. Nevertheless, all these variants can be used in a combined form to monitor the security threats.

9.5.2.2 DPI

In addition to the overall traffic flow analysis, it is also possible to analyse in more detail the traffic type and contents. The aim of DPI is to collect information and, if necessary, take action based on the inspected information or what can be inferred from the content of the communication [1]. The challenge of the DPI is that the IP traffic is increasingly dynamic and distributed both from multiple sources as well as to various receivers. Also, the traffic typically involves various protocols with varying syntax presentations, and many kinds of packet types and respective ports. DPI needs to adjust to this highly dynamic environment in real time to, e.g., prevent access to unauthorized zones and contents, or to limit throughput per communications type, while not interfering with legitimate traffic.

DPI solutions share the same type of requirements for all stakeholders, including MNOs, service providers and governments. These requirements include reliability for correct and timely analysis, fault tolerance, and adequate capacity and ability to distribute the analysis over different elements in a parallel fashion. As an example, the Advanced Telecommunications Computing Architecture (ATCA) complies with typical industry requirements for DPI.

There are many types of DPI applications for the various entities involved. The focus of DPI can thus be adjusted accordingly. Some of the practical roles for DPI are policy enforcement, network security, subscriber analytics, traffic monitoring, legal interception, content optimization, billing volume metering, content caching, load balancing and content modification. Table 9.1 summarizes some of the typical roles and their practical examples.

9.6 Virus Protection

As soon as the first smart devices were introduced into consumer markets, security threats began to emerge. As the smart devices are based on applications, viruses are becoming an important threat, as they have been in the fixed IT environment for many years. In fact, along with the enhanced protection mechanisms of the network infrastructure, the threats are moving towards the application layer.

The app store concept aims to minimize the incidences caused by malicious code in the apps. There is a prerequisite to pass the testing and certification processes prior to the introduction of the app to the store. Nevertheless, there are certain aspects that cannot completely eliminate the malicious intentions. One aspect is related to the privileges the app may require upon installation in order to function. It is logical that, e.g., a camera app requests permanent permission from the user to access the photo gallery. The situation gets slightly complicated if the app requests privileges that do not sound necessary for the app to work properly, such as a flashlight which wants access the microphone of the device. Even if the app developer does not use these types of extended privileges at the time of the delivery of the app except for perhaps testing purposes, they may expose a severe security threat in the later phase of the app lifetime if the app communication is used for illicit activities such as eavesdropping on the user's conversations, either by the app developer or someone hacking the app. One way to prevent and minimize such incidences is to evaluate before installing such apps if all the access rights are really necessary compared to the benefits the app brings.

Among other security-related apps the smart devices may have, there also are plenty of virus protection apps which may embed other useful protection tools. Some of these cost money whereas others are free of charge, and they are designed to protect against malware, adware, spyware and other malicious code. They may also have functionalities that clean up

Table 9.1 Key roles of DPI

PDI focus	Description	Examples
Policy enforcement	Functions for traffic shaping, prioritization, access control, admission, content filtering	Traffic management provides fair utilization of the resources between the users and enhances user experience
Network security	Firewalls, network-based antivirus applications, intrusion detection/prevention, data leak prevention, anti-spam, spam Internet telephony, spam instant messaging	Web application firewalls for protection and to support less real-time upgradeable endpoint SW. Aids network-based security solutions when user devices lack security, e.g., M2M applications
Network analytics	Reveals the state of network functioning via performance and capacity analysis	Provides information to MNOs and service providers about network quality
Subscriber analytics	Provides understanding about subscriber base behaviour	Provides information to service providers and MNOs about the typical utilization of resources and services, which optimizes marketing efforts
Traffic monitoring	Network diagnostics	Ensures information for prompt trouble shooting
Legal interception	Regulatory traffic monitoring	Provides contents for legal purposes
Content optimization	Proxying and content modification	Reduction of image and video quality to reduce needed capacity. Web page reformatting to optimize the bandwidth. The measures allow increased number of simultaneous users to share the resources
Billing volume metering	Traffic volume monitoring	Multiple schemes can be applied, e.g., varying the data rate based on consumed amount of bits or contents. Also division of the payment between different service providers
Content caching	Storing popular contents closer to the end users	Allows service provider to select cached content via intercepting the traffic
Load balancing	By investigating the content, redirection of the packets to different destination address	The traffic can be offloaded depending on the overall utilization, and by fine-tuning the redirection based on DPI results
Content modification	Examining and modifying content	Packet content modification provides the means to insert tracking IDs, modify packet headers, rewrite or add packets

unnecessary files, optimize the device's power utilization, manage the apps and provide additional privacy via encryption. There may also be anti-theft tools available, which wipe the contents of the device remotely if it is lost or stolen. The network or service provider may also offer virus protection and related tools as part of the service packet for the end-user.

Thus, virus protection is an elemental service in smart devices. It can be a real-time service based on the device's application layer, or a network-based service offered by the MNO or service provider. Selection of the most feasible options depends on the needs, remembering that the free-of-charge tools are probably equally useful but may be limited as for the protection functionalities, lack technical support and may display advertisements.

9.7 Legal Interception

Legal/Lawful Interception (LI) has been designed for authorized access to the communications of commercial, government and military environments. LI provides the means for mobile and fixed network operators and service providers to collect traffic and identification information of private or organizational communications for post-analysis for law enforcement officials. This method has been available for a long time in the mobile communications networks. As an example, LI was included as part of the first GPRS networks based on the 3GPP Release 97 specifications, under the term LIG (Legal Interception Gateway) to allow MNOs to mirror the traffic delivered via the GPRS nodes. LI can only be applied in accordance with national and regional laws and technical regulations.

The EPS of the LTE/LTE-A networks supports the interception of the IP layer's Content of Communication (CoC) data flows. The LTE/LTE-A voice connections also represent IP data flows via VoIP solutions. If a fall-back type of functionality is applied during the LTE voice call to switch over to circuit-switched technology, the respective 2G/3G network also contains the LI. In addition to the user plane interception, the LI solution of EPS can generate Intercept Related Information (IRI) records in the control plane messages which identifies the called parties, the location of the LTE terminal and other call-related information.

The functional architecture of the EPS lawful interception is comparable to the functional architecture of the packet-switched domain of 3G networks of 3GPP. Figures 9.5, 9.6 and 9.7 depict the configurations for the MME, Home Subscriber Server (HSS), Serving Gateway (S-GW) and Packet Data Network Gateway (PDN-GW), respectively, as defined in the 3GPP

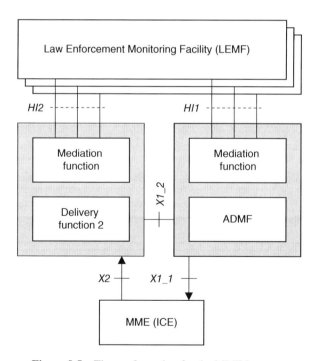

Figure 9.5 The configuration for the MME intercept

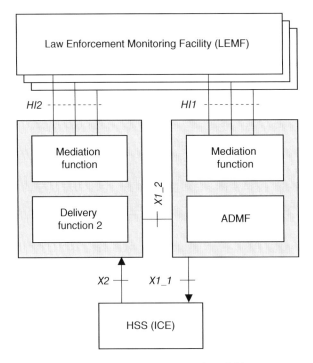

Figure 9.6   The configuration for the HSS intercept

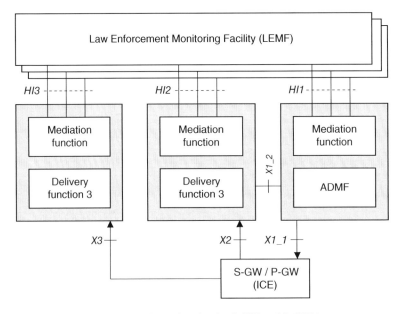

Figure 9.7   The configuration for the S-GW and P-GW intercept

standards for the EPS lawful interception [8]. The key identities for the interception are IMSI, MSISDN and IMEI.

The MME element manages the control plane while the HSS handles signalling. The interception of CoC is thus applicable only via the S-GW and P-GW elements of the LTE/LTE-A. In the Figures 9.5, 9.6 and 9.7, the Administration Function (ADMF) refers to a functionality that interfaces with the Law Enforcement Monitoring Facilities (LEMF) of the Law Enforcement Agencies (LEA) that may request the interception. The ADMF functionality has a direct interface with the network elements that are intercepted while it keeps the interception-related activities of each LEA separated from each other. The ADMF, together with the delivery functions of the intercepted information, is hidden from the Intercepting Control Element (ICE), even in the case of various simultaneous activations on behalf of separate LEAs related to the same subscription.

The physical ICE of the LTE/SAE network is connected to the ADMF via an *X1_1* interface which delivers the intercepted information from each ICE. Each ICE carries out the interception, i.e., the activation, deactivation, interrogation and invocation procedures independently. The *HI1* interface of the ADMF is defined towards the requester of the LI. For the communication between independent delivery functions and LEA, there are *HI2* and *HI3* interfaces. The delivery functions distribute the IRI and CoC to the relevant LEA.

Some use cases for the activation of the LI may be triggered when there is a change in the location information of the subscriber, or the terminating or originating short message transfer is initiated by the target. Also, when the terminating or originating circuit switched call is being initiated by the target, or the terminating or originating packet data service is initiated by the target, the LI can be set to activate.

The CoC can be intercepted from the media plane entities via the LI concept. In addition, various identities related to the intercepted communications can be stored. Some examples of the IRI that can be intercepted from the subscribers are: MSISDN, IMSI, Mobile Equipment Identifier (ME ID), event type, event time and date, (Network Element Identifier (NE ID) and location.

Ref. [12] contains updated requirements for the LTE Release 10 phase and beyond. As an example, it emphasizes that the intercept function shall only be accessible by authorized personnel and that the interception must take place without the knowledge of either party to the communication. Thus, decryption must also take place without either party being aware that it is happening so no indication shall be given to any person except authorized personnel that the intercept function has been activated on a target.

Ref. [13] contains a more detailed description of the LTE interception as of Release 10 and beyond. It includes additional items such as the interception of the MBMS and IMS conference services. Ref. [13] is thus one of the most relevant and up-to-date sources of information for those interested in learning more about the legal intercept in the 3GPP networks.

9.8   Personal Safety and Privacy

9.8.1   CMAS

The Commercial Mobile Alert System (CMAS) was introduced to the LTE networks as of 3GPP Release 9. It is capable of delivering multiple, concurrent warning notifications. The CMAS warning notification is broadcastd in *SystemInformationBlockType12*. The paging is used in order to inform CMAS-capable UE about the message in both the RRC_*Idle* and RRC_*Connected* state. Upon the UE receiving the paging message with the CMAS indication, it starts receiving CMAS notifications based on the scheduling information list which is found in *SystemInformationBlockType1*. In order to comply with the requirements of replacing and cancellation of the notifications, additional procedures are included in the LTE between MME and eNodeB. The respective CMAS signalling is presented in Figures 9.8 and 9.9. In this case, the MME initiates the Write-Replace procedure via a Write-Replace Warning Request message which contains message identifier, warning area list, instruction of the broadcasting and the contents. The eNodeB acknowledges via the Write-Replace Warning Response message and initiates the broadcasting. Please note that the ETWS and CMAS are independent services, and that the ETWS and CMAS messages are differentiated over *S1* for different handling.

The broadcasting of a Public Warning System (PWS) message is stopped via the Kill procedure. The MME initiates the procedure via the Kill Request message which contains the message identifier, serial number of the message and the warning area list where the broadcasting will be killed. Upon receiving the request, the eNodeB acknowledges it via the Kill Response message and stops broadcasting.

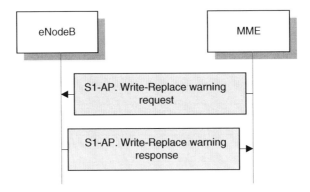

Figure 9.8   Write-Replace warning procedure

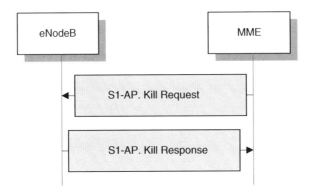

Figure 9.9   Kill procedure

9.8.2   Location Privacy

Along with the considerably growing popularity of the LBSs, with enhancements of the supporting technologies such as satellite positioning (GPS, GALILEO, GLONASS, and variety of other international and national variants), integrated mobile network services such as cell ID, arrival of time from multiple cells and assisted location service and location tracking methods of other wireless systems such as via Wi-Fi, there may also arise concerns about the privacy protection of the individual users – especially as the location information provided by current user devices may be extremely accurate, in the order of some metres.

The physical location tracking may not be the only issue as long as the information stays in proper hands, the location data may also be embedded automatically to the user's photos and other contents by default until such functionality is deactivated, e.g., from the smart device's camera app settings. The issue becomes even trickier when these photos are shared in social media, so it may make the illicit tasks of burglars easier knowing that the family publicly shares photos from a distant holiday resort with location and time stamp tagged automatically to the metadata of the pictures. The issue is two-fold: it is logically nice to remember the locations and times of such instances among family members and friends, but it may be wise to think carefully about uploading such detailed information for everyone's eyes in real time.

In addition to the user's communication devices, there may also be an increasing number of location tracking technologies embedded into other daily objects such as cars. The tracking devices are useful for maintenance purposes and for providing auxiliary measures, e.g., in the event of traffic accidents. Nevertheless, it may be good idea to read the respective privacy conditions of such objects to fully understand how such private data may be observed and by whom.

As Ref. [10] reasons, the trace information may reveal a surprisingly lot about individuals' habits, interests, activities and relationships, as well as personal or corporate secrets. Despite of the good intentions of some service providers to deliver focused announcements for individuals, the location tracking may trigger also unwanted advertisements and location-based spams which in turn may have a negative impact on social reputation or even cause economic damage via exploitation of such information by criminals.

There is plenty of literature related to the respective trends, legal aspects and the general pros and cons of location tracking. As an example, Ref. [9] discusses the counter-measures an individual may be able to take. The high-level principle is to change pseudonyms of the users frequently, even while they are being tracked. The principle is based on users adopting a series of new, unused pseudonyms for each application with which they interact.

On the other hand, avoiding such tracking by applying anonymous communications – which could be argued to especially interest the illicitly behaving entities wanting to hide themselves – would also prevent the positive impacts such as sufficiently accurate location information embedded to the emergency call. Nevertheless, it is not only about the exposure or revealing of such location information mapped to individuals for trusted parties, but also some important questions are, how accurate this information is, how to trust that the information is not exposed to criminals, and in the case of information monitored by the legal entities, what are the means to ensure the location data is authentically correct in order not to make wrong conclusions of an individual's movements? Some examples of the legislation of location tracking can be studied from Ref. [11], which discusses the principles of commercially available GPS devices with communication or recording features.

In order to tackle the respective attacks, Ref. [10] discusses the Location-Privacy Protection Mechanisms (LPPMs), and reckons that their assessment and comparison is problematic because of the lack of systematic methods to quantify them. Furthermore, assumptions about the attacker's model tend to be incomplete, with the risk of a wrong estimation of the user's location privacy. Ref. [10] provides a framework for the analysis of LPPM variants by capturing the prior information that might be available to the attacker as well as possible attacks. It also presents a simple model to formulate all types of location-information disclosure attacks, and by formalizing the adversary's performance, it proposes and justifies proper metrics to quantify location privacy.

9.8.3 Bio-effects

A distantly wireless security-related item is the RF radiation. The RF radiation is non-ionizing which means that it does not cause genetic alterations as can be the case with, e.g., excess ionizing X-ray exposure. The consensus of the scientifically relevant investigation bodies and industries is that the only measurable effect of RF radiation is the temperature rise in human cells. If there is too much radiation, the temperature can increase over healthy limits – which can be seen by observing the effects of microwave heating. This can happen on the non-licensed 2.45 GHz frequency band which is optimal for warming up the water atoms thanks to their resonance peak that results in friction on that specific band. Nevertheless, the radiating power level of a microwave oven being hundreds up to over a thousand watts is far beyond the user devices of cellular systems which typically use power levels up to 1–2 watts. As the topic is related to the health aspects of human beings, it is fruitful ground for debate. To understand the topic in a reliable way, it is recommended to refer to high quality, repeatable scientific results that rely on professional methods and equipment.

The bio-effects as such do not belong directly within the scope of this book – except when speculating scenarios that involve cyber-attacks with an aim to deliberately re-direct high RF power sources such as flight radar antenna systems which are dangerously close to a population. Another theoretical case may involve a hacking effort to increase the radiating power level of the mobile device to its maximum, or to overload the electrical circuits or short-circuit the device, which could warm up the mobile phone or battery out of the permitted limits. However, these topics would require another cyber-attack related book. As for the overall mobile communications, there are many guidelines from official entities, and information about limits by, e.g., national frequency regulators. Relevant and scientifically proved information about the effects of RF exposure can be found from various official sources.

Ref. [3] states the COST study results, which also coincide with the common understanding of the field from the last few years of research, have not identified adverse health effects due to exposure to electromagnetic fields at the low levels occurring in most occupational or environmental settings. Nevertheless, Ref. [3] also states that a number of uncertainties still exist about existing exposure situations, and new applications of electromagnetic fields due to emerging technologies may also motivate further research activities. The European Council has recommended its member nations to closely follow further development, and to facilitate research at the national level. For those interested, more details about the EU-funded studies can be found in Ref. [2], and the COST 244bis study detailing the biomedical effects of electromagnetic fields can be found in Ref. [3].

References

[1] Radisys. DPI: Deep packet inspection motivations, technology, and approaches for improving broadband service provider ROI. White paper, September 2010.

[2] Health and electromagnetic fields. EU-funded research into the impact of electromagnetic fields and mobile telephones on health. http://ec.europa.eu/health/ph_determinants/environment/EMF/brochure_en.pdf (accessed 5 December 2015).

[3] COST 244bis. Biomedical effects of electromagnetic fields, 3 November 2000. ftp://ftp.cordis.europa.eu/pub/cost/docs/244bisfinalreport.pdf (accessed 5 December 2015).

[4] Check Point. Next generation security for 3G and 4G LTE Networks. White paper, November 2013. https://www.checkpoint.com/downloads/product-related/whitepapers/wp-ng-mobile-network-security.pdf (accessed 5 December 2015).

[5] Morrie Gasser. *Building a Secure Computer System*. Van Nostrand Reinhold, 1988.

[6] Ulrich Wimböck. Securing your M2M business M2Mission Possible. Giesecke & Devrient, Belgrad, 26 September 2013. https://m2m.telekom.com/upload/Event_Presentation_2013_BS_Ulrich_Wimboeck_4276.pdf (accessed 30 December 2015).

[7] Embedded UICC Protection Profile, Version 1.0. GSMA, 22 September 2014.

[8] J. Penttinen. The Telecommunications Handbook. John Wiley & Sons, Inc., Hoboken, NJ, 2015.

[9] Alastair R. Beresford and Frank Stajano. Location privacy in pervasive computing. *Pervasive Computing*, January-March 2003.

[10] Reza Shokri, George Theodorakopoulos, Jean-Yves Le Boudec, and Jean-Pierre Hubaux. Quantifying location privacy. IEEE Symposium on Security and Privacy, 2011.

[11] GPS location privacy in the USA. http://www.gps.gov/policy/privacy/ (accessed 8 January 2016).

[12] ETSI TS 133 106, V10.0.0 (2011-05). Technical Specification, Universal Mobile Telecommunications System (UMTS); LTE; Lawful interception requirements, Release 10.

[13] ETSI TS 133 107, V10.4.0 (2011-06). Technical Specification, Universal Mobile Telecommunications System (UMTS); LTE; 3G security; Lawful interception architecture and functions, Release 10.

10

Future of Wireless Solutions and Security

10.1 Overview

This chapter discusses future views of wireless security, including the development of trends and the impact of key enablers such as Big Data. Related security threats and solutions are discussed, emphasizing the importance of balancing proper security mechanisms while providing fluent user experience and avoiding performance degradation even with extensive data transmission, which can be expected as the number of IoT devices grows. It also summarizes the evolution of sensor networks and their security. Finally, this chapter gives an introduction to the mobile communications systems of 5G and beyond, including items under preparation for standardization and security challenges of future wireless technologies.

10.2 IoT as a Driving Force

As the GSMA has identified, proper preparation for future IoT networks is essential. The GSMA, among other industry initiatives, works on establishing common capabilities among mobile operators. It enables networks, providing the possibility for value creation for all stakeholders in the IoT environment. The GSMA has further concluded that the essential capabilities include security, billing and charging and device management, all of which can enhance the IoT environment by enabling the development of new services. Through the provision of these value-added services, MNOs can move beyond connectivity and act as a trusted partner for their customers. For the remote M2M provisioning, the GSMA embedded SIM specification is expected to accelerate the growth and operational efficiency of the M2M solutions via respective technical specifications to enable the remote provisioning and management of eSIMs for OTA provisioning of an initial operator subscription and the subsequent change of subscription from one operator to another.

The GSMA recognizes that the growing IoT environment provides a range of socio-economic benefits, but it requires businesses to be incentivized to develop devices, applications and

Wireless Communications Security: Solutions for the Internet of Things, First Edition. Jyrki T. J. Penttinen.
© 2017 John Wiley & Sons, Ltd. Published 2017 by John Wiley & Sons, Ltd.

services while consumers need to trust the securing of data. This is a good moment for the industry and standardization organizations to create a common base as the IoT environment will continue to develop very quickly in the forthcoming years.

Some building blocks for a cost-efficient IoT environment include Big Data, so the respective threats need to be understood and solutions for the security mechanisms are required in order to avoid performance degradation of wireless technologies along with extensive data transmission. Another important element in this environment are the sensor networks, including smart grids, which are equally vulnerable if proper actions are not taken in the very initial phase.

10.3   Evolution of 4G

Release 8 LTE entered the markets soon after the 3GPP standards were announced. As some examples, Verizon Wireless launched LTE in the United States in December 2010, and the Verizon Wireless LTE network covered over a 250 million population area by November 2012. By that time, AT&T had covered a 150 million population area. Other major US operators advanced with LTE deployment in an exceptionally fast time schedule, and the global deployment figures are also growing quickly.

As an example of the speed of development of equipment and services, Nokia Networks together with Sprint achieved 2.6 Gb/s downlink throughput demo. This demo was done in a single sector case and 120 MHz aggregated bandwidth, and exceeds by far the strict requirements of the ITU-R for the most complete 4G performance. At the same time, NSN released a public announcement about the Flexi Multiradio 10 Base Station product family capable of supporting up to 5 Gb/s peak downlink throughput per single site. These activities indicate the fast pace of the development towards the next big step for 5G networks that may be a reality in standardized form as of 2020.

As for the 4G era, so far, the ITU-R has approved two systems under the 4G umbrella, which are the LTE-A and the WiMAX2. They share some important high-level principles as both are designed to transfer packet data for all the services, including voice traffic. Furthermore, both systems are based on Orthogonal Frequency Division Multiplexing (OFDM) technology in the downlink.

The WiMAX and its evolution are defined in IEEE 802.16 standard set. Similarly, as is the case of the previous generation of the IEEE-defined Wi-Fi, it is an open standard with extensive revisions by the engineering community prior to having been ratified as a standard. This provides the means to introduce WiMAX equipment in such a scale that the cost for the end-user is relatively low. Similarly, the 3GPP standard is a gateway for LTE/SAE equipment interoperability, and thus for the scale of economics.

As for the mobile operators, the most important difference between the 3GPP and IEEE approaches is that LTE/SAE can be deployed as a continuum within the existing GSM and/or UMTS infrastructure, as depicted in Figure 10.1. WiMAX and its development path, in turn, requires a new network. The main benefit of LTE and LTE-A is that the already deployed and interoperable 2G/3G infrastructure is widely utilized globally. The LTE/LTE-A user equipment typically includes 2G and/or 3G, which provides seamless LTE/LTE-A services for end-users while the operators can still benefit from the existing infrastructure.

IEEE 802.16m, or WiMAX2, is a set of additional definitions to the IEEE 802.16-2009 standard. IEEE 802.16m and LTE Advanced are the two systems that comply with the

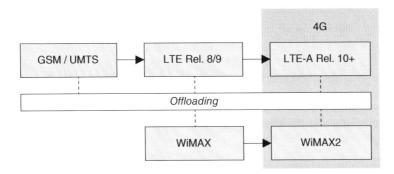

Figure 10.1 LTE-A and WiMAX2 are the result of their own evolution paths, but can be used in a cooperative environment via data offloading and inter-working

requirements for International Mobile Telecommunications 4G, i.e.IMT-Advanced systems. IEEE 802.16 m was submitted to the ITU in October 2009 in order to answer the requirements for IMT-Advanced. As defined by the ITU-R, the IMT-Advanced requirement is a maximum data rate of 1 Gb/s in low mobility scenarios, which is compatible with IEEE 802.16 m and LTE-Advanced [10].

In addition to the highly increased data rates of 4G (up to 1 Gb/s for static environment), the main requirements are, e.g., support of the 100 MHz bandwidth and a round trip time value of 1 ms. The focus of IEEE 802.16 m has been on achieving the values defined by the ITU-R yet maintaining the backwards compatibility with the legacy IEEE 802.16 system. The IEEE 802.16 m standard was released at the end of 2010, and the commercial launch in 2012.

IEEE 802.16 m includes several features to reach the IMT-Advanced goals. It increases the data rate accordingly, and provides enhanced FDD support as of Release 1.5 of WiMAX. WiMAX2 supports a frequency bandwidth of 5 to 20 MHz per carrier, and includes the possibility for carrier aggregation up to 100 MHz. The carrier aggregation can be contiguous or non-contiguous in IEEE 802.16 m. As a comparison, the previous IEEE 802.16e does not have carrier aggregation.

IEEE 802.16 m, like IEEE 802.16e, defines MIMO antenna configuration options of 2×2, 2×4, 4×2, 4×4, 4×8 and 8×8 in downlink, and 1×2, 1×4, 2×4 and 4×4 for uplink. In IEE 802.16 m, the support of 2x2 MIMO is mandatory. The duplex scheme of IEEE 802.16 m, as well as for IEEE 802.16e, is TDD, FDD and Hybrid-FDD. The cell range of IEEE 802.16 m is up to 5 km for achieving optimal performance, while service with graceful degradation can be achieved up to 30 km, and basic connectivity up to 100 km. The speed of the IEEE 802.16 m user device may be up to 10 km/h for optimal performance while 120 km/h is supported for the vehicular environment with graceful degradation, and up to 350 km/h for high speed vehicular environment with basic connectivity.

The targeted frequency bands of IEEE 802.16 m are the following:

- 450–470 MHz (also included in IMT-2000)
- 698–960 MHz (also included in IEEE 802.16e R1.0 target)
- 1710–2025 MHz (also included in IMT-2000)
- 2110–2200 MHz (also included in IMT-2000)
- 2300–2400 MHz (also included in IEEE 802.16e R1.0 target)

- 2500–2690 MHz (also included in IEEE 802.16e R1.0 target)
- 3400–3600 MHz (also included in IEEE 802.16e R1.0 target)

The IEEE 802.16m standard also supports functionalities implemented in physical and MAC mechanisms. Some of these are femto base stations, multi base station MIMO, relay stations, SONs, LBSs, and Enhanced Multicast Broadcast Services (E-MBSs). It should be noted that some of the implementations are already specified in other standards such as IEEE 802.16e (Mobile WiMAX) and IEEE 802.16j (Relay Stations for Mobile WiMAX).

For more information about the LTE-A, please refer to Chapter 2 which further details the cellular systems.

10.4 Development of Devices

10.4.1 Security Aspects of Smartcards

The SE based on UICC is still highly useful, whether it is of traditional form factor or embedded into the device permanently. These tamper-resistant HW solutions remain a good base for providing at least TLS4 security level while it is much more challenging to achieve the same level with the SW-based security solutions. Along with the development of the eSE concept and respective subscription management, the final trends towards 5G are still to be seen. It may be possible that due to the fast development of the IoT environment, 5G may rely increasingly on solutions apart from the 'traditional' SIM/UICC concept.

10.4.2 Mobile Device Considerations

Some time ago, the logical base for the connectivity of the IoT/M2M devices was considered to be GSM. However, its role in the consumer voice service is reducing along with the deployment of 3G and 4G networks, yet it is still widely spread being able to serve devices such as those embedded into utilities and cars. At the same time, advanced systems such as LTE were seen as unnecessarily complicated to serve the IoT domain. The trend indicates that LTE can in fact be a very useful base for all kinds of IoT services. A concrete proof of this trend in the inclusion of the M2M category (Cat M), which is designed to support very low data rates (up to 1 Mb/s) compared to other categories which are in tens or hundreds of Mb/s. Furthermore, the 3GPP specifications divide the M2M categories into Cat M1 (which was originally the Cat-M) and Cat M2 (which was called previously Narrow Band-IoT, i.e. NB-IoT) in order to further emphasise the low bit-rate and narrowband characteristics of many IoT devices. The reason for this 'mind-set change' is that LTE/LTE-A is becoming the default standard at the global level which ensures rapid deployment and large coverage areas. As LTE/LTE-A provides better spectral efficiency, dynamic scalability and thus more capacity than any of the previous generations, it is able to support IoT devices such as wearables very efficiently.

These are some of the reasons why the 3GPP LTE-A Release 12 and 13 with an increasing number of definitions for M2M device support are relevant solutions for the IoT environment. The Cat-M1 and Cat-M2 re especially adjusted for Machine-Type Communications (MTC). Cat M1 provides the peak data rate of 1 Mb/s which is considered s a suitable base for IoT/ M2M devices that typically do not require more demanding data rates. Cat M2 specifies even narrower bandwidth and reduced complexity in order to reduce device cost and power consumption. This development in the standardization field could accelerate the lowering of importance of previous mobile generations and possibly shut down complete networks that were deployed prior to LTE/LTE-A quicker than anticipated, along with new ideas of re-farming the 2G and 3G frequencies for MTC. This development may start taking effect as of 2017 when Release 12 and 13 capable networks will be deployed.

Meanwhile, the chipset vendors are preparing the device base to better tackle the M2M environment, by providing Category 1 chips that allow data rates of up to 10 Mb/s. Even if this category provides more rapid connections than the cost-efficient and low-power M2M devices actually would require, it still better optimizes the device complexity compared to the other typically offered Categories 3 and 4 for smart devices.

While waiting for the Category M1/M2 chipsets, many of the current and near-future IoT devices will benefit from the Category 1 chipsets, e.g., wearable devices such as smart watches/ health equipment, and other devices that benefit from very low-power consumption with low-capacity and small-sized batteries, or remotely located devices such as utility meters which need to work autonomously with the same power source for a very long time.

10.4.3   IoT Device Considerations

Ref. [7] summarizes some of the key aspects for IoT device security. The main message for the industry is that the IoT devices need to include in-built security solutions instead of relying solely on the protection mechanisms of the network infrastructure. There is a big temptation in device manufacturing to leave the security level at its minimum, the reason being that many of the devices will be extremely low cost and simple – and unfortunately for the manufacturers as well as for the end-users, inclusion of advanced security solutions may increase the total cost of the device considerably.

Ref. [7] also divides the cyber-security threats against the IoT devices into application and system levels. The application level includes a huge amount of non-intentional as well as deliberately created security threats. As an example of the latter, there have been cases with backdoors left in the devices which have been revealed later – leaving many of the devices vulnerable tor cyber-attacks as it might be very challenging to update, e.g., routers after they have been deployed in locations that no one might be directly responsible for.

Ref. [7] further discusses that while application layer attacks are prominent in embedded devices, attacks against system layer services are also found. The system-level breaches during 2015 include, e.g., a largely reported security breach against Jeep, which was demoed for remote control 'hijacking' [8], and the Heartbleed vulnerability revelation [9] which showed the weakness of OpenSSL cryptographic libraries that are widely used in embedded devices. The Heartbleed bug can be used as a reference to potential IoT device vulnerabilities as such a wide community has been relying on it. It is a fundamental security threat as it allows anyone on the Internet to read the memory of the systems protected by vulnerable versions of the OpenSSL software. As a result of this vulnerability, the secret keys for identifying service providers and to encrypt the traffic, names and passwords of users and the actual content are compromised. As a consequence, knowledgeable attackers can exploit the vulnerability to eavesdrop on communications, and to access server and user data and copy it.

As IoT devices grow in number, it is essential to ensure a sufficiently high QoS as well as fluent user experience – along with a sufficiently good security level, including event-based fraud detection in real time. This requires strong end-to-end cryptography solutions that ensure end-users' privacy and identity as well as content protection.

Ref. [11] presents some potential solutions for tackling potential security breaches of IoT devices. One of the most feasible and highly logical solutions is the inclusion of the SE into the IoT device, in one or another form (traditional SIM/UICC card, eSE or external HW such as micro-SD). The benefit of the SE/eSE is the device agnostic, horizontal approach. Furthermore, SE/eSE provides robust, proven security mechanisms and management that rely on international standards. The SE/eSE is thus a future proofed concept that provides feasible end-to-end security. It can also support multiple applications while the trusted parties manage the identities.

Ref. [12] has concluded that the role of the eSE will be essential in the forthcoming years along with the increasing number of IoT devices. The reference notes, though, that the traditional SIM cards limit the number of retail outlets and sales. As a result of this, sales of cellular connected consumer electronics devices have not been developed internationally, and Wi-Fi connectivity has taken over cellular connectivity for that reason. Nevertheless, the reference also notes that the role of the eUICC in consumer electronics devices is crucial in changing this. The consumer electronics domain is expected to become a strong adapter of eSIM specifications from which GSMA has great potential as the preferred solution for new connected products, providing interoperable profile management during the lifecycle of IoT devices as well as consumer devices.

Totally new services are expected in the commercial market, which optimizes IoT communications, as well as converging existing technologies. A main building block for the highly dynamic utilization of the devices, in both M2M and consumer realms, is the renewed subscription management as described throughout this book. Also new adaptations related to optimal communications, such as advanced offloading methods, are seen as the Google Fi initiative indicates [14]. It is a programme that delivers a fluent and seamless wireless experience for end-users in close partnership with leading carriers and HW makers by optimizing the communications via selected cellular networks and Wi-Fi hotspots.

10.4.4   Sensor Networks and Big Data

Big Data refers to an extensive amount of data or complex data that 'traditional' data-processing applications are not capable of handling. Not only the amount of data itself but issues arise in harvesting and post-processing it when storing, analysing, correcting, searching, sharing, transferring and presenting the data. Interestingly, Ref. [20] states that there is such an excess of data that it often stays stored without further analysis.

A significant contributor of Big Data is expected to be the distributed Wireless Sensor Networks (WSNs). Even if the data generated in such an environment is not extensive per device, the accumulated data produced by all the sensors within a dense WSN may represent a considerable share of Big Data. An ongoing area of research is, thus, the energy-efficient methods and techniques in the field of Big Data gathering in the densely distributed sensor network [16]. There are significant business opportunities involved, although the proper handling of privacy is equally important.

One of the drivers in the development of Big Data technologies has been the huge increase of data inflows to central processing points, and the increasing demand for outputs from those processing points, i.e., server-based technologies. This trend continues as new equipment creates increasing amounts of data, including all kinds of devices such as accelerometers, cameras and GPS [17]. To tackle this overwhelming amount of data, new thinking and solutions are needed, such as stream processing. Also, optimized ways are required such as simpler deployment tools, programming interfaces and libraries in order to process and mine data sources.

As Ref. [18] notes, Big Data through cloud computing is a feasible way to offload a significant share of computation and data from data centres and terminal devices as it provides flexibility and scalability, as well as economic savings. Nevertheless, cloud computing may not be the optimal application because it requires real-time response time and mobility support. For the WSN, the operational areas need to be close to the physical world whereas the cloud makes it possible to manage part of data storage and computation at the edge of the network. This is another interesting area for the research community – to develop solutions for Big Data in the ubiquitous WSN environment including related computation, storage, data analysis, mining and distributed algorithms while ensuring adequate QoS and system integrity.

The technical advances in the generation of Big Data via current and innovative sensor technologies, as well as collecting and analysing the data, may also bring concerns for the privacy of individuals and the security of the data. As an example, Ref. [19] reports on a Google demonstration on wireless environmental sensors, emphasizing that these sensors may expose the potential for the use and misuse of relentless data collection. This setup consisted of some hundreds of wireless devices for detecting noise levels, humidity and temperature. The WSN was based on ZiGBee connectivity. The network produced data in 4000 continuous streams to the cloud platform as an entry point for the data, which was posteriorly processed using the Google Compute Engine, analysed using Google BigQuery and presented using an interactive Web application. The demo indicated the means for managing Big Data, and the benefits of a developed system based on the cloud. Nevertheless, Ref. [19] reminds us that if the sensors were replaced by cameras and motion sensors, the setup may become more intrusive.

As Ref. [21] states, there are social benefits in Big Data analytics, for example, in scientific and medical research. Properly handled, the data can contribute to new innovations which may lead to enhanced quality of life. Nevertheless, it is important to handle with care the direct measurement of the environment and the collection of Big Data and its post-processing. Thus, proper technologies and regulation are needed to ensure that the data is stored securely and handled accordingly in order not to jeopardize privacy rights. One example of the preparation for the next waves can be seen in Ref. [21] of the ICO which notes the interest in data protection and privacy risks posed by Big Data, and raises the need for understanding what data protection issues it raises with suggestions on how to comply with the Data Protection Act (DPA).

10.5 5G Mobile Communications

10.5.1 Standardization

5G represents the idea of a much more efficient system beyond 4G, as the 4G concept is defined by the ITU's IMT-Advanced requirements. The aim of the global 5G standard is to increase the data rate and speed up the response time. The overall concept of 5G is already generating large interest even if the actual form of it is still not clear. The focus of the discussions is on enabling a seamlessly connected society as of 2020. The idea of 5G is, via convergence systems and technologies, to bring together people, things, data, apps, transport systems, cities – in other words, everything that can be connected. 5G thus may function as an integral platform to ensure the smooth development of IoT and act as an enabler for smart networked communications.

The ITU-R has established a programme to develop 5G via the IMT for 2020 and beyond, i.e., 'IMT-2020', which is the next evolution step after the IMT-2000 of 3G and IMT-Advanced of 4G requirement descriptions of the ITU. This programme sets the stage for international 5G research activities, and the aim of the ITU-R is to finalize the vision of a 5G mobile broadband society which, in turn, is an instrumental base for the ITU frequency allocation discussions at the WRC. The WRC is an essential forum for the decisions on how best to reorganize the current frequency bands for the most efficient international use of the forthcoming 5G networks.

Most concretely, the ITU-R WP5D working group is actively driving for the information sharing related to the advances and requirements of 5G, including the vision and technical trends, requirements, RF sharing and compatibility, support for applications and deployments, and most importantly, the IMT specifications.

One of the active standardization bodies driving for 5G is the 3GPP, which is committed to submit candidate technology to the IMT-2020 process of the ITU-R. The initial technical submission is planned for ITU-R WP5D meeting #32 in June 2019, and the detailed specification submission is planned for ITU-R WP5D meeting #36 in October 2020 [15]. To align the technical specification work accordingly, the 3GPP has decided to submit the candidate proposal based on the specifications frozen by December 2019. As for the 3GPP specifications, 5G has an impact on several technology areas from which the radio interface is the most clearly visible as the aim would be to increase the theoretical 4G (and thus LTE-A) data rates considerably while the response time would dramatically reduce. Thus, the Radio Access Network Technical Specification Group (3GPP RAN TSG) is committed to identifying the requirements for the IMT-2020 as well as the scope and the 3GPP requirements for the new radio interface, working in a parallel fashion to enhance the ongoing LTE evolution belonging to the 4G phase which is the LTE-A of the 3GPP, complying with the IMT-Advanced requirements of the ITU.

10.5.2 Concept

5G refers to the 5th generation wireless systems. They belong to the next major phase of mobile telecommunications standards beyond the current 4G networks that will comply with the forthcoming IMT-2020 requirements of the ITU-R. The idea of 5G is to provide much faster data rates with less delays compared to the current systems up to 4G and thus to facilitate the adaptation of more advanced services in the wireless environment.

The industry seems to agree that 5G is a combination of novelty, yet to be developed and standardized solutions and existing systems including commercially available mobile generations, as well as other feasible wireless access technologies that together contribute to considerably increased data rates (at least 10-fold compared to the current LTE-A), lower latency (practically zero) and support of increased capacity demands (thousands of simultaneously connected consumer and M2M devices). As a result of the key enablers of 5G, some of the expected highly enhanced use cases include the support of tactile Internet and augmented (virtual) reality which provides completely new, fluent and highly attractive user experiences.

At present, there are many ideas about the more concrete form of 5G. Some major operators are driving the technology in practice via concrete demos and trials with the aim of contributing to the standardization and thus speeding up the system definitions. While these activities are beneficial for the overall development of 5G, they represent proprietary solutions until the international standardization ensures the jointly agreed 5G definitions which in turn facilitates global 5G interoperability.

5G is the result of a long development in the mobile communications area, the roots going back to the 1980s when 1G mobile communication networks began to be a reality. Since then, the new generations up to 4G have been based on earlier experiences and learning, giving the developers a base for designing enhanced security and technologies for the access, transport, signalling and overall performance of the systems. Regardless of the highly performing 4G that is still under deployment, the telecommunications industry has identified a great need for considerably faster end-user data rates as a result of the never-ending demands of the evolving multimedia. 5G would be capable of coping with extremely challenging capacity requirements to provide fluent user experiences that are suitable for practically all needs, up to the most advanced virtual reality applications. At the same time, the exponentially enhancing and growing IoT requires new security measures, including potential security breach monitoring and prevention.

5G will be an enabler for responding to the ever growing needs of the consumers in the advanced multimedia utilization, as well as tackling the vast needs of the exponentially growing IoT environment. Along with the new M2M/IoT applications and services, it can be expected that there will be role-changing technologies developed and new ideas to rely on, support and complement the existing ones. 5G is one of the most logical bases for managing this environment together with the existing systems in the markets.

As for the security assurance of the new 5G era, there can be impacts expected in the 'traditional' forms of the SIM/UICC and subscription types as the environment will be much more dynamic, with constantly changing devices which may be using a single user's subscription data and credentials. The ongoing efforts in developing interoperable subscription management solutions that respond in near real time for changing the subscriptions upon need illustrate a building block for this always connected, high-speed data society. It is still to be seen how the consumer and M2M devices will look like physically in the 5G era, but it can be taken for granted that there will be much more variety compared to previous mobile network generations, including multiple wearable devices per user, and highly advanced controlling and monitoring equipment taking part in our daily life in the connected society. Along with these completely new types of machines, the role of removable subscription identity modules such as the SIM/UICC will change; the much smaller personal devices require much smaller form factors from such elements meaning that the number of eSEs will grow. At the same time, techniques to tackle the constantly changing subscriptions need to be developed further as well as security solutions – both HW-based and SW-based. The feasibility of the cloud-based security solutions such as tokenization and HCE, and the further development of the device-based technologies such as TEE, will be highly relevant in the 5G era.

10.5.3   Industry and Investigation Initiatives

There are various MNOs and network equipment providers involved with practical field tests, and several investigative programmes have been established to study the feasibility and performance of new system ideas. As some examples, Verizon has established a 5G Technology Forum, and the EU coordinates 5G research programmes under various teams. More information about the latest EU-funded 5G research plans can be found in Ref. [13].

10.5.4   Role of 5G in IoT

Even if 5G is still undergoing brainstorming until the ITU officially dictates the respective requirements and selects the suitable technologies from the candidate systems, it is clear that 5G systems are aimed at tackling the vastly growing and developing IOT domain. Many novelty and future solutions can be expected, such as integrated wearable devices, household appliances, industry solutions, robotics, self-driving cars and other solutions that benefit from the 5G networks that can be assumed to be able to support growing amounts of simultaneously signalling, 'always-connected' devices.

The ongoing work on developing the next big step in mobile communications, i.e., 5G, includes IoT aspects as an integral part. Even if some of the most important key goals of 5G are to provide many-fold data rates compared to the 4G systems with close to zero delays, utilizing advanced technologies such as multi-antenna systems [6], they only represent one part of the complete picture. Another equally important aspect of the forthcoming 5G is the ability to manage huge amounts of IoT devices – which can be thousands under a single radio cell – which means that the overall data rate budget would be divided between these typically relatively low bit rate machines.

In addition to the 'traditional' type of IoT devices such as wearable watches with integrated mobile communications systems, car communications systems and utility meters, there are also emerging technology areas such as self-driving cars and drones that require high reliability for their functionality as well as secure communications [4]. More information about IoT development can be found, e.g., in Refs. [1–3,5].

References

[1] IoT white papers of GlobalPlatform. http://www.globalplatform.org/mediawhitepapers.asp

[2] IoT descriptions of RedBite. http://www.redbite.com/the-origin-of-the-internet-of-things/

[3] IoT definitions of WordPress. https://iotomorrow.wordpress.com/origin-definition/

[4] BBC. Drones: http://www.bbc.com/news/technology-34088404

[5] Nokia. LTE-M: Optimizing LTE for the Internet of Things. White paper. http://www.gsacom.com/downloads/pdf/Nokia_lte-m_-_optimizing_lte_for_the_internet_of_things_white_paper_2015.php4

[6] Nokia. LTE Multi-antenna Optimization. White paper. http://www.gsacom.com/downloads/pdf/Nokia_multi-antenna_optimization_in_LTE_white_paper_2015.php4

[7] Alan Grau. Security framework for IoT devices, 1 December 2015. http://www.embedded.com/design/safety-and-security/4440943/Security-framework-for-IoT-devices (accessed 30 December 2015).

[8] *Wired.* http://www.wired.com/2015/07/hackers-remotely-kill-jeep-highway/ (accessed 30 December 2015).

[9] Heartbleed bug report. http://heartbleed.com/ (accessed 30 December 2015).

[10] Rohde & Schwarz. IEEE 802.16m technology introduction. White paper, 2010.

[11] Ulrich Wimböck. Securing your M2M Business; M2Mission possible. Giesecke & Devrient, Belgrad, 26 September 2013. https://m2m.telekom.com/upload/Event_Presentation_2013_BS_Ulrich_Wimboeck_4276.pdf (accessed 30 December 2015).

[12] Beecham Research Ltd. Benefits analysis of GSMA embedded SIM specification on the mobile enabled M2M industry, 2014.

[13] European Union, 5G initiatives. http://ec.europa.eu/digital-agenda/en/towards-5g (accessed 30 December 2015).

[14] Google Fi project, 1 January 2016. https://fi.google.com/about/ (accessed 1 January 2016).

[15] Tentative 3GPP timeline for 5G. 17 March 2015. http://www.3gpp.org/news-events/3gpp-news/1674-timeline_5g (accessed 1 January 2015).

[16] Daisuke Takaishi, Hiroki Nishiyama, Nei Kato and Ryu Miura. Towards energy efficient Big Data gathering in densely distributed sensor networks. IEEE, 2014. http://ieeexplore.ieee.org/xpl/articleDetails. jsp?tp=&arnumber=6800057 (accessed 6 January 2016).

[17] Big Data now. Current perspectives from O'Reilly Media, January 2015. http://www.oreilly.com/data/free/files/big-data-now-2014-edition.pdf (accessed 6 January 2016).

[18] Fu Xiao, Chongsheng Zhang and Zhijie Han. Big Data in ubiquitous wireless sensor networks. *International Journal of Distributed Sensor Networks*, March 2014. http://www.hindawi.com/journals/ijdsn/2014/781729/ (accessed 6 January 2016).

[19] Google's wireless Sensors: Big Data or Big Brother? *Network Computing*, 22 May 2013. http://www.networkcomputing.com/wireless-infrastructure/googles-wireless-sensors-big-data-or-big-brother/a/d-id/1234212? (accessed 6 January 2016).

[20] SAS. Big Data. http://www.sas.com/en_us/insights/big-data/what-is-big-data.html (accessed 6 January 2016).

[21] Big Data and data protection. Data protection act, Information Commissioner's office, Version 1, June 2014.

Index

A5, 252
ABS, 154
access card, 174
ACP, 168
ADF, 181–182
ADMF, 281
AES, 35
AICD, 139
AID, 142
ALC, 91
AMI, 146
Android, 53
ANDSF, 107
ANSI, 29–32
APDU, 142, 184, 246
application, 50–55
   management, 22
   provider, 145
ARIB, 16, 29
ASME, 74
ASN.1, 158
ATIS, 16
ATR, 156
AuC, 61–62
authentication, 255
   AuC, 61–62
   LTE, 78

   network, 67
   subscriber, 66
   user, 66
   Wi-Fi, 97
authorization, 62, 78
automotive, 145

backhaul, 48
BAN, 146
bandwidth, 49
barcode, 135, 139, 189
BGA, 174
Big Data, 293
bio-effect, 286
biometric, 31
BIP, 168, 210, 231
BLOCK1/2, 64-65
Bluetooth, 134–137
broadcast, 94

Calypso, 204
capacity, 49
CAT, 23, 185, 210, 230
CBEFF, 31
CC, 91, 159
CCM, 184
CCSA, 16

Wireless Communications Security: Solutions for the Internet of Things, First Edition. Jyrki T. J. Penttinen.
© 2017 John Wiley & Sons, Ltd. Published 2017 by John Wiley & Sons, Ltd.

Printed and bound by CPI Group (UK) Ltd, Croydon, CR0 4YY